Spreadsheets in structural design

Spreadsheets in structural design

S. R. Davies
B.Sc., Ph.D., C.Eng., M.I.C.E.
Formerly Senior Lecturer
Dept. of Civil Engineering
University of Edinburgh

Longman Scientific & Technical

TO
Mary
Wendy, Susan
Megan, Radha, Heather

Longman Scientific & Technical,
Longman Group Limited,
Longman House, Burnt Mill, Harlow,
Essex, CM20 2JE, England
and Associated Companies throughout the world.

Copublished in the United States with
John Wiley & Sons, Inc., 605 Third Avenue, New York, NY 10158

First published 1995

Notice of liability
Neither the author nor the publishers accept any responsibility for loss occasioned to any person using the disk supplied with this book or acting or refraining from acting as a result of material contained in the disk.

British Library Cataloguing in Publication Data
A catalogue entry for this title is available from the British Library.

ISBN 0-582-22698-8

Library of Congress Cataloging-in-Publication Data
A catalog entry for this title is available from the Library of Congress.

ISBN 0-470-23459-8 (USA only)

Set by 4 in 10/11 pt Times
Printed and Bound by Bookcraft (Bath) Ltd

Contents

Preface

Although many engineers are familiar with the elementary concepts and uses of spreadsheets their wider application to more difficult or tedious design problems does not appear to have been generally accepted. Spreadsheets offer a simple method of obtaining solutions to a variety of problems and are very efficient with regard to the amount of preparation required to obtain these solutions. This efficiency arises from the fact that the spreadsheet already contains routines for data input, data or graphical output and a range of algebraic and logic functions and that there is no requirement for a programmer to rewrite these routines.

The text contains a number of examples of spreadsheet layouts which have been developed for solving specific design problems using a variety of techniques. Examples are drawn from different areas of design using a range of materials and a number of codes of practice. Most design routines can be simplified by use of a spreadsheet and the aim of this text has been to illustrate the techniques that are available so that the methods can then be applied to other problems. Structural engineering design consists of a mixture of solving equations, ensuring that limiting conditions are obeyed, selecting values from tables or graphs and a study of the effects of the use of trial values. All these processes can be handled by the use of a suitable spreadsheet function or formula.

A number of the design routines used by structural engineers are both laborious and time consuming because trial designs may have to be repeated a number of times before an acceptable solution is obtained. The spreadsheet approach reduces this labour and the effect of any changes in the value of any parameter can be studied without difficulty. This not only leads to more economical designs but also makes the method a useful educational tool. Using the spreadsheet students are able to explore the 'What if' situation and gain a better insight into the influence of the key variables.

It will be shown that many of the standard charts used by engineers can be replaced by a spreadsheet. For example all the charts used for the design of reinforced concrete columns, short or slender, subject to either uniaxial or biaxial bending, can be replaced by a single spreadsheet. Charts for beam and slab design are treated similarly and also replaced by spreadsheets. These spreadsheets require less preliminary calculation than the usual design approach since there is no requirement to group parameters together in order to determine values of the chart co-ordinates.

Many of the examples in the text use the equations and restrictions quoted in the relevant British Codes but it must be stressed that the method is not related to a particular code of practice. In fact it is not even confined to structural engineering and could easily be extended to other disciplines where similar processes are required.

The time taken to develop a spreadsheet layout suitable for a particular problem will obviously depend on the complexity of the problem but most of the layouts provided throughout the text were developed in a few hours. Simple functional solutions can usually be obtained quite quickly but it is then possible to produce more elegant solutions using cosmetic devices to improve the appearance. Some of these devices are illustrated in the examples.

Two artificial restrictions have been placed on the solutions provided in the text.

(a) In order to simplify presentation a limit was generally placed on the number of columns to be used in obtaining solutions so that output from the printer would be confined to a page width. (This limit was exceeded when the extension beyond a page width contained tabular values of a repetitive nature, or to house the operations of matrix inversion and multiplication in Chapters 7 and 8.)

(b) No Macros were to be used for obtaining the solutions since these are not always transportable from one package to another and their use would have resulted in additional and side tracking explanations.

Solutions developed by readers are not bound by these restrictions and in particular Macros can be used to make the whole process interactive.

The 3½″ high density disk supplied with this book contains a copy of each of the spreadsheets described in the text. It is suitable for use on IBM compatibles and also Apple Macintosh that includes an Apple File Exchanger. To use any of the spreadsheet files the user must have access to one of the standard packages such as Lotus 1-2-3, Quattro Pro or Excel, or any package which is capable of importing Lotus 1-2-3 files.

S.R.D.
November 1993

Acknowledgements

The methods described in this book have been illustrated by their application to different materials and different structural elements and references to many publications are scattered throughout the text. A list of the sources of these references are included at the end of the book and the author wishes to record his indebtedness to the authors of all these publications. Many other colleagues and authors have also contributed indirectly, either by discussion or education, and this assistance is also acknowledged.

Thanks are due to Mr P. Aspinall and Mr A. Ohri for assistance with the checking of the spreadsheets and to Mary, both for checking the manuscript and also her general encouragement throughout.

The author is also grateful to the following for granting permission to reproduce copyright material:

The Institution of Civil Engineers for Fig. 3 of the paper by B. Stafford-Smith and J. R. Riddington from the *Proceedings of the Institution of Civil Engineers,* **63**, 1977, 377–91; and for Figs 5 and 19 by R. H. Wood from the *Proceedings of the Institution of Civil Engineers,* **65**, June 1978, 381–411.

The Institution of Structural Engineers, 11 Upper Belgrave Street, London, SW1X 8BH, for permission to reproduce various extracts from the *Manual for the Design of Reinforced Concrete*.

Ellis Horwood Ltd, Chichester, for Figs 8.1, 8.2, 8.4 to 8.6 and 10.6 from *Load Bearing Brickwork Design — Second Edition* (1987) by A. W. Hendry, B. P. Sinha and S. R. Davies.

Extracts from British Standards, BS 8110, BS 5628 and BS 5950 are reproduced with the permission of the British Standards Institute from whom complete copies can be obtained by post from BSI Sales, Linford Wood, Milton Keynes, MK14 6LE.

Notations

A = Constant of integration. Eqn 8.7
A_c = Area of concrete
A_m = Cross-sectional area of masonry
A_{ps} = Area of prestressing tendons
A_s = Area of tension reinforcement
A_s' = Area of compressive reinforcement
A_{s1} = Area of steel near upper surface
A_{s2} = Area of steel near lower surface
A_{sc} = Total area of longitudinal reinforcement
A_{sv} = Area of reinforcement resisting shear
A_x = Area of reinforcement at level h_x
$[A^{-1}]$ = Inverse of matrix [A]
[A] = Matrix
a = Distance to point load from left
a_1,a_2,a_3,a_4 = Area of bars
a_s = Shear span
B = Slab width
B_1,B_2,B_3 = Horizontal dimensions of cross section
{B} = Vector
b = Width of beam or column (Chapter 3)
b = Distance, point load or end of UDL from right (Chapter 8)
b' = Distance defined in Figs 4.6 and 4.12
b_c = Length in compression
C_f = Force correction factor
C_i = Shape correction factor
c = $L-a$ (Chapter 8)
d = Effective depth of beam (Chapter 3)
d = Depth to tendon centroid (Chapter 6)
d = $L-b$ (Chapter 8)
d = Depth of beam in wallbeam (Chapter 9)
d' = Depth to compressive reinforcement
d_1 = Concrete cover to upper steel
d_2 = Concrete cover to lower steel
d_3,d_4 = Cover to steel parallel to XX (Fig. 4.6)
d_c = Depth of masonry in compression
d_m = Distance of centroid to neutral axis
E = Modulus of elasticity
E_b = Elastic modulus for beam
E_c = Modulus of elasticity for concrete
E_s = Modulus of elasticity of steel
E_w = Elastic modulus for wall
e_m = Eccentricity of resultant load
e_{min} = Lesser of 0.5 h and 20 mm
e_s = Distance from centroid to P_o (Fig. 6.1)
e_x = Eccentricity M/N
e_y = Eccentricity in y direction
F = Floor load, KNm^{-2}
$F_{(frame)}$ = Design strength of frame
$F_{(wall)}$ = Design strength of wall
F_a = Constant $\sqrt{(f_{cu})}/17.5$
F_o = Limiting sliding friction
f = Representative stress
f_b = Compressive strength of unit
f_{cu} = Characteristic cube strength of concrete
f_k = Characteristic compressive strength of masonry
f_m = Max vertical stress in wall beam
f_{ms} = Mortar strength
f_{pb} = Stress in tendon at beam failure
f_{pe} = Effective prestress
f_{pu} = Characteristic strength of tendons
f_s = Steel stress. Eqn 5.6
f_s' = Stress in compressive reinforcement
f_{s1} = Critical stress in service (top)
f_{s2} = Stress in steel near lower surface (Chapter 4)
f_{s2} = Critical stress in service (bottom) (Chapter 6)
f_{sc} = Limiting stress in service, compression
f_{st} = Limiting stress in service, tension
f_{sx} = Stress at level h_x
f_{t1} = Critical stress at transfer (top)
f_{t2} = Critical stress at transfer (bottom)
f_{tc} = Limiting stress at transfer, compression
f_{tt} = Limiting stress at transfer, tension
f_v = Characteristic shear strength of masonry
f_y = Characteristic tensile strength of steel
f_{yv} = Characteristic strength of links
G_k = Characteristic dead load
H_1,H_2,H_3 = Vertical dimensions of cross-section (Chapter 6)
H_1,H_2 = Storey heights (Chapter 7)
h = Height of frame, column, wall or storey
h' = Distance defined in Figs 4.6 and 4.12
h_1,h_2,h_3 = Horizontal forces (Fig. 7.8)
h_{eff} = Effective height
h_{effx} = Effective height perpendicular to XX
h_{effy} = Effective height perpendicular to YY
h_x = Representative depth to reinforcement

h_{x1},h_{x2} = Depths to reinforcement (Fig. 4.10)
I = Floor level counter from top (Appendix 1)
I = Second moment of area
IL = Imposed load kNm^{-2}
i = Node point
J_1 = Dimensionless constant
J_2 = Dimensionless constant
K = $M/f_{cu}bd^2$
K' = $M_u/f_{cu}bd^2$
K_a = Axial stiffness parameter (Davies and Ahmed, 1978)
K_b = Material constant. Eqn A.17
K_d = Flexural stiffness parameter (Davies and Ahmed, 1978)
K_m = M/bd^2
K_r = Reduction factor — Eqn 4.29
K_s = Flexural stiffness parameter (Stafford-Smith and Riddington, 1977)
k = Coefficient used by Wood and Simms (1969)
k_1,k_2,k_3 = Stress block ratios
k_5 = Stiffness of beam 5
k_A = Total joint stiffness at joint A
k_B = Total joint stiffness at joint B
L = Span of member, beam, slab, wall or frame
L_1 = Length of shorter wall (Appendix 1)
L_1,L_2 = Bay lengths (Chapter 7)
L_2 = Length of longer wall (Appendix 1)
L_x = Length of slab side
L_y = Length of slab side
l_e = Effective length
l_o = Distance between restraints
M = Design bending moment
M_1,M_2 = Support moments fixed beam
M1x = Smaller moment about XX
M1y = Smaller moment about YY
M2x = Larger moment about XX
M2y = Larger moment about YY
M_A,M_E = Bending moment at A, E
M_{An} = Moment at joint A in member n
MBS = Moment at bottom of wall due to imposed load
M_{Bn} = Moment at B in member n
MD = Function of α. Eqn A.14
MTS = Moment at top of wall due to imposed load
M_a = Increased design moment for Case (c)
M_{addx} = Additional moment about XX due to slenderness
M_{addy} = Additional moment about YY due to slenderness
M_{as} = Moment in slender columns
M_{asx} = Moment perpendicular to XX — slender columns
M_{asy} = Moment perpendicular to YY — slender columns
M_{ax} = Increased design moment about XX. Case (c)
M_{ay} = Increased design moment about YY. Case (c)
M_c = Moment at centre of beam
M_d = Design moment of resistance
Mix = Combined moment (0.41M1x + 0.6M2x)
M_m = Maximum moment in wall beam
M_p = Plastic moment of beam or column
M_{pb} = Plastic moment of beams
M_{pc} = Plastic moment of columns
M_s = Bending moment in service
M_{sx} = Mid-section moment XX
M_{sy} = Mid-section moment YY
M_t = Bending moment at transfer
Mtx = Total moment XX including slenderness effect
Mty = Total moment YY including slenderness effect
M_u = Ultimate moment of resistance
Mx = Design bending moment perpendicular to XX
Mx′ = Increased moment due to biaxial effect
M_{xx} = Bending moment at x
My = Design bending moment perpendicular to YY
My′ = Increased moment due to biaxial effect
m_1,m_2 = Yield moments
m_d = Parameter for moment used by Wood (1978)
m_e = Framed-panel coefficient
N = Number of rows or columns in matrix (Chapter 1)
N = Design axial load (Chapter 4)
N = Number of strips (Hillerborg method) (Chapter 5)
N_d = Design axial load resistance
N_{dz} = Ultimate axial load with zero moment
N_s = Function of α. Eqn A.10
n = Design ultimate load
P_o = Prestressing force
P_s = Effective prestressing force
P_t = Prestressing force at transfer
Q_k = Characteristic imposed load
q = Uniformly distributed load
q_k = Imposed load intensity
R = Roof load kN m^{-2}
R_1,R_2 = Support reactions fixed beam
R_{An} = Reaction at A in member n
R_{Bn} = Reaction at B in member n
R_l = Reaction at left support
S = Strip number
S_R = Slenderness ratio
S_c = Wall spacing
S_d = Number of discontinuous edges
S_v = Spacing of shear reinforcement
T = Tie force in beam
t = Wall thickness (Appendix 1)
t = Overall width of column (Chapter 4)
t_e = Effective wall thickness
t_f = Flange thickness
t_i = Thickness of inner wall
t_s = Slab thickness
u = Interval between node points
V = Shear force
V_2 = Wind speed on longer wall
V_s = Wind speed

v	= Shear stress
v_1,v_2	= Vertical forces (Fig. 7.8)
v_c	= Limiting value of shear stress
W	= Wall weight kNm^{-2} (Appendix 1)
W	= Point load (Chapter 8)
W	= Total load on wall beam (Chapter 9)
W_k	= Characteristic wind load
w	= Uniformly distributed load
w_n	= Uniformly distributed load in member n
$X(m)$	= Load reduction factor
X_1,X_2,X_3	= Coefficients related to K_d (Davies and Ahmed, 1978)
$\{X\}$	= Vector of unknowns
x	= Depth to neutral axis (Chapter 3)
x	= Distance in x direction (Chapter 5)
x_1,x_2	= Values of variable x
y	= Distance in y direction (Chapter 5)
y	= Deflection at x from left (Chapter 8)
y_1,y_2	= Values of variable y
Z	= Section modulus
Z_1	= Section modulus referred to bottom of section
Z_2	= Section modulus referred to top of section
z	= Lever arm depth
α	= Slab/wall ratio (Appendix 1)
α	= Coefficient in BS 5628 related to N/N_{dz} (Chapter 4)
α	= Initial prestress coefficient (Chapter 6)
α	= Coefficient related to h/L (Davies and Ahmed, 1978) (Chapter 9)
α_{sy}	= Moment coefficient
α_x	= Moment coefficient
β	= Capacity reduction factor (Appendix 1)
β	= Coefficient related to end conditions (Table 4.3)
β	= Coefficient related to h/L (Davies and Ahmed, 1978) (Chapter 9)
$\beta_1,\beta_2,\beta_3,\beta_4$	= Moment coefficients (Chapter 5)
β_1,β_2	= Bond factors (Chapter 6)
β_a	= Coefficient related to slenderness (Table 4.2)
β_b	= Moment redistribution ratio
β_y	= Moment coefficient
γ	= Coefficient related to h/L (Davies and Ahmed, 1978)
γ_m	= Safety factor; shear strength with no steel
γ_{mm}	= Partial safety factor for material
γ_{ms}	= Partial safety factor for steel strength
γ_{mv}	= Partial safety factor for shear strength
Δ	= Material constant. Eqn A.17
Δ_ϕ	= Correction factor for framed panel design
δ_p	= Parameter related to m_d
δ_x	= Deflection at x from left
ϵ_s	= Representative strain
ϵ_1	= Strain in upper reinforcement
ϵ_2	= Strain in lower reinforcement
ϵ_A	= Strain at A
ϵ_B	= Strain at B
ϵ	= Strain
ϵ_{cu}	= Ultimate compressive strain in concrete
ϵ_e	= Prestress strain in concrete at tendon level
ϵ_m	= Maximum strain in masonry
ϵ_{pa}	= Total strain at tendon level due to ultimate load
ϵ_s	= Strain in compressive reinforcement
$\epsilon_{s'}$	= Strain in compressive reinforcement
ϵ_u	= Ultimate strain in masonry (Chapter 3)
ϵ_u	= Concrete strain at tendon level — ultimate loads
ϵ_x	= Strain at level h_x
ϵ_y	= Yield strain in reinforcement
η	= Ratio of P_s/P_t
η_s	= Ratio of P_s/P_o
μ_p	= Ratio of M_{pb}/M_{pc}
θ	= Stress loss at transfer coefficient (Chapter 6)
θ	= rotation due to bending at end of member (Chapter 7)
θ_E	= Rotation due to bending at end E of member
θ_F	= Rotation due to bending at end F of member
ϕ	= Yield line angle to horizontal
ϕ_s	= Parameter related to m_e
ρ	= Steel ratio
ρ_B	= Steel ratio for balanced section
σ	= Arbitrary wind stress at base
τ_m	= Maximum shear stress

1

Spreadsheets: application to design

1.1 Introduction

There are quite a number of different spreadsheet packages available and most engineers will be familiar with the use of at least one. Fortunately these packages do not differ greatly from each other in method of application and, with most, similar tasks can be performed. In the main the differences arise from the use of different words or symbols to describe similar functions. For example the words **block** and **range** and the symbols $ and + might be used differently.

The aim of this chapter is to define the basic operations which will be required in later chapters. Some of the terms used might differ from those with which the reader is already familiar and it is essential that any differences are noted at the outset.

1.2 Basic concepts

One of the basic principles of a spreadsheet is that it consists of a number of **cells** each of which can contain a **number** or a **label** or a **formula**. Each cell has a unique address such as C9, meaning the cell at the intersection of column C and row 9, and the contents of each cell can be interacted with the contents of other cells.

For example in the spreadsheet shown below:

Cell B1 contains the number **6**

Cell A2 contains the number **9**

In cell C3 the unseen formula (**B1*A2**) has been entered into the spreadsheet (* = ×) and on the screen the result of the calculation is shown as the number 54 (6 × 9). This value has been calculated from the formula and represents the product of the contents of the two cells B1 and A2.

Similarly cell D2 contains the unseen formula (**B1+A2**) so that the sum of the contents of cells B1 and A2 is shown in cell D2 on the screen.

	A	B	C	D	E
1		**6**			
2	**9**			**15**	
3			**54**		

Spreadsheet as displayed

	C	D
1		
2		(B1+A2)
3	(B1*A2)	

Underlying formulae

This simple example could be expanded so as to include more cells and more complicated algebraic and logical operations since the principles are the same whatever the complexity. Additionally, any change made to a number in one cell results in appropriate changes to other cells which are connected by formulae to the first cell.

	A	B	C	D	E
1		**10**			
2	**9**			**19**	
3			**90**		

Spreadsheet as displayed

	C	D
1		
2		(B1+A2)
3	(B1*A2)	

Underlying formulae

In the above spreadsheet the contents of cell B1 has been changed to a number **10** and this has resulted in changes to both cell C3 and cell D2. This principle can be extended so as to include **labels** and **formulae**. (The word **value** is used to define the contents of the cell irrespective of type.)

For example if cell B1 contained the label **STRESS** and cell C3 contained the formula **+B1** then the resulting spreadsheet would be as follows:

	A	B	C	D	E
1		**STRESS**			
2	**9**			**19**	
3			**STRESS**		

Spreadsheet as displayed

	C
1	
2	
3	+B1

Underlying formulae

Note that the entry **+B1** placed in cell C3 implies the formula:

Make contents of C3 = contents of B1

A similar interaction can exist for formulae as shown below:

	A	B	C	D	E
1	**6**	**10**	**(A1+B1)**		
2	**9**	**8**	**(A2+B2)**		
3	**7**	**5**	**(A3+B3)**		

In this example the formulae themselves are shown in cells C1, C2, and C3 instead of the results obtained by use of the formulae. It is assumed that the formula in cell C1 was entered from the keyboard and the formulae in C2 and C3 obtained by use of the COPY command, see section 1.6. The spreadsheet automatically adjusts the formulae so as to address the correct cells so that A1 and B1 referred to in cell C1 automatically change to A2, B2 and A3, B3 in cells C2 and C3 respectively.

It can be seen from the above that the benefits derived from the use of spreadsheets arise from two basic principles:

(a) Values in cells can be connected by arithmetic, algebraic or logical operations.
(b) Changes to any value are transmitted throughout the spreadsheet resulting in appropriate changes to any connected cells.

Other more complicated operations are usually available but in the main these are labour-saving devices which are included to simplify input and clarify output but do not add to the basic concepts described above.

1.3 Contents of cells

At least one of the cells of the spreadsheet will be highlighted by the cursor on the screen. For this case of a single highlight, any keyboard input will be directed automatically to this highlighted cell. If a range of cells are highlighted then input to these cells only results automatically by means of one of the editing commands or from the application of a formula. The location of the highlight can be changed by using the arrow keys. As stated above a cell can contain a number, a label or a formula and usually the design of the spreadsheet program is such that the program itself is able to recognise the input mode. Different spreadsheets may use different methods of effecting this recognition but the usual method is illustrated in Table 1.1.

Note that:

(a) The first entry for a **number** must be a digit, or a decimal point, or an addition or subtraction sign, an opening bracket, or a currency symbol.
(b) These entries may also be the first entry for a **formula** but the list is now extended so as to include the @ and # symbols. If the first part of a formula begins with a cell address then the input must begin with a + sign.

Table 1.1 Types of cell numbers, labels and formulae

Example	*Type*
654.32	Number
-456	Number
£23.69	Number
55.2%	Number
5.32E+02	Number
'SALES	Label (' is the label prefix character which alerts the cell to expect a label)
'-456	Label (because of the ')
+8-4	Formula
+C2+E3	Formula
@sum(A2..A9)	Formula

It may seem strange that both numbers and formulae can start with a digit and still be recognised but in reality a number can be considered as a simple formula. Consider the number **10** and the formula **10+0**. One is only a simpler version of the other. (More complicated formulae would begin with the @ sign.)

If the first entry is not one of those listed above then the program will assume that it is the beginning of a **label** or **text** entry and automatically attach a prefix to the entry. Normally the prefix is the apostrophe (') although other label prefixes are also used in order to justify the entry to the centre or the right of the cell. Sometimes the user might wish to override the program and force a number or a formula to be a label. This can be achieved by starting with the prefix (') so that, for example, **625** is a number whilst **'625** is a label.

1.4 Command menus

The user can control input in a number of ways. The number of decimal places displayed, the alignment of numbers and labels within the cells, the width of a particular column so as to allow more input per cell and the format of the display can all be changed at the user's discretion.

Control of the various editing, filing, graphics and printing facilities are effected by means of a command tree which can be summoned by means of one of the keyboard keys. The usual key for this operation is the forward slash / and the use of this key will be assumed in this text for calling up the commands.

Use of / will result in two command lines appearing either at the top or the bottom of the spreadsheet. The first of these command lines represents the **main menu** which lists the main commands available. One of these commands, usually the first, will be highlighted. The second command line is a **submenu** containing a further list, or description, related to the highlighted command in the main menu time.

For example the main menu might be:

Worksheet, Range, Copy, Move, File, Print, Graph, Date, System, Quit and the submenu:

Global, Inset, Delete, Column-width, Erase, Titles, Window

To select one of these commands it must first be highlighted, using the arrow keys, and then 'entered' by pressing RETURN. Errors can be rectified by using the ESC key. Some commands are **global** in that they operate on the whole spreadsheet whilst others only affect a part, in which case the part to be operated upon must be defined, i.e. a **range** or **block** must be defined.

1.5 Editing

It is essential that the contents of each cell can be edited and each spreadsheet program will include the means of copying, moving, erasing, changing, aligning, formatting, etc. This applies to the contents as seen on the screen, such as numbers or labels, or unseen, such as formulae. For example some of the attributes, such as width and type of display, of the cells themselves can be changed so that the output to the screen is the formula contained in the cell rather than the result obtained by application of the formula.

Although some differences may exist in the way different packages handle the required editing facilities these differences are slight and readers will be familiar with the rules which apply to their preferred package. Use of most of the commands is fairly straightforward but since the COPY command is used a great deal throughout this text a more detailed description of this command is given below.

1.6 The COPY command

To copy a single label or number entry

Using the appropriate key, such as the forward slash /, call up the COMMAND mode and highlight the COPY command. Press the return key to bring up the SUBMENU of the COPY command. The SUBMENU will now request the address of the FROM block which has to be copied. For this case only the contents of one cell is to be copied and this cell must be highlighted, or alternatively the cell address entered. Either way pressing return will inform the program of the FROM address. The SUBMENU will now request the address of the TO block and this can be entered by using the arrow keys to move the highlight to the required cell and pressing return.

To copy a block of label or number entries

Highlight the first cell of the block containing the entry and use /C to issue the COPY command. The SUBMENU will now request the address of the FROM block which has to be copied. For this case the addresses of all cells in the block must be defined and this can be achieved by **pointing** the first cell with the full-stop key and using the arrow keys to spread the highlighted area over the entire block. Pointing with the full-stop can be considered as anchoring the first position of the highlight before using the arrow keys to increase the highlighted area. A block of column entries or row entries or both can be defined in this way and once the FROM block is defined it can be entered by using RETURN. The SUBMENU will now request the address of the TO block and this can be defined by repeating the above process. First move the highlight to the start position of the required TO block and POINT it. Now spread the highlight over the required area and enter. Note that if the first highlight has been anchored in the wrong cell this can be cancelled using the ESC key.

To copy a single formula entry

Although the method for copying a formula from one cell to another is similar to that described for the case of copying labels or numbers there is a difference in that cell addresses may be copied in different ways. These addresses may be relative, absolute or mixed and the program must be able to recognise which the user requires. The symbol usually used to define the required type of address is the dollar sign, $, and this is the symbol used in this text. Other application packages may use a different symbol. Examples of the different types of addresses are given below.

Relative Address. Assume that addresses such as A1 or B5 in a formula refer to addresses in the FROM block. When copied to the TO block these addresses change so that they refer to the cells in the same relative position.

	A	B	C	D	E
1	**EXPENSES**			**EXPENSES**	
2	**FOOD**	**£60.00**		**FOOD**	**£60.00**
3	**RENT**	**£320.00**		**RENT**	**£320.00**
4	**CAR**	**£100.00**		**CAR**	**£100.00**
5	**TOTAL**	**£480.00**		**TOTAL**	**£480.00**

Spreadsheet as displayed

	A	B	C	D	E
1					
2					
3					
4					
5		@SUM(B2.B4)			@SUM(E2..E4)

Underlying formulae

In this example the FROM block is (A1 to B5) and the TO block is (D1 to E5). If cell B5 in the FROM block contains the formula **@sum(B2..B4)** then this formula would become **@sum(E2..E4)** in cell E5 of the TO block.

Absolute Address. There are occasions when the relative change is not required and a method is required to suppress this change. If two dollar signs ($ $) are included in the address, one for the column letter and the other for the row number then this is an instruction to the program not to change the address during the copying process.

	A	B	C	D
1	**BREADTH**	**200**	**200**	
2	**DEPTH**	**100**	**150**	**200**
3	**AREA**	**20000**	**30000**	**40000**

In this example the FROM block is the single cell B3 which contains the formula **($B$1*B2)**. The TO block are the cells C3 and D3. When copied the cell C3 will contain the formula **($B$1*C2)** and the cell D3 the formula **($B$1*D2)**. Copying in this case has been partly absolute and partly relative. Note that many spreadsheets do not accept commas to denote 1000's.

Mixed Address. On occasion it may be necessary to make one direction (row or column) absolute and the other direction relative. This can be achieved by the use of one dollar sign:

(a) The use of **$B6** would keep the same column value but use relative row values.
(b) The use of **B$6** would keep the same row value but use relative column values.

The use of the dollar sign, for copying, can be illustrated by considering the four possible cases, i.e. **B1, $B$1, $B1** and **B$1**. In the examples shown below the block of values shown in rows 1 to 3 are initial values. Cell B4 contains one of the above four cases as indicated by the label in A4. Cells C4 to E6 contain the results of copying cell B4 to the block C4..E6.

	A	B	C	D	E
1		**1**	**2**	**3**	
2		**4**	**5**	**6**	
3		**7**	**8**	**9**	
4	**B1**	**1**	**2**	**3**	**0**
5			**5**	**6**	**0**
6			**8**	**9**	**0**

In the above example copying the formula **+B1** from cell B4 to the block C4 to E6 would place the formulae **+C1, +D1,** and **+E1** in cells C4 to E4. Similarly **+C2, +D2** and **+E2** would be placed in cells C5 to E5 etc. Copying is relative.

	A	B	C	D	E
1		**1**	**2**	**3**	
2		**4**	**5**	**6**	
3		**7**	**8**	**9**	
4	**$B$1**	**1**	**1**	**1**	**1**
5			**1**	**1**	**1**
6			**1**	**1**	**1**

In this example copying formula **+$B$1** from cell B4 results in all the values being the same, i.e. equal to the value in cell B1. Copying is absolute.

	A	B	C	D	E
1		**1**	**2**	**3**	
2		**4**	**5**	**6**	
3		**7**	**8**	**9**	
4	**$B1**	**1**	**1**	**1**	**1**
5			**4**	**4**	**4**
6			**7**	**7**	**7**

In this example copying formula **+$B1** from cell B4 has the effect of anchoring the column, column B in this case, so that copying results in all columns containing the same values and only the row values change. Copying is mixed.

	A	B	C	D	E
1		**1**	**2**	**3**	
2		**4**	**5**	**6**	
3		**7**	**8**	**9**	
4	**B$1**	**1**	**2**	**3**	**0**
5			**2**	**3**	**0**
6			**2**	**3**	**0**

In this example copying formula **+B$1** from cell B4 has the effect of anchoring the row, row 1 in this case, so that copying results in all rows containing the same values and only the column values change. Copying is mixed.

To copy a block of formulae entries

This is similar to that described above bearing in mind the importance of the different types of address. First highlight the first cell of the FROM block. POINT this cell in order to anchor the first cell and then use the arrow keys to define the block. Press RETURN. Move to the first cell of the TO block and POINT it. Define the TO block with the arrow keys and ENTER.

1.7 Graphs

A number of different types of graph can be created from the data on the spreadsheet. Usually five types are available:

pie chart;
bar graph;
stacked bar chart;
line graph;
XY graph.

Probably the most useful for the purposes of this text is the XY graph.

The XY graph is a plot between at least two sets of variables X and Y. In some applications more than one set of Y values can be plotted against a common set of X values. In the examples that follow in later chapters it will be shown that the XY graph can be a useful device for testing the accuracy of data and also for helping to select data to give the best solution.

To use the graphics facility input /G and ENTER. The blocks which contain the X data and at least one set of Y data must be defined and the resulting graph(s) can then be viewed on screen. Other refinements, such as entering titles and using different line displays are possible, but these are not usually necessary for the simple checking or convergence processes described in the following chapters.

1.8 Other useful facilities and functions

The description given above is only a brief outline of the facilities that are available within each spreadsheet package. Many other useful facilities are provided but the number and use of these may vary according to the software package. The manuals, for the spreadsheet package used by the reader, will describe the other facilities which are available.

In particular it is essential that the user can SAVE and RETRIEVE files. It is also useful to be able to PRINT blocks of data or formulae so that hard copies are available for purposes of storage or checking. The WINDOW facility is useful for enabling two widely-separated parts of the screen to be viewed simultaneously and other facilities may be required from time to time.

Spreadsheet packages contain a set of built-in functions for which a particular operation is defined. For example the operation of summation can be performed using the function **@sum(B5..E5)** to produce the sum of all values in cells B5 to E5. This example assumes that the @ prefix has been defined as the prefix required to inform the computer that a built-in function is being input. Other packages might use a different symbol for this function prefix.

Although a large number of functions are available there are two, in particular, which are of special interest in this text. These are the @IF and @LOOKUP functions. Other functions may also be used in the chapters that follow but their use is fairly straightforward whereas the @IF and @LOOKUP functions merit special attention. They can be used in special ways which can be very useful to an engineer.

1.9 @VLOOKUP and @HLOOKUP

The @HLOOKUP function is used for a horizontal search of a table of values whilst the @VLOOKUP function is used for a vertical search. The format of the vertical table function is:

@VLOOKUP(X,Block,Column)

where:

X = A string or numeric value.
Block = The address of a block.
Column = A positive integer.

The horizontal table function is similar except for the third argument which would become 'Row'. Most of the descriptions given below refer to the @VLOOKUP function but these references could generally be replaced by the @HLOOKUP function.

In addition to being a string or a number, X can be the address of a cell which contains one of these values or an expression which results in one of these values. The following would be acceptable values of X:

5, 10.54, "FEB", B6, (C8*3), 8*4/3

(assuming B6 contains a string or a number and C8 contains a number).

The block consists of a number of adjacent column entries in which the column on the extreme left (the X column) contains possible values of X and no blank cells. If X is numeric then the numbers must be in ascending order. The block can consist of any number of columns but there must be no blank columns and columns to the right of the X column are numbered 1,2... etc. Column X is numbered 0. The coordinates of the block are entered by the upper left and lower right cell addresses.

An example of a block in which there are a total of four columns, i.e. one X column and three others, is shown below. This block would be referred to as A2..D5 assuming that the three columns to the right of the X column were to be used. The names in row 1 are not included in the block reference and if only part of the table is required the reference can be reduced accordingly, e.g. A2..B5, using only one column.

	A	B	C	D
1	**X**	**C1**	**C2**	**C3**
2	**1.5**	**0.6**	**1.5**	**2.7**
3	**2.7**	**0.81**	**2.6**	**4.5**
4	**3.8**	**0.76**	**4.2**	**0.4**
5	**4.9**	**0.56**	**3.7**	**0.9**

The above example also illustrates the use of the argument (column) in @VLOOKUP. In this example, the 'column' entry would be an integer, 0 to 3, depending on which column is to be referenced. The first, or X column, is numbered 0 and the others 1 to 3. Alternatively it could be a cell address which contained one of the integers 0 to 3.

In use the @VLOOKUP function uses the X value to find a matching string or number in the X column and then returns the value from the column, specified by the column argument, on the same row as the matching X value. If X is a number and no exact match is found then the previous number is used.

Using the spreadsheet shown above:

@VLOOKUP(2.7,A2..D5,3) would return the value 4.5

@VLOOKUP(3.7,A2..D5,3) would return the value 4.5

@VLOOKUP(3.9,A2..B5,1) would return the value 0.76

@VLOOKUP(0.6*4,A2..B5,1) would return the value 0.6

	A	B	C	D	E
1	**DAY**	**C1**	**C2**	**C3**	
2	**MON**	**0.6**	**1.5**	**2.7**	
3	**TUES**	**0.81**	**SUN**	**4.5**	
4	**WED**	**0.76**	**4.2**	**0.4**	
5	**THU**	**0.56**	**3.7**	**JULY**	

And from the above spreadsheet:

@VLOOKUP("MON",A2..D5,3) would return the value 2.7

@VLOOKUP("WED",A2..D5,3) would return the value 0.4

@VLOOKUP("TUES",A2..C5,2) would return the value SUN

@VLOOKUP("MON",A2..B5,1) would return the value 0.6

Further examples of this function are included later in this chapter when the application of the function to design is illustrated.

1.10 The @IF function

The format of this function is:

@IF(CONDITION,TRUE,FALSE)

where:

CONDITION = a logical expression.
TRUE = a numeric or string value to be used if the condition is true.
FALSE = a numeric or string value to be used if the condition is false.

In simple form CONDITION represents any logical expression that can be evaluated as being TRUE or FALSE, for example **C9>10** or **C9/3=20**. However, the argument can be compounded by using the logical operators #AND# or #OR#. The operator #NOT# is also available. Examples of these are:

(a) The condition **A4>12#AND#A4<16** implies that if A4 is between 12 and 16 then it is true. Both conditions must be met in order for the compound condition to be true.
(b) The condition **A4>12#OR#A4<8** will be true for all values of A4 less than 8 and greater than 12 since the compound condition is true if either condition is true. Values of A4 less than 8 will be true because of condition 2, values greater than 12 will be true because of condition 1. Values between 8 and 12 will result in the condition being false.
(c) The condition **#NOT#B6<16** implies that the condition is true if B6 is greater than 16.

The second and third arguments can be either:

numeric values;
expressions resulting in numeric values;
text (expressed in double quotes);
cell addresses containing any of the above.

Examples (in order) are:

@IF(A4=6,12.6,4.5)
@IF(A4=6,12.6*3/4.5,4.5/(12.6*3))
@IF(A4=6,"TOM","DICK")
@IF(A4=6,C1,D3)

The main power of this function lies in the fact that each of the second and third arguments can be expanded by using a series of other @IF functions, i.e. 'nested' functions. Using this facility quite powerful logical operations can be constructed within any block of cells. As an example of nesting consider the function:

@IF(A4=6,@IF(B4=12,@IF(C4=16,C9,C6),C6),C6)

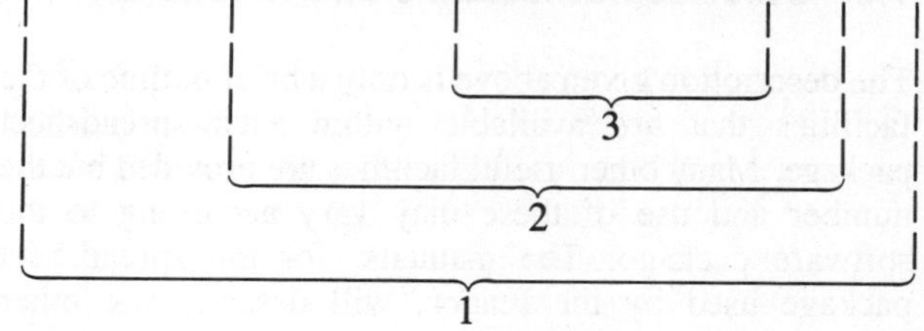

The @IF construction marked 3 has the three arguments **(C4=16,C9,C6)** with C9 and C6 being the cell addresses of the true and false results respectively. The second @IF construction has a normal CONDITION argument **(B4=12)**, but uses the third @IF construction for the TRUE argument and C6 as the cell address of the FALSE argument. The first @IF construction uses **A4=6** for the CONDITION argument, the second @IF for the true argument and C6 as the cell address of the false argument.

The above example, which is used only to demonstrate the procedure for nesting, represents the statement:

> 'If the values in cells A4, B4 and C4 are 6, 12 and 16 respectively then the current cell is to have the same value as that in cell C9. If any of the values are different then the current cell is to have the same value as that in cell C6. The current cell is the one in which the @IF construction has been entered.'

Alternative methods for achieving the same result are usually available and the user has to decide on the most suitable approach. For this example the same result can be obtained without nesting:

@IF(A4=6#AND#B4=12#AND#C4=16,C9,C6)

1.11 Applications to aid design

Although the design process involves the application of many different theoretical and practical decisions it is found that a common set of problems/tasks can be identified. These problems appear over and over again irrespective of the particular element which is currently under design scrutiny.

Three such common problems can be identified:

1. Having calculated or assumed the value of a required quantity the engineer has to refer to a set of tables in order to determine the value of a dependent variable. This process usually requires some interpolation to find the exact match.
2. Starting from the assumed or calculated values the engineer has to refer to a graph in order to determine the value of a dependent variable.

3. Quite often the equations that describe the problem have no direct solution and the engineer has to obtain a solution by trial. This can be a very cumbersome process.

It can be shown that the spreadsheet is ideal for dealing with these problems and, in addition to helping to obtain a solution, enables the engineer to study the influences of changes in the values of the variables so that a better understanding of the problem can be obtained.

1.12 Reading from tables

As described in section 1.9, if a numeric value of X is entered into a table using the @VLOOKUP function and this value of X is not found in the table then the results obtained from the @VLOOKUP function will be the values from the columns for the next lowest X value. No interpolation between tabular values is performed automatically and it is necessary for the user to allow for this if an exact interpolated match is required.

This is illustrated using the two-column spreadsheet table shown below:

	B	C
1	**Variable_X**	**Variable_Y**
2	x_1	y_1
3	x_2	y_2
4	x_3	y_3
5	x_4	y_4

If a value between x_2 and x_3 is entered into the table using the @VLOOKUP function then the value returned would be y_2 and not the interpolated value which would be between y_2 and y_3.

This can be corrected by inserting an additional column as shown below:

	B	C	D
1	**Variable_X**	**Variable_Y**	**Difference_Ratio**
2	x_1	y_1	$(y_2-y_1)/(x_2-x_1)$
3	x_2	y_2	$(y_3-y_2)/(x_3-x_2)$
4	x_3	y_3	$(y_4-y_3)/(x_4-x_3)$
5	x_4	y_4	

The interpolated value (Y) for any value X is now given by:

@VLOOKUP(*X*,B2..D5,1)
+(*X*−(@VLOOKUP(*X*,B2..D5,0)))
***(@VLOOKUP(*X*,B2..D5,2))**

or expressed in algebraic form (assuming X is between x_2 and x_3):

$$Y = y_2 + (X-x_2) \times (y_3-y_2)/(x_3-x_2) \quad [1.1]$$

The above example assumes a linear interpolation but other forms of interpolation could be allowed for by a similar approach if it was found necessary.

Note that in the example given above the difference ratio is left blank in row 5 because there is no point (x_5,y_5) and therefore no requirement to define the slope of the curve beyond point (x_4,y_4). The difference ratio for any row represents the slope of the line joining the coordinates of the value in that row to the coordinates of the value in the next higher row, for example, the difference ratio in row 4 represents the slope of the line connecting (x_3,y_3) and (x_4,y_4).

Sometimes a table of values is represented in a slightly different form such that the value x_4 is not a terminal value, beyond which the curve is not defined, but rather a limited value, beyond which all subsequent y values remain the same. This case can be dealt with as shown below:

	B	C	D
1	**Variable_X**	**Variable_Y**	**Difference_Ratio**
2	x_1	y_1	$(y_2-y_1)/(x_2-x_1)$
3	x_2	y_2	$(y_3-y_2)/(x_3-x_2)$
4	x_3	y_3	$(y_4-y_3)/(x_4-x_3)$
5	x_4	y_4	
6	$x_4+?$	y_4	

In row 5, $x = x_4$ and $y = y_4$. Assume that y remains equal to y_4 for all subsequent values of x higher than x_4. To deal with this case the difference ratio remains blank for both rows 5 and 6 and the value of y in row 6 is made the same as the value in row 5.

The question mark after x_4 in row 6 indicates that any suitable value can be inserted depending on the range of values to be included. Examples of this technique are given later in Sheets 4.3 to 4.5.

1.13 Reading from graphs

This problem can be treated in a similar way to that outlined above for tables if the graph is first linearised and the results put into tabular form on the spreadsheet.

As examples of this consider the simplified stress/strain relationships assumed for the reinforcing steel in the *British Codes for Reinforced Concrete* (BS 8110) and *Reinforced Masonry* (BS 5628). The assumed forms differ slightly in the two codes in that the concrete code assumes a bilinear form of the stress/strain graph and the masonry code a trilinear form. The two cases are considered below.

Case A: bilinear analysis. The stress/strain relationship assumed for the reinforcement in the current British code for reinforced concrete, is as shown in Fig. 1.1 (BS 8110, *Code of Practice for the Use of Concrete, Part 2: Structural Use of Reinforced and Prestressed Concrete*).

In order to produce a table from this graph the values of the strains at A and B must be determined. Using the values supplied in the BS code the expressions for these strains are:

Strain at A:

$$\epsilon_A = f_y/(200 \times 1000 \times \gamma_{ms})$$

Strain at B:

$$\epsilon_B = -f_y/(200 \times 1000 \times \gamma_{ms})$$

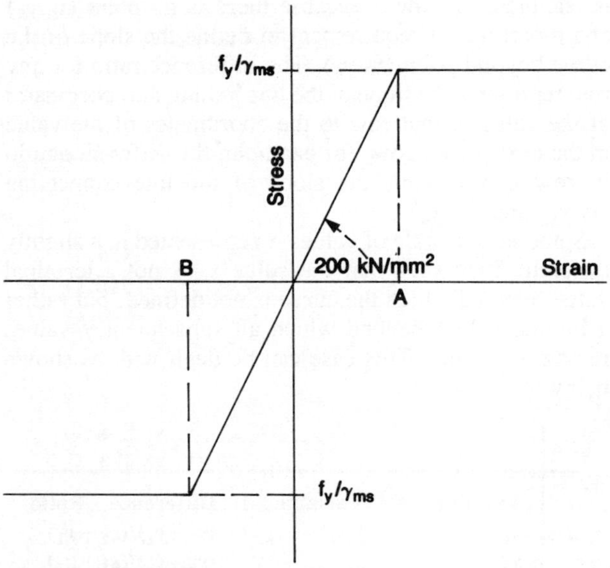

Figure 1.1 Stress/strain curve for reinforcement to BS 8110

Table 1.2

B	*C*
STRAIN	STRESS
−0.003	$-f_y/\gamma_{ms}$
ϵ_B	$-f_y/\gamma_{ms}$
ϵ_A	f_y/γ_{ms}
+0.003	f_y/γ_{ms}

So that the table of BS 8110 stress–strain values could be written as in Table 1.2.

The table has been extended on either side using arbitrary values of the strains, of ±0.003, to allow for strains less than ϵ_B and greater than ϵ_A.

The design of a spreadsheet which uses this table in order to determine the value of the stress for any given value of the strain is illustrated on Sheet 1.1.

The initial values of f_y and γ_{ms}, the characteristic strength and partial safety factor of the reinforcing steel, are placed in cells C8 and C9 respectively and these values are then used to calculate the strains at B and A in cells F8 and F9. The values in C8 and C9 can be changed as

Sheet 1.1

```
1/A      B         C          D          E          F          G         H
   2  Stress Interpolation for Given Value of Strain
   3  Using Assumed Stress/Strain Relationship from BS8110
   4  ------------------------------------------------------------------
   5
   6  Initial Values                       Preliminary Calculations
   7  ----------------------------         ------------------------------
   8  fy                   460             Strain B      -0.002
   9  Gms                  1.15            Strain A       0.002
  10  ------------------------------------------------------------------
  11            Table 1                              Table of values
  12  Strain    Stress      Difference               which represents the
  13  -0.003    -400.00           0.0                Stress - Strain curve
  14  -0.002    -400.00      200000.0                for reinforcement
  15   0.002     400.00           0.0                according to BS8110
  16   0.003     400.00
  17  ------------------------------------------------------------------
  18  Col 0     Col 1       Col 2      Col 3
  19  Results obtained using the @VLOOKUP                     Final
  20  function for each column and strains                    Interpolation
  21  taken from Table 1.                                     for
  22  Strain    Lookup_0    Lookup_1   Lookup_2               Stress
  23  -0.003    -0.003      -400.00         0.0               -400.00
  24  -0.002    -0.002      -400.00    200000.0               -400.00
  25   0.002     0.002       400.00         0.0                400.00
  26   0.003     0.003       400.00         0.0                400.00
  27  Showing that calculated stresses are the same as those in
  28  the Table.
  29  If any arbitrary value of strain is used (say 0.001)
  30   0.001    -0.002      -400.00    200000.0                200.00
  31  the stress is calculated to be 200N/mm2
```

required to allow for other types of steel. The formulae placed in cells F8 and F9 are:

F8: **−C8/(200*1000*C9)**
F9: **−F8**

The values from Table 1.2 together with the difference ratio required for interpolation are entered in block B13 to D16. Some typical examples of the formulae placed in this block are:

B13: **−0.003** (the arbitrary strain)
C13: **−C8/C9** (the stress due to arbitrary strain)
D13: **(C14−C13)/(B14−B13)** (difference ratio)
B14: **+F8** (the strain at B)
C14: **−C8/C9** (the stress at B)
D14: **(C15−C14)/(B15−B14)** (difference ratio)
B15: **+F9** (the strain at A)
C15: **+C8/C9** (the stress at A)
D15: **(C16−C15)/(B16−B15)** (difference ratio)

To illustrate the method used for interpolating for the value of the stress for any given value of the strain the method is shown step-by-step in rows 23 to 26. Although any value of strain could have been placed in cells B23 to B26, values taken from Table 1.2 have been used since the stresses are known for these values and this enables the process to be checked.

Using these values of strain the @VLOOKUP function is applied in turn in columns C, D and E. For example the formulae in the Lookup_0 column takes the adjacent value of the strain from column B and then searches the 0 column of the table to find a value either equal to the strain value or failing this the next value below the given value. For the present example an exact match is found so that the values in columns B and C will be identical.

Similarly the Lookup_1 formulae use the adjacent strain to find either the exact match or the value below this from the 0 column and having located the appropriate value outputs the value in the adjacent 1 column. Examples of the formulae in the cells of row 23 and 26 are:

B23: **−0.003**
C23: **@VLOOKUP($B$23,$B$13..$D$16,0)**
D23: **@VLOOKUP($B$23,$B$13..$D$16,1)**
E23: **@VLOOKUP($B$23,$B$13..$D$16,2)**
G23: **+D23+(B23−C23)*E23**
B24: **+F8**
C24: **@VLOOKUP($B$24,$B$13..$D$16,0)**
D24: **@VLOOKUP($B$24,$B$13..$D$16,1)**
E24: **@VLOOKUP($B$24,$B$13..$D$16,2)**
G24: **+D24+(B24−C24)*E24**

The formulae in cells G23 and G24 shown above represent the interpolation equation developed previously in the section dealing with tables. In row 30 of Sheet 1.1 an example is given of the stress calculation for an arbitrary strain value. The formulae required in the cells of Row 30 are similar to those given above for rows 23 and 24.

It must be stressed that this particular spreadsheet which outputs each of the required @VLOOKUP functions in rows C, D and E was designed for illustrative purposes. The actual spreadsheet required to output the stress for any given strain would be much simpler. It would consist of rows 1 to 16, the value of the strain in cell B30 and the resulting stress in cell G30 which could be calculated using the full interpolation formula without the intermediate steps taken in C30 to E30.

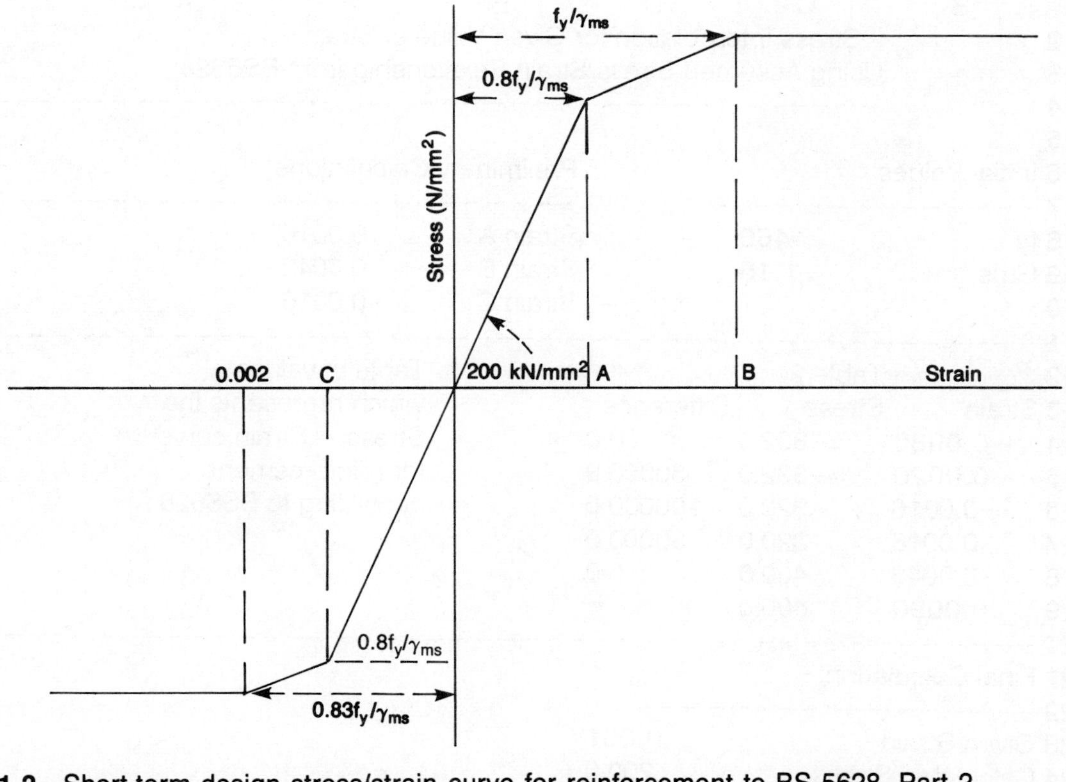

Figure 1.2 Short-term design stress/strain curve for reinforcement to BS 5628, Part 2

Case B: trilinear analysis. The stress/strain relationship assumed for the reinforcement in the current British Standard code for reinforced masonry, is as shown in Fig. 1.2 (BS 5628, *Code of Practice for the Use of Masonry, Part 2: Structural Use of Reinforced and Prestressed Masonry*).

In order to produce a table for this graph the values of the strains at A, B and C must be determined. Using the values supplied in the BS code for reinforced masonry the expressions for these strains are:

Strain at A:

$$\epsilon_A = 0.8 \times f_y/(200 \times 1000 \times \gamma_{ms})$$

Strain at B:

$$\epsilon_B = \epsilon_A + 0.2 \times (0.002 - \epsilon_A)/0.03$$

Strain at C:

$$\epsilon_C = -\epsilon_A$$

So that the table of BS 5628 reinforcement values can be written as in Table 1.3.

The table has been extended on either side using arbitrary values of the strains of −0.003 and 0.009 in order to allow for strains less than −0.002 and greater than $+\epsilon_B$.

The design of a spreadsheet which uses this table in order to determine the value of the stress for any given value of the strain is illustrated in Sheet 1.2.

The initial values of f_y and γ_{ms}, the characteristic strength and partial safety factor of the reinforcing steel, are placed in cells C8 and C9 respectively and these values are then used to calculate the strains at A, B and C in cells F8, F9 and F10. The values in C8 and C9 can be changed as required to allow for other types of steel. The formulae placed in cells F8 to F10 are:

F8: **0.8*C8/(200*1000*C9)** (the strain at A)
F9: **+F8+0.2*(0.002−F8)/0.03** (the strain at B)
F10: **−0.8*C8/(200*1000*C9)** (the strain at C)

The values from Table 1.3 together with the difference ratio required for interpolation are entered in block B14 to D19. Some typical examples of the formulae placed in this block are:

B14: **−0.003** (the arbitrary strain)
C14: **−0.83*C8/C9** (the stress due to arbitrary strain)
D14: **(C15−C14)/(B15−B14)** (the difference ratio)
B15: **−0.002** (the strain at changeover (Fig. 1.2))
C15: **−0.83*C8/C9** (the stress at C)
D15: **(C16−C15)/(B16−B15)** (difference ratio)
B16: **+F10** (the strain at A)
C16: **−0.8/C8/C9** (the stress at A)
D16: **(C17−C16)/(B17−B16)** (difference ratio)

Table 1.3

B	*C*
STRAIN	STRESS
−0.003	$-0.83^*f_y/\gamma_{ms}$
−0.002	$-0.83^*f_y/\gamma_{ms}$
ϵ_C	$-0.8^*f_y/\gamma_{ms}$
ϵ_A	$0.8^*f_y/\gamma_{ms}$
ϵ_B	f_y/γ_{ms}
0.009	$0.83^*f_y/\gamma_{ms}$

Sheet 1.2

```
1/A    B              C           D          E              F              G        H
 2            Stress Interpolation for Given Value of Strain
 3          Using Assumed Stress/Strain Relationship from BS5628
 4          ----------------------------------------------------
 5
 6 Initial Values                          Preliminary Calculations
 7 ---------------------                   -------------------------
 8 fy                     460              Strain A       0.0016
 9 Gms                    1.15             Strain B       0.0043
10                                         Strain C      -0.0016
11 -----------------------------------------------------------------
12               Table 2                            Table of values
13 Strain        Stress       Difference            which represents the
14   -0.0030     -332.0           0.0               Stress - Strain curve
15   -0.0020     -332.0       30000.0               for reinforcement
16   -0.0016     -320.0      200000.0               according to BS5628
17    0.0016      320.0       30000.0
18    0.0043      400.0           0.0
19    0.0090      400.0
20 -----------------------------------------------------------------
21 Final Calculation
22 ---------------------------
23 Given Strain                 0.001
24 Calculated Stress            200.0
```

The final calculation is performed by the formula in cell D24. Given the value of strain placed in cell D23 then the formula in cell D24 interpolates the value of the strain using the values from Table 1.3 and the @VLOOKUP functions given previously. The contents of cells D23 and D24 are:

D23: **0.001** (the given value of the strain)
D24: **@VLOOKUP(D23,B14..D19,1)**
+(D23 − @VLOOKUP(D23,B14..D19,0))
***@VLOOKUP(D23,B14..D19,2)**

Note that the intermediate stages of the calculation have been omitted for Case B and the stress calculated directly. Other methods of determining the stress are possible, for example using a nested @IF function, and examples of the use of these will be given later.

1.14 Solution of equations

The sets of equations derived to explain the behaviour of certain elements under a given load system usually have no simple direct solution. This arises, not from any complexity in the equations themselves, but from additional relationships that exist between the unknowns. For example the equation for beam bending can be expressed in terms of two functions (F_1,F_2) of a number of parameters:

$$M = F_1(b,d,A,f_y,f_m,\gamma_{ms},\gamma_{mm}) \qquad [1.2]$$

$$M = F_2(b,d,f_m,\gamma_{mm}) \qquad [1.3]$$

where:

b,d are the breadth and depth of the section;
A is the area or reinforcement;
f_y is the characteristic strength of the steel;
f_m is the characteristic strength of concrete or masonry;
γ_{ms}, γ_{mm} are partial safety factors.

Usually Eqn 1.3 is used as a limiting equation to ensure that in the limit failure would be by compression of the concrete or masonry rather than tensile failure of the steel. The solution must therefore be obtained from Eqn 1.2 and this solution checked using Eqn 1.3, to ensure that it is within the limit. In practice both A and f_m are unknown and suitable values that satisfy the equations must be determined by trial. Design curves to simplify this calculation have been produced (Kong and Evans, 1987 and Hendry, *et al.*, 1987).

The spreadsheet is ideal for this type of problem since the limiting equation can be expressed as a condition in an @IF function and the effect of changing a variable can be studied by solving for a range of values. It is therefore possible to produce more economical sections by selecting variables that result in solutions which are closer to the limiting values.

Additionally, since the spreadsheet solution is independent of the assumed values of the parameters it is found that the solution can be applied to both symmetrical and unsymmetrical sections. In Hendry (1991) a set of charts is provided for the direct solution of columns subjected to uniaxial bending. These charts only apply to symmetrical sections whereas the spreadsheet solution applies also to unsymmetrical sections.

The values required to plot the design curves in Hendry, *et al.* (1987) and the charts in Hendry (1991) were originally produced by the author using a spreadsheet. In the following chapters examples of these spreadsheets for the solution of a number of different cases will be given.

Many spreadsheet packages also contain commands for multiplying and inverting matrices and these can be used directly to solve simultaneous equations. If the coefficients of the unknowns are placed in the [A] matrix, N rows by N columns, and the constants placed in the {B} vector, N rows by 1 column, then the equations can be expressed by:

$$[A]\{X\} = \{B\}$$

where {X} represents the vector of the unknowns, N rows by 1 column. The solution to these equations is given by:

$$\{X\} = [A^{-1}]\{B\}$$

where $[A^{-1}]$ is the inverse matrix of [A].

Thus to obtain a solution of the equations the inverse command of the spreadsheet is used to produce $[A^{-1}]$ and this inverse multiplied by {B} to obtain {X}.

The example in Sheet 1.3 shows a layout for obtaining the solution of the five equations with five unknowns shown in rows 4 to 8.

The [A] matrix has been entered in block B12 to F16 and the {B} vector in the range H12 to H16. The inverse of the A matrix has been output to the block B20 to F24, using the inverse command, and the {B} vector premultiplied by $[A^{-1}]$ to obtain the solution shown in I20 to I24.

Note that the layout of Sheet 1.3 is such that the inverse and multiplication operations used are not automatic so that for each new set of equations the operations would have to be repeated by the user. It will be shown in Chapters 7 and 8 how the required matrix operations can be included in the spreadsheet layout as an automatic process so that user interaction with the spreadsheet is reduced.

Sheet 1.3

1/A	B	C	D	E	F	G	H	I
2		Solution of Equations						
3		------------------------------						
4	4x1 + 3x2 + 2x3 + 5x5 = 0							
5	3x1 + 3x2 − 1x3 − 4x4 + 2x5 = −10							
6	2x1 − 1x2 + 2x3 + 3x4 + 6x5 = 20							
7	−4x2 + 3x3 + 1x4 − 5x5 = 24							
8	5x1 + 2x2 + 6x3 − 5x4 + 4x5 = 0							
9	--							
10		A Matrix					B Matrix	
11		----------					----------	
12	4	3	2	0	5		0	
13	3	3	−1	−4	2		−10	
14	2	−1	2	3	6		20	
15	0	−4	3	1	−5		24	
16	5	2	6	−5	4		0	
17								
18		Inverse Matrix					Results	
19		--------------------					----------	
20	0.16	0.33	0.09	0.27	−0.16		x1 =	4.79
21	0.33	−0.25	−0.29	−0.12	0.00		x2 =	−6.11
22	0.09	−0.29	−0.12	−0.05	0.15		x3 =	−0.59
23	0.27	−0.12	−0.05	0.06	−0.12		x4 =	1.69
24	−0.16	0.00	0.15	−0.12	0.06		x5 =	0.07

2 Illustrative examples

2.1 Introduction

In this chapter some further examples are given in order to illustrate some of the techniques that can be used during the design of a spreadsheet. Some of these techniques are used simply to improve the appearance of the spreadsheet whilst others play a more essential role in that they assist in obtaining more direct solutions. Later chapters refer to the design of particular elements and, rather than interrupt the thread of the argument with a discussion on techniques, it is convenient to collect descriptions of the problems that arise into a single chapter and to outline here the method used for obtaining solutions by the use of spreadsheets.

2.2 Example 1: suppression of unwanted output

For many engineering designs some of the initial data is first grouped into a set of limiting parameters and, depending on the magnitude of these parameters, a decision made as to whether the design is to proceed along one path or another. A typical example of this occurs with the design of reinforced concrete beams. If certain parameters are less than some limiting factor then the design path is for a singly reinforced beam, if greater than the limiting factor then design is for a doubly reinforced section. The design process is often one of deciding which of alternative paths must be followed and there are a number of examples of the use of this technique throughout the rest of this text.

An improvement in the appearance of the spreadsheet can be achieved by ensuring that only the relevant design path appears on the spreadsheet. If the limiting factors are such that design must be in accordance with a set of equations B, rather than set A, then it is better to suppress spreadsheet output which relates to design A and to show only the output for design B. To achieve this improvement in appearance use is made of the null string " " to produce a blank cell or "---" to produce a broken line. This can be applied not only to the numerical calculations but also the labels which define them.

Consider the following example in which 'Path 1' design is to be followed if 'Limit' is ≤ 10 and 'Path 2' design if 'Limit' is > 10. If no cosmetic supression is in place then part of the spreadsheet would appear as:

A	B	C	D	E
1	**Limit**	**9**		
2				
3				
4	**Path_1**		**Path_2**	
5	**Calc_1**	**22**	**Calc_1**	**46**
6	**Calc_2**	**25**	**Calc_2**	**55**

However since 'Limit' is < 10 and this information has already been entered, or been calculated, in the spreadsheet all that is required are the details of Path_1 design and information relating to Path_2 can be suppressed as shown below:

A	B	C	D	E
1	**Limit**	**9**		
2				
3				
4	**Path_1**			
5	**Calc_1**	**22**		
6	**Calc_2**	**25**		

Examples of some of the formulae required to achieve this are:

B4: @IF(C1>10," ","Path_1")
D4: @IF(C1>10,"Path_2"," ")
B5: @IF(C1>10," ","Calc_1")
C5: @IF(C1>10," ",Insert the formula for Calc_1)

Only one design path appears on the spreadsheet. A similar approach can be used for designs with any number of paths.

Note that if the same principle is used for the value calculated in C1 so that the output that appears in cell C1 could be " " then the method will not work for earlier versions of some spreadsheet packages. This occurs in Sheet 3.2 (cells G25 to G27) and Sheet 8.1 (cells D65 to D69). For example the formula used in cell D69 in Sheet 8.1 could be written as either Case 1:

D69: @IF(H11>0,5," ")

or Case 2:

D69: @IF(H11>0,5,0)

For the first case, if the condition is false, the null string would be output whilst for the second case a zero value would appear. However since the formula in cell E69 uses the value in D69 as a limiting condition it will be found that the Case 1 formulation will lead to an error in cell E69 if a null string appears in cell D69. This difficulty does not arise in later versions of most software packages.

2.3 Example 2: multiple 'trial and error' solutions

A standard approach for starting the design process for many engineering problems is to assume a trial value for some unknown quantity and then check if the subsequent calculations, based on this trial value, satisfy the limiting equations. If the limits are not satisfied then a new trial value must be tried until a satisfactory solution is found. A typical example of this occurs in the design of reinforced concrete beams where a trial value is assumed for the depth of concrete in compression.

With a spreadsheet solution many trial values can be used at the same time so that a range of solutions is available to the designer who must then select the one which is most appropriate. Usually this selection process is based on the calculation of the out-of-balance resulting from a known function using a series of trial values and then determining the trial value that results in a zero out-of-balance value of the function. Practical examples of the use of this process are given in Chapters 3 and 6 using different methods.

Consider the following example of trial values of x and the out-of-balance that results from a known function which should be zero for the correct value of x.

	Trial Value (x)	*Out-of-Balance*
	A	B
1	0.0	10.0
2	0.2	8.0
3	0.4	4.0
4	0.6	−2.0
5	0.8	−6.0
6	1.0	−8.0

The value of the out-of-balance changes from positive to negative and an aproximate value of the trial value, which would result in a near zero value of the out-of-balance, can be obtained by interpolation. As shown in Fig. 2.1 the required value of x can be determined graphically.

From Fig. 2.1:

$$x = [\{4/(4-(-2))\}(0.6-0.4)] + 0.4$$
$$= 0.5333$$

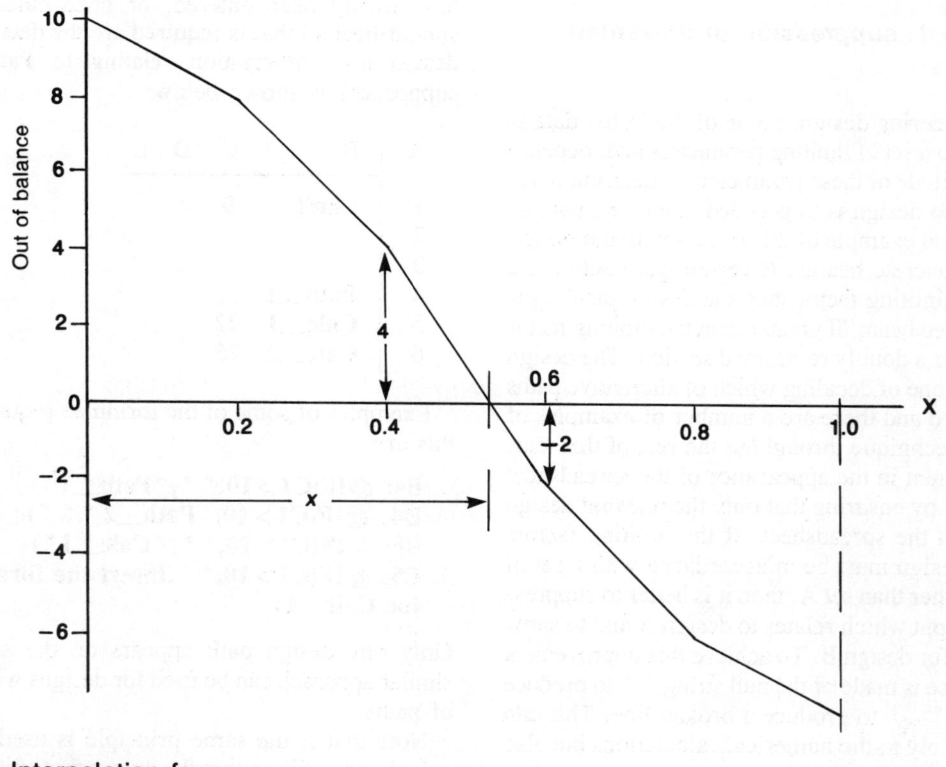

Figure 2.1 Interpolation for x

Although this simple interpolation is fairly straightforward some additional factors must be considered when it is applied to a spreadsheet solution. These are:

(a) The out-of-balance function may be nonlinear, in which case the value found for x will only be approximate.
(b) The location of the cells where the out-of-balance changes sign have to be found. Visually this is easy but in the spreadsheet a formula is required to define the point of changeover.
(c) Having located the cells where the changeover occurs it is necessary to extract the four key values required for the interpolation. In the above example these key values are 0.4 in A3, 0.6 in A4, 4 in B3 and −2 in B4.

A convenient method for determining the four key values that are required for the interpolation is to add four extra columns as shown below:

Trial Value (x) *Out-of-Balance*

	A	B	C	D	E	F
1	**0.0**	**10.0**	**0.0**		**0.0**	
2	**0.2**	**8.0**	**0.0**	**0.0**	**0.0**	**0.0**
3	**0.4**	**4.0**	**4.0**	**0.0**	**0.4**	**0.0**
4	**0.6**	**−2.0**	**0.0**	**−2.0**	**0.0**	**0.6**
5	**0.8**	**−6.0**	**0.0**	**0.0**	**0.0**	**0.0**
6	**1.0**	**−8.0**		**0.0**	**0.0**	**0.0**

The formulae entered in these extra columns are of the form:

C1: **@IF(B1>0#AND#B2<0,B1,0)**
D2: **@IF(B2<0#AND#B1>0,B2,0)**
E1: **@IF(C1<>0,A1,0)**
F2: **@IF(D2<>0,A2,0)**

In columns C and D a check is first made to determine if a change in sign has occurred and if not output 0, otherwise output the value of the out-of-balance. In column E (or F), a check is first made to determine if the value in cell C (or D) is zero and if not output the value of x found in that row. Cells D1, F1 and C6 are left blank since their inclusion would be meaningless. Note that if equal intervals of x are used then only three columns are required since the second value of x is known once the first has been determined. Note also that the above example assumes that the out-of-balance value runs from + to − and the formulae have been entered using this assumption. If the out-of-balance runs from − to + then the > and < signs in the formulae in columns C and D would have to be reversed.

To gather the interpolation factors from the four columns the @SUM function can be used. For example **@SUM(C1..C6)** would result in the value 4.0 and the summation of the other columns the remaining interpolation factors.

The error which results when the out-of-balance function is non-linear can be reduced by use of two additional rows as shown below (rows 7 and 8):

Trial Value (x) *Out-of-Balance*

	A	B	C	D	E	F
1	**0.0**	**10.0**	**0.0**		**0.0**	
2	**0.2**	**8.0**	**0.0**	**0.0**	**0.0**	**0.0**
3	**0.4**	**4.0**	**4.0**	**0.0**	**0.4**	**0.0**
4	**0.6**	**−2.0**	**0.0**	**−2.0**	**0.0**	**0.6**
5	**0.8**	**−6.0**	**0.0**	**0.0**	**0.0**	**0.0**
6	**1.0**	**−8.0**		**0.0**	**0.0**	**0.0**
7	**0.533**	**−0.03**	**4.0**	**−2.0**	**0.4**	**0.6**
8	**0.512(Trial)**	**0.00**				

In cells C7 to F7 the formulae for the summation of each column defines the interpolation factors and these factors are used in cell A7 which contains the interpolation formula. Cell B7 is the out-of-balance which results from the use of the interpolated value of x calculated in cell A7, and for the present example it is assumed that the out-of-balance is −0.03. Using the values in A7 and B7 as a guide, new trial values are input to A8 until the resulting out-of-balance is zero. With this method the user *must* enter a new trial value in cell A8 because all subsequent calculations are based on the value of x in A8 and not the value in A7. The method is quite effective since only a small number of trials are usually required.

The above method is used in Chapter 6 whilst an alternative method, using @VLOOKUP functions, is used in Chapter 3.

2.4 Example 3: controlling the start/end values of a series

It is sometimes necessary to control the start value of a series so that the subsequent calculations which use the series are within a particular range. An example of this occurs in Chapter 4 where the range of the series for a parameter (*dc/t*) should be selected so that the plotted values of coordinates M,N all lie in the first quadrant. Without such a control some of the coordinates, calculated from the series, might be negative so that the curve would continue into a negative quadrant.

In the examples that arise in this text this control has been achieved by using a trial value to start the series and then using this trial value to continue the series. For example if cells A2 to A4 contain formulae of the type:

A2: **+A1+0.1**
A3: **+A2+0.1**

where the value 0.1 has been selected as the series interval, then the values of the series will depend on the trial value placed in cell A1.

If the values of the series, calculated from this initial trial value, do not satisfy the required conditions then a new trial value is selected and the conditions rechecked until they are satisfied.

Controlling the end of a series can usually be achieved by use of the @MAX or @MIN functions. Using defined upper, or lower, limits of a parameter these functions can be used to control output to cells. Combining this technique with that described in Example 1 above, output to a cell can either be a value within the required range or a blank, represented by " ".

2.5 Example 4: other useful cosmetic techniques

The design of a spreadsheet layout for the solution of a particular problem can usually be achieved in a number of ways for there are quite a number of choices available to the designer both in the way a particular equation is solved and also in the way the layout is structured. If the spreadsheet is to be used by others then the descriptive labels would have to reflect this, but even for personal layouts, there should always be sufficient information for the spreadsheet to be understandable after a lapse of time. Fortunately descriptive notes can be inserted at any point in the spreadsheet and it is usually worth taking the extra time to do this.

Other techniques which have been used in this text are summarised below:

(a) A comprehensive range of graphical facilities are available and these can sometimes be incorporated into the spreadsheet design process. An example of this technique occurs in Chapter 4, with reinforced column design, where the interaction diagram is used to decide on the acceptability of the design parameters.

(b) Some spreadsheet options require additional input from the user so that, in addition to inserting the initial design values, the user has to interact with the output and respond to a set of questions. This occurs for example with operations on databases and matrices. This interruption to the flow affects the elegance of the solution and it may be more convenient not to use the built-in spreadsheet commands but to achieve the same result using direct spreadsheet solutions requiring no additional user input. An example of this technique is given in Chapter 7 for matrix multiplication and inversion.

(c) Later versions of some spreadsheet packages contain excellent drawing facilities and these can be used to output diagrams of relevant parts, such as the cross-section and location of reinforcement, as the design proceeds. Such diagrams help to identify errors and to show if any changes are needed. A variation of this technique is used in Chapter 4 where a diagram of a rectangular column cross-section is included in the spreadsheet so that the user is aware of the location and diameters of bars being used for the current design.

2.6 Example 5: triple interpolation from tabular values

In Example 2 above it was shown how an interpolated value could be determined for a function of one variable x. In Chapter 9 an interpolation will be required for a function of three variables and it is convenient, at this stage, to describe the mechanics for achieving this using a spreadsheet. It is assumed that the equation can be expressed as $y = F(u,v,w)$ and that tabular values of y are available in terms of values of u, v and w. The problem is one of determining y, from the tables, for any given values u, v and w. Note that the actual notation used below is different from the simplified version in this paragraph.

In Wood's design method for Framed Panels (Wood, 1978), a correction factor Δ_ϕ must be determined using a design chart supplied, see Fig. 2.2. Basically the design chart is a plot of Δ_ϕ, along the Y-axis, against m_e along the X-axis and if there was only one curve on the chart then interpolation for different values of m_e would be straightforward and in line with the procedure outlined in Chapter 1. However this is not the case and it is found that the relationship between Δ_ϕ and m_e also depends on two other parameters μ_p and L/h. This is shown in Fig.

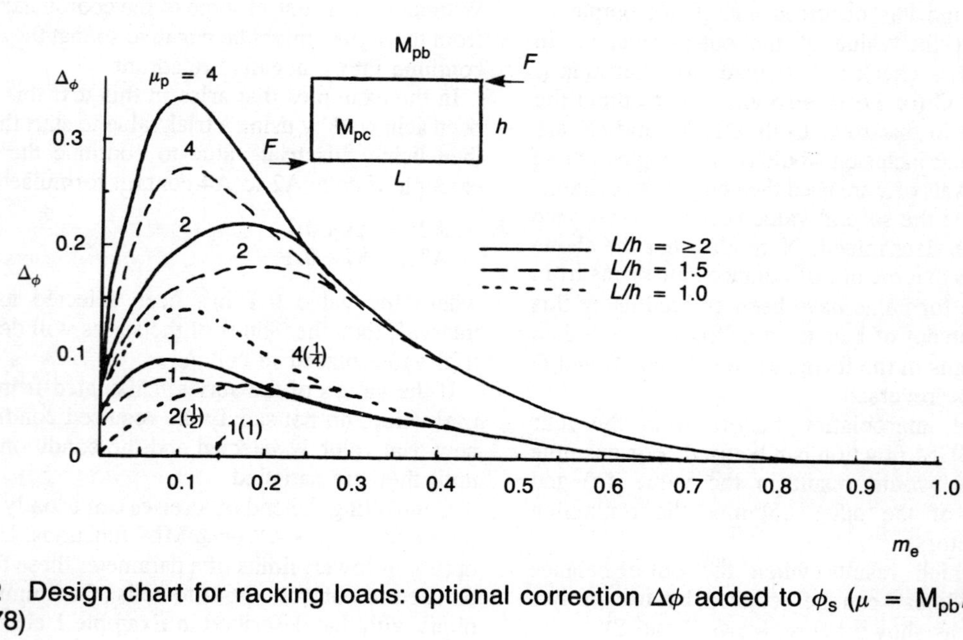

Figure 2.2 Design chart for racking loads: optional correction $\Delta\phi$ added to ϕ_s ($\mu = M_{pb}/M_{pc}$) from Wood (1978)

2.2 which shows that there are a number of curves related to the values of μ_p and L/h.

Although only eight curves are shown in Fig. 2.2 three more are implied because two of the curves represent a duplicate set of values of the parameters and also the X axis represents the curve for $\mu_p = 1$ and $L/h = 1.0$. The conditions set by Wood for use of the charts are:

(a) If $\mu_p \geq 1$ then use the chart directly.
(b) If $\mu_p < 1$ and $L/h = 1$ then use μ_p values in brackets.
(c) If $\mu_p < 1$ and $L/h > 1$ then use $\mu_p = 1$ curve.

The problem can be stated quite simply by: 'given the values of three parameters μ_p, L/h and m_e, find the value of Δ_ϕ'.

The solution to this problem is quite easy if a curve exists with the same values as the given values of the parameters but becomes more difficult if the value of Δ_ϕ has to be determined by interpolating between curves.

For example if $m_e = 0.2$, $L/h = 1.5$ and $\mu_p = 2$ then it is fairly simple to read off the value of Δ_ϕ from the Y axis because a curve exists for these values of L/h and μ_p.

However if $m_e = 0.25$, $L/h = 1.75$ and $\mu_p = 3$ then the designer would have to locate the required position along the Y axis by first interpolating between the curves representing $\mu_p = 4$ and 2 in order to obtain the $\mu_p = 3$ set of curves and then interpolate between $L/h = 1.5$ and 2 for the final curve to be used for the given value of m_e.

Direct use of the design chart for these interpolations is not always straightforward but it can be shown that by use of a spreadsheet a direct solution can be obtained for any appropriate set of m_e, μ_p and L/h values.

2.7 Use of spreadsheet for design chart interpolation

A spreadsheet which enables the designer to output the value of Δ_ϕ from any given values of the three parameters μ_p, L/h and m_e is shown in Sheet 2.1.

Initial values

The initial values are the given values of μ_p, L/h and m_e and these are entered in cells C6 to C8. Note that although the spreadsheet shows the final interpolated value of Δ_ϕ in cell G7 this is placed in this location only for convenience and is a copy of the final result for Δ_ϕ calculated at the end of the spreadsheet in cell D77.

Tabular values of Δ_ϕ against m_e

From the given design chart it is first necessary to determine the values of the coordinates of the curves and to achieve this use can be made of the graphical facility of the spreadsheet. As a first step it is usually convenient to enlarge the diagram, using a photocopier, and to then draw a suitable grid on the enlargement so that values of the coordinates can be read for each curve. These coordinates will be approximate but can be corrected by plotting the resulting curves and smoothing out any obvious kinks.

The ordinates along the X axis are chosen so that they reflect the changes in the shape of the curves so that the intervals are closer together for rapid changes and further apart for linear portions. For this particular problem the X axis starts with intervals of 0.025 up to a value of m_e equal to 0.2 because of the rapid changes in the Y ordinate over this portion. This interval increases, first to 0.05 and later to 0.1, as the curves become more linear.

Only a portion of the tabular values are shown on Sheet 2.1. The actual spreadsheet contains values of Δ_ϕ for three values of L/h (1, 1.5 and ≥ 2.0), and five values of μ_p, ($\frac{1}{4}$, $\frac{1}{2}$, 1, 2, 4). This represents 15 columns of which only the first six are shown on Sheet 2.1.

Interpolation for m_e

In Chapter 1 it was shown that in order to interpolate between specified values it was convenient to set up difference columns which represented the appropriate ratios of the Y intervals to the X intervals. This procedure becomes cumbersome if the table consists of a large number of columns, as in the present example, since a difference column would have to be prepared for each column. To overcome this an alternative approach is possible.

In general the initial value of m_e will be between two specified values in the first column of the table and therefore only the two rows which contain these values are of importance for the purpose of interpolation. Generally, one of the tabular values of m_e will be lower than the initial value and the other higher. The @VLOOKUP function is used to locate the row containing the lower value but to locate the row containing the next higher value requires some additional information. This can be obtained from the fact that the interval between consecutive tabular values of m_e is known and that increasing the initial value of m_e by a value equal to the interval would place the new value of m_e with the next interval. This is illustrated in Fig. 2.3.

Figure 2.3 illustrates the above principle that if the initial value of m_e is in the interval between A and B then adding the value of the interval results in a new value of m_e placed within the next interval (B – C). Applying appropriate @VLOOKUP functions to these two values of m_e would locate A and B. Allowance has to be made for the fact that the interval between consecutive values of m_e in the table is not constant.

The formulae required for this operation are entered in cells E36 and E37:

E36: **+C8** (the given value of m_e)
E37: **@IF($E$36>=0#AND#$E$36<0.2,**
$E$36+0.025001,
@IF($E$36>=0.2#AND#$E$36<0.4,
$E$36+0.05001,$E$36+0.1001))

where the nested @IF's allow for the change in the interval values of m_e. Note that the interval has been increased slightly to allow for the case when the given

Sheet 2.1 (Page 1)

1/A	B	C	D	E	F	G	H
2	Design Chart Interpolation for Framed Panels						
3	--						
4 Initial Values							
5 -------------							
6 mup		3.00		Interpolated Value of			
7 L/h		1.75		DeltaPhi =		0.201	
8 me		0.23					
9 --							
10	Tabular Values of DeltaPhi against 'me'						
11		For range of L/h and 'mup'					
12	------------------------------------						
13	L/h	1.0	1.5	>2.0	1.0	1.5	>2.0
14	mup	1/4	1/4	1/4	1/2	1/2	1/2
15	me		DeltaPhi				
16	0.000	0.000	0.000	0.000	0.000	0.000	0.000
17	0.025	0.100	0.060	0.070	0.025	0.060	0.070
18	0.050	0.120	0.070	0.080	0.040	0.070	0.080
19	0.075	0.130	0.075	0.090	0.050	0.075	0.090
20	0.100	0.140	0.080	0.095	0.060	0.080	0.095
21	0.125	0.135	0.080	0.092	0.065	0.080	0.092
22	0.150	0.130	0.078	0.085	0.070	0.078	0.085
23	0.175	0.120	0.073	0.081	0.077	0.073	0.081
24	0.200	0.105	0.070	0.075	0.075	0.070	0.075
25	0.250	0.085	0.065	0.068	0.073	0.065	0.068
26	0.300	0.065	0.060	0.060	0.065	0.060	0.060
27	0.350	0.055	0.055	0.055	0.055	0.055	0.055
28	0.400	0.044	0.044	0.044	0.044	0.044	0.044
29	0.500	0.025	0.025	0.025	0.025	0.025	0.025
30	0.600	0.013	0.013	0.013	0.013	0.013	0.013
31	0.700	0.000	0.000	0.000	0.000	0.000	0.000
32	0.800	0.000	0.000	0.000	0.000	0.000	0.000
33	0.900	0.000	0.000	0.000	0.000	0.000	0.000
34	1.000	0.000	0.000	0.000	0.000	0.000	0.000
35 --							
36		Value of me =		0.230			
37 Increased value of me =				0.280			
38							
39		Selected Set of above Tabular Values					
40		----------------------------------					
41	L/h	1	1.5	2	1	1.5	2
42	mup	0.25	0.25	0.25	0.5	0.5	0.5
43	0	1	2	3	4	5	6
44	me		DeltaPhi				
45	0.200	0.105	0.070	0.075	0.075	0.070	0.075
46	0.250	0.085	0.065	0.068	0.073	0.065	0.068
47	0.230	0.093	0.067	0.071	0.074	0.067	0.071
48 --------							

Sheet 2.1 (Page 2)

1/A	B	C	D	E	F	G	H
55							
56	Tabular Values of DeltaPhi against 'mup'						
57	For range of 'L/h' and given 'me'						
58	--						
59		L/h	1	1.5	2	Diff1	Diff1.5
60		mup	DeltaPhi				
61		0.25	0.093	0.067	0.071	−0.077	0.000
62		0.5	0.074	0.067	0.071	−0.148	0.000
63		1	0.000	0.067	0.071	0.074	0.107
64		2	0.074	0.174	0.208	0.010	0.011
65		4	0.093	0.196	0.224	0.000	0.000
66		5	0.093	0.196	0.224		
67	mup =	3.000	0.083	0.185	0.216		
68	--						
69	Tabular Values of DeltaPhi against 'L/h'						
70	For given 'mup' and 'me'						
71	--						
72		L/h	DeltaPhi	Diff			
73		1	0.083	0.203			
74		1.5	0.185	0.062			
75		2	0.216	0.000			
76		4	0.216				
77	L/h =	1.75	0.201				
78	--						
79	Interpolated Value of DeltaPhi =					0.201	
80	--						
81							

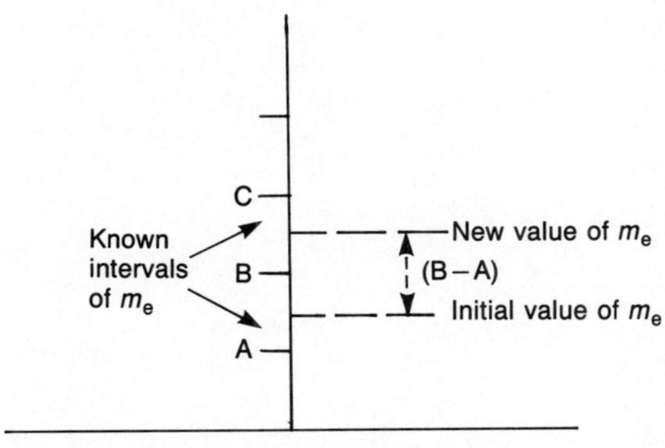

Figure 2.3 Illustration of tubular entry values

value of m_e is at one of the tabular values since rounding can lead to errors for this case.

Rows 41 and 42 are copies of rows 13 and 14. Row 43 contains a counter which is required in subsequent @VLOOKUP functions to identify the columns. The values in cells E36 and E37 can now be used to output the two rows of the table which embrace the initial value of m_e, namely rows 45 and 46. The interpolated values for each column of the table and for the intial value of m_e, are calculated in row 47. Examples of the formulae placed in the cells of rows 45, 46 and 47 are:

B45: @VLOOKUP($C$8,$B$16..$R$34,B$43)
C45: @VLOOKUP($C$8,$B$16..$R$34,C$43)
B46: @VLOOKUP($E$37,$B$16..$R$34,B$43)
C46: @VLOOKUP($E$37,$B$16..$R$34,C$43)
B47: +C8
C47: +C$45+(($B$47−$B$45)*(C$46−C$45)/($B$46−$B$45))

This method results in values of Δ_ϕ for each of the specified values of μ_p and *L/h* in the design chart and for the given value of m_e. It avoids the necessity for preparing columns showing the difference ratios and is quite useful when there are a large number of columns.

Interpolation for μ_p

To continue with the calculation it is now necessary to interpolate for the given value of μ_p using the values of Δ_ϕ previously calculated for the given m_e in row 47.

A table is prepared for the three values of *L/h* (1, 1.5, 2), and the five values of μ_p (0.25, 0.5, 1, 2, 4). An extra value of $\mu_p = 5$, using values of Δ_ϕ equal to those for $\mu_p = 4$, is included in the table since this enables the difference ratios to be calculated for $\mu_p = 4$.

Cells C61 to C66 contain the values of μ_p and cells D59 to F59 the values of *L/h*. The values of Δ_ϕ for the given value of m_e are placed in the block D61 to F66 and the relevant difference ratios in block G61 to I65. The I column is not shown on Sheet 2.1. Examples of the formulae in these cells are:

D61: **+C47**
E61: **+D47**
F61: **+E47**
G61: **(D62 – D61)/($C62 – $C61)**
H61: **(E62 – E61)/($C62 – $C61)**
D62: **+F47**
E62: **+G47**
F62: **+H47**
G62: **(D63 – D62)/($C63 – $C62)**
H62: **(E63 – E62)/($C63 – $C62)**

The given value of μ_p can now be inserted in cell C67 and the values of Δ_ϕ calculated in cells D67 to F67 for the three values of *L/h*. The formulae in these cells are:

C67: **+C6** (the value of μ_p)
D67: **@VLOOKUP($C$67,$C$61..$I$65,1)+ ($C$67 – (@VLOOKUP($C$67,SC$61..$I$65,0))) *(@VLOOKUP($C$67,$C$61..$I$65,4))**
E67: **@VLOOKUP($C$67,$C$61..$I$65,2)+ ($C$67 – (@VLOOKUP($C$67,SC$61..$I$65,0))) *(@VLOOKUP($C$67,$C$61..$I$65,5))**
F67: **@VLOOKUP($C$67,$C$61..$I$65,3)+ ($C$67 – (@VLOOKUP($C$67,SC$61..$I$65,0))) *(@VLOOKUP($C$67,$C$61..$I$65,6))**

so that row 67 contains values of Δ_ϕ for the given values of m_e and μ_p and for the three values of *L/h*.

Interpolation for *L/h*

Following a similar procedure to that outlined in the previous section, values of *L/h* are placed in cells C73 to C75 and the corresponding values of Δ_ϕ from row 67 placed in cells D73 to D75. An extra value of $L/h = 4$, using values of Δ_ϕ equal to those for $L/h = 2$, is included in the table in cell C76 since this enables the difference ratios to be calculated for $L/h = 2$.

Examples of the formulae entered in the block C72 to E75 are:

D73: **+D67**
E73: **(D74 – D73)/($C74 – $C73)**
D74: **+E67**
E74: **(D75 – D74)/($C75 – $C74)**
D76: **+D75**

Finally the given value of *L/h* is entered in cell C77 and the value of Δ_ϕ calculated from the table in cell D77. These cells contain:

C77: **+C7**
D77: **@VLOOKUP($C$77,$C$73..$E$76,1)+ ($C$77 – (@VLOOKUP($C$77,SC$73..$E$76,0))) *(@VLOOKUP($C$77,$C$73..$E$76,2))**

The value shown in cell D77 represents the value of Δ_ϕ for the given values of μ_p, *L/h* and m_e by interpolation from the original design curves. This is the value copied to cell G7, at the start of the spreadsheet, for ease of reference.

3
Design of reinforced beams

3.1 Introduction

The design of reinforced beams, of either concrete or masonry, can be based on either a linear elastic theory or an ultimate limit state theory but only the latter will be considered in the examples described in this chapter.

Reinforced concrete beams are made up of two materials and the load carrying capacity is based on the interaction and the stress/strain properties of these materials, concrete and the steel reinforcement. It is generally assumed that the stress/strain relationship for the reinforcement can be idealised into the bilinear form shown in Fig. 1.1

Reinforced masonry beams are made up of three materials and the load carrying capacity is based on the interaction and the stress/strain properties of these materials, masonry, steel reinforcement, and the concrete grout surrounding the bars. It is generally assumed that the grout has the same properties as the masonry and also that the stress/strain relationship for the reinforcement can be idealised into the trilinear form shown in Fig. 1.2.

If design is to be in accordance with BS 5628 then there is no requirement to define the stress/strain relationship for masonry since the equations that define the behaviour of the beam in flexure are given in the BS code. If however the designer wished to work from first principles then the stress/strain relationship would have to be defined. The relationship might be considered as parabolic or simplified even further by idealising the parabola into an equivalent rectangle.

Although the direct use of the equations given in the BS code represents the simplest design approach it will be shown in this chapter that the spreadsheet solution can incorporate designs which are based on alternative forms of the stress/strain relationship. It will be shown that this can be achieved without much increase in the required input or in the complexity of the solution. A comparison can therefore be made between the results obtained by using the assumptions of the code with those assuming some other form of stress distribution.

REINFORCED CONCRETE

3.2 Design for flexure in concrete, rectangular sections

Reinforced concrete beams may be either singly or doubly reinforced and any design method must be capable of dealing with both cases. The notation used in the case of a doubly reinforced beam is illustrated in Fig. 3.1.

The failure mode in bending may be either by yielding of the tensile reinforcement or by crushing of the concrete in compression. Designers usually prefer the first case since the extensive yielding, which results in large deflections and cracking, gives a warning of failure whilst failure by crushing does not. The first case is referred to as under-reinforced and the second as over-reinforced and, in order to give these terms meaning, it is necessary to define the transition from one case to the other. This transition state results from an assumed balanced section in which the amount of reinforcement is such that, at failure, concrete crushing and yielding of the reinforcement occurs simultaneously. It will be found that the determination of the amount of reinforcement for a balanced section plays a key role in the design process.

In the BS code the design procedure is simplified by the provision of a number of design charts which relate the parameter M/bd^2 to the steel ratio A_s/bd, where M represents the design bending moment. Each chart is drawn for a particular set of values of the characteristic material strengths, f_{cu} and f_y, and for a doubly reinforced section the geometrical ratio d'/d. The designer can obtain results for other intermediate values by interpolation from the charts.

A spreadsheet approach results in a direct solution for any set of values and the procedure for setting up such a spreadsheet is described below. Initially an assumption, based on the stress/strain diagram for concrete, has to be made with regard to the form of the distribution of the stresses in the compression zone of the section. Two assumed stress distributions, or stress blocks, are con-

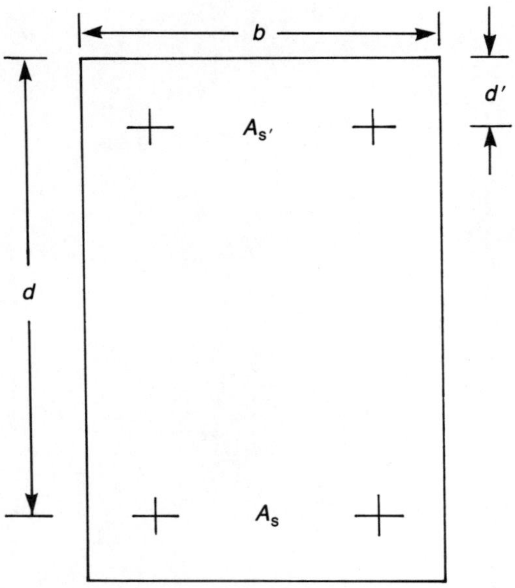

Figure 3.1 Beam cross-section

sidered in the British code and details of these are given in the next section.

The stress block

The stress/strain curve for concrete in compression is usually assumed to be partly parabolic and partly rectangular as shown in Fig. 3.2.

In Fig. 3.2, f_{cu} represents the characteristic cube strength of concrete and γ_{mm} is the partial safety factor.

In the example which follows γ_{mm} is taken as 1.5, so that the maximum stress will be written as $0.45f_{cu}$ instead of $0.67f_{cu}/\gamma_{mm}$ as shown in Fig. 3.2, and the maximum strain ϵ_{cu} taken as 0.0035. Similarly, the value of γ_{ms} will be taken as 1.15 so that the maximum stress in the reinforcement, Fig. 1.1, is given by $0.87f_y$. The equations which define the problem could have been written in more general terms, so that any values of γ_{mm}, γ_{ms} and ϵ_{cu} could have been used, but this approach has not been followed for the present example.

The strain ϵ_1 is given by the equation:

$$\epsilon_1 = (\sqrt{f_{cu}})/5000\ldots \quad [3.1]$$

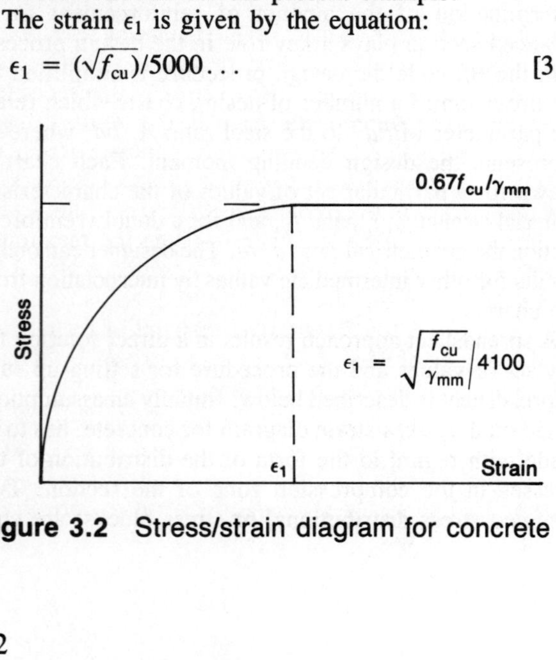

Figure 3.2 Stress/strain diagram for concrete

The two forms of the stress block assumed in the BS code are shown in Fig. 3.3.

Comparing the parabolic + rectangular form shown in Fig. 3.3(a) with the stress/strain diagram shown in Fig. 3.2 it is evident that the diagrams are the same except that the strain axis has been replaced with a distance from the neutral axis. A simplified rectangular form of the stress block is shown in Fig. 3.3(b) in which the compression zone is assumed to extend for a depth of 0.9 times the neutral axis depth x.

In the analysis which follows both of the above stress blocks will be considered and compared although, in practice, there is no advantage in introducing the simplified rectangular stress block with a spreadsheet solution.

Note that both the above stress blocks are further simplified by assuming they can be represented by an equivalent rectangle of width k_1f_{cu} and depth x. The values of the constants k_1 and k_2, shown in Fig. 3.3, can be obtained by equating the area of the stress block to the area of the equivalent rectangle, for k_1, and taking the moments of area of the two areas about the top of the section for k_2. This leads to the equations:

$$k_1 = 0.45(1-(\mathrm{F_a}/3)) \quad [3.2]$$

$$k_2 = ((2-\mathrm{F_a})^2 + 2)/4(3-\mathrm{F_a}) \quad [3.3]$$

where $\mathrm{F_a} = (\sqrt{f_{cu}})/17.5$.

It can be shown that values of k_1 and k_2 for the rectangular stress block shown in Fig. 3.3(b) are given by the equations:

$$k_1 = 0.405 \quad [3.4]$$

$$k_2 = 0.45 \quad [3.5]$$

Having decided which stress block to use, the appropriate design equations can now be developed in terms of the values of k_1 and k_2 and this procedure is outlined below.

Basic equations for singly reinforced sections

A balanced section can be defined as the section in which the maximum strain in the concrete, taken as 0.0035, is reached simultaneously with a strain in the reinforcement of ϵ_y. This state is illustrated in Fig. 3.4.

The equivalent rectangular stress block shown in Fig. 3.4(c) must not be confused with the simplified rectangular stress block shown in Fig. 3.3(b). The rectangular stress block in Fig. 3.4(c) represents both the parabolic/rectangular stress block and the simplified rectangular stress block but the values used for k_1 and k_2 would differ for the two cases as shown above by equations 3.2 to 3.5. By equating the forces in tension and compression it can be shown that the equation for the steel ratio at balance is given by:

$$\rho_B = (k_1f_{cu}0.0035)/(0.87f_y(0.0035 + \epsilon_y)). \quad [3.6]$$

where ϵ_y is represented by $0.87f_y/200$ as illustrated by the initial slope of the stress/strain curve for the reinforcement (Fig. 1.1).

If the section is assumed to be underreinforced then design is based on the assumption that the strain in the

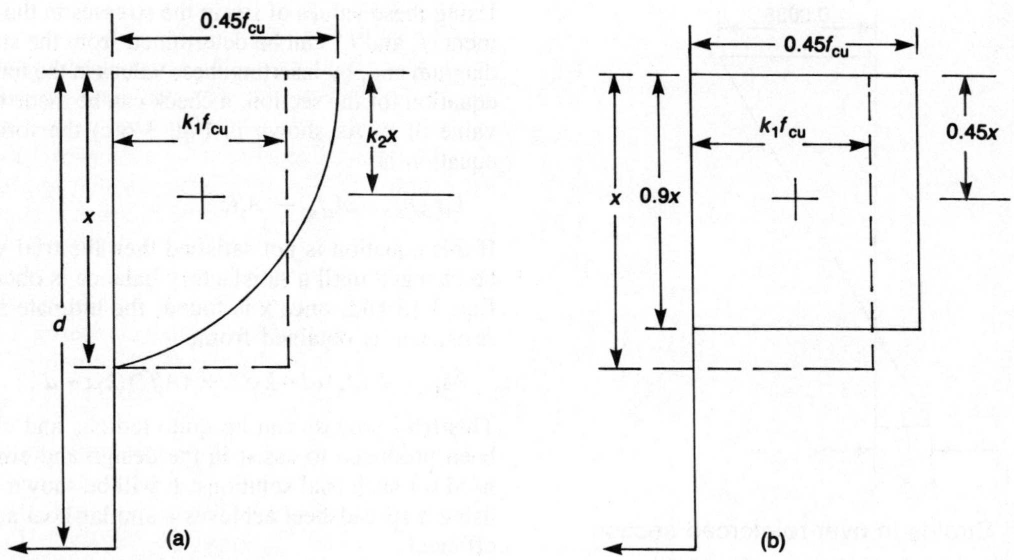

Figure 3.3 Alternative stress blocks

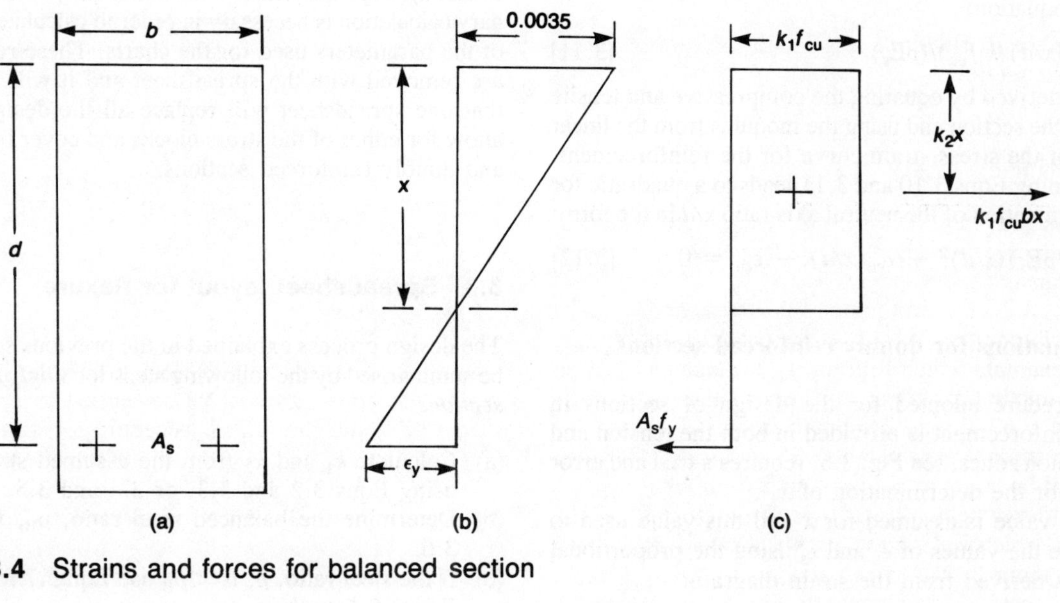

Figure 3.4 Strains and forces for balanced section

reinforcement reaches the value ϵ_y and remains at this value until the strain in the concrete increases to 0.0035. The ultimate moment for this case, which can be obtained by taking moments about the top of the section, is represented by:

$$M_u = A_s 0.87 f_y [1 - (\rho k_2 0.87 f_y / k_1 f_{cu})] d \qquad [3.7]$$

An important parameter which often helps in the interpretation of the results obtained from a particular problem is the neutral axis ratio x/d. For an underreinforced section this ratio is defined by the equation:

$$x/d = (0.87 f_y \rho)/(k_1 f_{cu}) \qquad [3.8]$$

which can be derived by equating the forces in Fig. 3.4(c) and using the resulting expression to determine x/d. Note that although Fig. 3.4 was drawn to represent a balanced section it also represents an underreinforced section. The only difference is that for the balanced section the maximum concrete strain and the yield strain in the reinforcement are attained simultaneously whilst for the underreinforced section the strain in the reinforcement reaches yield before the maximum concrete strain of 0.0035 and then remains at ϵ_y until ϵ_{cu} reaches 0.0035.

If the section is overreinforced then the strain in the reinforcement will be less than ϵ_y when the maximum concrete strain reaches 0.0035, as illustrated in Fig. 3.5, so that at failure the strains are 0.0035 and ϵ_s.

An expression for the ultimate moment of resistance, for the overreinforced case, can be derived by taking moments about the reinforcement giving:

$$M_u = k_1 f_{cu} bx(d - k_2 x) \qquad [3.9]$$

where all the terms on the left-hand side, except x, are known. To find x, use is made of the ratio represented in Fig. 3.5:

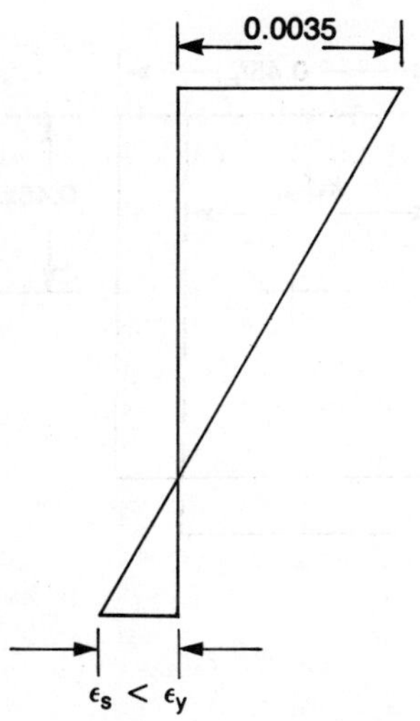

Figure 3.5 Strains in over-reinforced section

$$x/d = 0.0035/(\epsilon_{cu}+\epsilon_s) \quad [3.10]$$

and the equation:

$$\epsilon_s = (x/d)(k_1 f_{cu})/(\rho E_s) \quad [3.11]$$

which is derived by equating the compressive and tensile forces in the section and using the modulus from the linear portion of the stress strain curve for the reinforcement.

Combining Eqns 3.10 and 3.11 leads to a quadratic for the determination of the neutral axis ratio x/d in the form:

$$(k_1 f_{cu}/\rho E_s)(x/d)^2 + \epsilon_{cu}(x/d) - \epsilon_{cu} = 0 \quad [3.12]$$

Basic equations for doubly reinforced sections

The procedure adopted for the design of sections in which reinforcement is provided in both the tension and compression zones, see Fig. 3.6, requires a trial and error process for the determination of x.

A trial value is assumed for x and this value used to determine the values of ϵ_s and ϵ_s' using the proportional equations derived from the strain diagram:

$$\epsilon_s = (d-x)0.0035/x \quad [3.13]$$

$$\epsilon_s' = (x-d')0.0035/x \quad [3.14]$$

Using these values of strain the stresses in the reinforcement, f_s and f_s', can be determined from the stress/strain diagram and, by inserting these values in the force balance equation for the section, a check can be made on the trial value of x. As shown in Fig. 3.6(c) the force balance equation is:

$$k_1 f_{cu} b_x = A_s f_s - A_s' f_s' \quad [3.15]$$

If this equation is not satisfied then the trial value must be changed until a satisfactory balance is obtained from Eqn 3.15 and, once x is found, the ultimate moment of resistance is obtained from:

$$M_u = (A_s f_s)(d-k_2 x) + (A_s' f_s')(k_2 x - d') \quad [3.16]$$

This trial process can be quite tedious and charts have been produced to assist in the design and eliminate the need for such trial solutions. It will be shown below that using a spreadsheet achieves a similar goal and is more efficient.

Since only a limited number of design charts are available the designer must interpolate between charts to obtain appropriate values and additionally some preliminary calculation is necessary in order to calculate the value of the parameters used for the charts. These restrictions are removed with the spreadsheet and it will be shown that one spreadsheet will replace all the design charts, allow for either of the stress blocks and cover both singly and doubly reinforced sections.

3.3 Spreadsheet layout for flexure

The design process explained in the previous section can be summarised by the following steps for *singly reinforced sections*:

(a) Calculate k_1 and k_2 from the assumed stress block using Eqns 3.2 and 3.3, or 3.4 and 3.5.
(b) Determine the balanced steel ratio, ρ_B, from Eqn 3.6.
(c) If the steel ratio, ρ, is $<\rho_B$ use Eqn 3.7 for M_u and Eqn 3.8 for x/d.
(d) If the steel ratio is $>\rho_B$ use Eqn 3.12 to determine x/d and Eqn 3.9 for M_u.

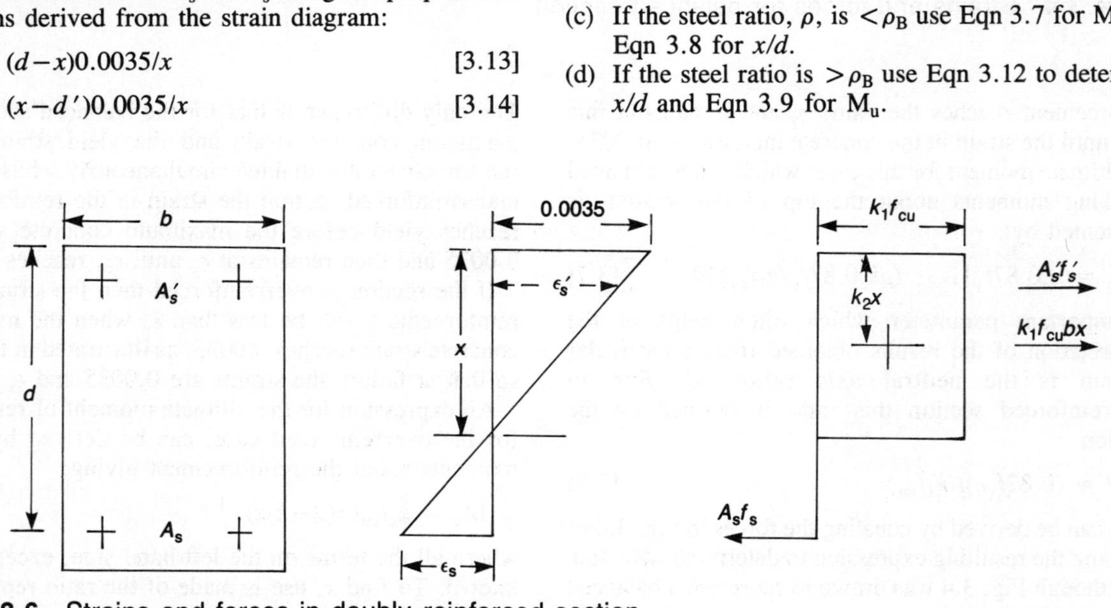

Figure 3.6 Strains and forces in doubly reinforced section

For *doubly reinforced sections*:

(1) Assume a value for x and calculate ϵ_s and ϵ_s' from Eqns 3.13 and 3.14.
(2) Determine f_s and f_s' from the calculated strains using either the tabular interpolation method described in Chapter 1 or more directly by use of nested @IF functions.
(3) Check Eqn 3.15 and adjust x accordingly until balance is obtained.
(4) Use this balanced value of x to calculate M_u from Eqn 3.16.

Constants and initial data

Values of the constants ϵ_{cu}, γ_{mm} and γ_{ms} are placed in cells C8 to C10 in Sheet 3.1 and the initial values required to define the section geometry, with the material constants and information about the reinforcement placed in cells C13 to C19. The value of the design moment is placed in cell C20. As far as the design of singly reinforced sections is concerned the initial values entered in cells C13 to C20 together with the selection of the stress block in cell G15 are the only entries required.

If a doubly reinforced section is required then the user is made aware of this and additional information must be entered giving details about the top and bottom reinforcement, the depth to the top reinforcement and the value of x/d entered in cell C60.

Preliminary calculations

The effective depth, the product $0.87f_y$, the area of steel and the steel ratio A_s/bd are calculated in cells F8 to F11. These cells contain the formulae:

F8: **(C14−C15)**
F9: **0.87*C17**
F10: **@PI*C19*C19*C18/4**
F11: **(F10/(C13*F8))**

Choice of stress block

The user has to enter either 1 or 2 in cell G15 in order to make a choice with regard to the required stress block and the values, 1 and 2, already entered in cells G13 and G14 are used as memoranda for this selection process.

If the spreadsheet is being designed for another user then it is useful to provide a reminder that only the values 1 or 2 should be entered in cell G15 and this can be achieved by placing the following formula in cell H15:

H15: **@IF(G15=1#OR#G15=2," ",**
"Must be 1 or 2")

which tests if cell G15 contains the value 1 or 2 and, if true, outputs the null string and, if false, outputs the warning "Must be 1 or 2".

Values of k_1 and k_2 and ρ_B

Values of k_1 and k_2 can be calculated from equations 3.2 and 3.3 or equations 3.4 and 3.5 depending on which stress block has been selected in cell G15. To simplify the entries to the spreadsheet for calculating these values it is convenient to make the preliminary calculation of the value of $\sqrt{f_{cu}}/17.5$ and this formula has been entered in cell F16.

The contents of cells F16 to F18 are:

F16: **(@SQRT(C16))/17.5**
F17: **@IF(G15=1,0.45*(1−(F16/3)),0.405)**
F18: **@IF(G15=1,**
(((2−F16)^2)+2)/((4*(3−F16))),0.45)

The steel ratio for a balanced section can now be determined using equation 3.6 and the formula for this calculation has been entered in cell F19 in the form:

F19: **(F17*C16*0.0035)/(F9*(0.0035+(F9/200000)))**

This value of the steel ratio for a balanced section in cell F19 can now be compared, using the spreadsheet, with the design steel ratio shown in cell F11 and the appropriate action taken following the guidelines outlined in the introduction to this section. If ρ is less than ρ_B then the spreadsheet calculation continues in block B22 to C27. If ρ is greater than ρ_B then the calculation continues initially in block E22 to F28 and then moves on to page 2 of the spreadsheet since it is probable that the section will have to be doubly reinforced. To simplify the reading of the output this particular spreadsheet has been designed so that only one of the above blocks will appear on the spreadsheet at a time depending on whether the value of ρ is $<$ or $>$ ρ_B. The procedure for achieving this is described below.

Values of $\rho < \rho_B$

If the number of bars and the diameter of the bars entered in cells C18 and C19 are such that the steel ratio is less than that required for a balanced section then values of M_u, M_u/bd^2, x/d and x will appear in cells C24 to C27 respectively. To obtain a satisfactory design the ultimate moment of resistance M_u must be greater than the design moment M and the user must change the number of bars and/or the bar diameter, in C18 and C19, to ensure that this is the case. On Sheet 3.1 the design moment is 350 kNm and the ultimate moment of resistance 380 kNm so the above condition is satisfied whilst ρ still remains less than ρ_B. If the value of M is such that it is not possible to make M_u satisfy the condition $M_u > M$ without making $\rho > \rho_B$ then the spreadsheet would indicate this by flipping over from block B22−C27 to block E22−F28. An example of this will be given on Sheet 3.2.

To achieve this control on the output use is made of an @IF function which checks the condition $\rho < \rho_B$ and then outputs either the null string, " ", or the required label or formula depending on whether the condition is true or false. Some examples of the formulae in block B22 to C27 are:

Sheet 3.1 (Page 1)

```
 1/A          B          C          D          E          F          G          H
  2       Design of Reinforced Concrete Beams
  3   ------------------------------------------------
  4       Singly Reinforced – Rectangular
  5   ------------------------------------------
  6 Constants                         Preliminary Calculations
  7 ----------                        ------------------------------
  8 ecu               0.0035          Eff.Depth          630 mm
  9 gmm                  1.5          0.87fy           400.2
 10 gms                 1.15          As                 1810 mm2
 11 Initial Data                      Rho              0.0131
 12 -------------                     ------------------------------
 13 Breadth           220 mm          Parabolic Block                1
 14 Depth             680 mm          Simple Block                   2
 15 Cover              50 mm          Select?                        1
 16 fcu                35 N/mm2       Prelim.          0.338 F
 17 fy                460 N/mm2       K1               0.399
 18 No Bars             4             K2               0.447
 19 Dia                24 mm          Rho(Bal)         0.022
 20 Design M          350 kNm
 21 -------------------------------------------------------------
 22 Rho less than Rho(Bal)
 23 ---------------------
 24 Mu                  380 kNm
 25 Mu/bd2             4.35
 26 x/d               0.374
 27 x              235.54 mm
 28 ---------------------
 29
```

```
B22: @IF(F11<F19,"Rho less than Rho(Bal)"," ")
B24: @IF(B22<>" ","Mu"," ")
B25: @IF(B22<>" ","Mu/bd2"," ")
C24: @IF(B22<>" ",(F10*F9*F8
     *(1-((F11*F18*F9)/(F17*C16))))/1000000," ")
C25: @IF(B22<>" ",
     (C24*1000000)/(C13*F8*F8)," ")
C26: @IF(B22<>" ",
     (F9*F11)/(F17*C16)," ")
C27: @IF(B22<>" ",(C26*F8)," ")
B28: @IF(B22<>" ","----------------"," ")
```

Values of $\rho > \rho_B$

If the number of bars and the diameter of the bars entered in cells C18 and C19 as in Sheet 3.2 are such that the steel ratio is greater than that required for a balanced section then values of x/d, M_u, M_u/bd^2 and x will appear in cells F25 to F28 respectively. A preliminary calculation has been made in cell F24 in order to simplify the input required for some of the other cells in this block. The ultimate moment of resistance M_u must be greater than the design moment M and the user must change the number of bars and/or the bar diameter to ensure that this is the case. On Sheet 3.2 the design moment is 1000 kNm and the ultimate moment of resistance 735 kNm so the above condition is not yet satisfied. More steel or a change in some of the other initial data is required but, as indicated in the statement alongside the block, the value of x/d already exceeds 0.5 and increasing the amount of bottom steel will increase x/d still further. To overcome this problem compression steel must be introduced and the method of determining appropriate values is given in the next section.

Some examples of the formulae entered in the cells of block E22 to F28 are:

```
E22: @IF(F11>F19,"Rho>than Rho(Bal)"," ")
E24: @IF(E$22<>" ","Prelim."," ")
E25: @IF(E$22<>" ","x/d"," ")
F24: @IF(E22<>" ",(700*F11)/(F17*C16)," ")
F25: @IF(E22<>" ",(-F24+@SQRT((F24^2)
     +(4*F24)))/2," ")
F26: @IF(E22<>" ",(F17*C16*C13*F25*F8*
     (F8-(F18*F25*F8)))/1000000," ")
F27: @IF(E22<>" ",
     (F26*1000000)/(C13*F8*F8)," ")
F28: @IF(E22<>" ",+F25*F8," ")
E29: @IF(E22<>" ","----------------"," ")
```

Sheet 3.1 (Page 2)

	A	B	C	D	E	F	G	H
35		B	C	D	E	F	G	H
36		Doubly Reinforced – Rectangular						
37		------------------------------------						
38	Additional Data			Preliminary Calculations				
39	----------------			--------------------------				
40	d'		50 mm	Strain B		−0.0020		
41				Strain A		0.0020		
42				d'/d		0.079		
43	Top Steel			Bottom Steel				
44	----------			------------				
45	No Bars		0	No Bars		4		
46	Dia		0 mm	Dia		24 mm		
47	As'		0 mm2	As		1810 mm2		
48	Rho'		0.0000	Rho		0.0131		
49			x/d Values	es	es'	fs	fs'	Check
50			0.1	0.032	0.001	400.200	144.444	−3.83
51			0.2	0.014	0.002	400.200	400.200	−2.43
52			0.3	0.008	0.003	400.200	400.200	−1.03
53			0.4	0.005	0.003	400.200	400.200	0.37
54			0.5	0.003	0.003	400.200	400.200	1.76
55			0.6	0.002	0.003	400.200	400.200	3.16
56			0.7	0.001	0.003	300.000	400.200	5.87
57			0.8	0.001	0.003	175.000	400.200	8.90
58			0.9	0.000	0.003	77.778	400.200	11.56
59	Calc. x/d		0.374	0.006	0.003	400.200	400.200	0.00
60	Trial x/d		**0.374**	0.006	0.003	400.200	400.200	0.00
61								
62	Doubly Reinforced			Interpolation for x/d				
63	Final Calculation			-----------------------				
64	------------------				0.30	−1.03		
65	Mu/bd2		4.35		0.4	0.37		
66	Mu		380 kNm		x/d	0.374		

To complete this part of the spreadsheet a reminder has been placed in block G25 to G27 to the effect that if the value of *x/d* in cell F25 is greater than 0.5 then the user should introduce compression steel into the analysis and an illustration of how this can be achieved is given in the next section. This limit of 0.5 on the value of *x/d* is introduced in the British Standard code to prevent a possible brittle failure of the beam although this limit varies with the degree of moment redistribution which can occur in continuous beams (Kong and Evans, 1987).

The formulae placed in cells G25 to G27 to effect this reminder take the form:

G25: **@IF(F25>0.5,"Note. x/d** > 0.5**"," ")**
G26: **@IF(E22<>" "#AND#G25<>" ",**
"Use compression"," ")
G27: **@IF(E22<>" "#AND#G25<>" ",**
"Reinforcement"," ")

Note that for *Lotus 1-2-3 (Release 2)* the formula for G25 is changed to:

G25: **@IF(E22<>" ","Note. If x/d > 0.5",**
" ")

since the condition F25>0 would result in an error if the value in cell F25 is the null string " ".

Doubly reinforced section

Having decided to introduce compression steel it is necessary to enter additional data on the spreadsheet. The depth d′ of the top reinforcement is entered in cell C40 and the number and diameter of the bars for the top and bottom reinforcement in cells C45, C46, F45 and F46. An entry must also be placed in cell C60 but details about this entry will be given later. (Page 2 of Sheet 3.2 refers)

During the analysis it is necessary to calculate the stress in both the top and bottom reinforcement from predetermined strains and this can be achieved either by using the @VLOOKUP function as illustrated in Chapter 1 or alternatively by using a nested @IF function. In the present example the nested @IF function will be used and, as a preliminary, the strains at B and A are calculated from the stress/strain diagram for the reinforcement in cells F40 and F41 using:

Sheet 3.2 (Page 1)

1/A	B	C	D	E	F	G	H
2	Design of Reinforced Concrete Beams						
3	------------------------------------						
4	Singly Reinforced – Rectangular						
5	------------------------------------						
6 Constants			Preliminary Calculations				
7 ----------			------------------------------				
8 ecu		**0.0035**	Eff.Depth		640 mm		
9 gmm		**1.5**	0.87fy		400.2		
10 gms		**1.15**	As		4825 mm2		
11 Initial Data			Rho		0.0314		
12 ------------			------------------------------				
13 Breadth		**240** mm	Parabolic Block			1	
14 Depth		**720** mm	Simple Block			2	
15 Cover		**80** mm	Select?			**1**	
16 fcu		**40** N/mm2	Prelim.		0.361 F		
17 fy		**460** N/mm2	K1		0.396		
18 No Bars		**6**	K2		0.444		
19 Dia		**32** mm	Rho(Bal)		0.025		
20 Design M		**1000** kNm					
21 ------------			------------------------------				
22			Rho > than Rho(Bal)				
23			------------------------				
24			Prelim.		1.39		
25			x/d		0.673 Note. x/d > 0.5		
26			Mu		735 Use Compression		
27			Mu/bd2		7.47 Reinforcement		
28			x		431.02		
29			------------------------				

F40: **−F9/(200*1000)**
F41: **+F9/(200*1000)**

Other preliminary calculations for the ratio d'/d, the areas of the reinforcement and the steel ratios for the compressive and tensile reinforcement are made in F42, C47, F47, C48 and F48 respectively and examples of the type of formulae required in these cells have been given previously.

To continue the solution it is now necessary to find the value of x/d which will satisfy Eqn 3.15 and, by using the spreadsheet, it is possible to avoid the usual tedious process of trial and error. This is achieved by placing values of x/d from 0.1 to 0.9 in intervals of 0.1, in block C50 to C58 and calculating ϵ_s, ϵ_s', f_s and f_s' in the adjacent columns D to G. The calculated values of f_s and f_s' can now be inserted into Eqn 3.15 and any out-of-balance residual value determined. This is carried out in the 'check' column of cells H50 to H58.

Examples of the formulae used for the above are:

C50: **0.1**
D50: **(1−C50)*0.0035/C50** (using equation 3.13)
E50: **(C50−$F$42)*0.0035/C50** (using equation 3.14)
F50: **@IF(D50<$F$40,−$F$9,@IF(D50>$F$41, +$F$9,(($F$9*D50)/($F$41))))**
G50: **@IF(E50<$F$40,−$F$9,@IF(E50>$F$41, +$F$9,(($F$9*E50)/($F$41))))**
C51: **0.1+C50**
D51: **(1−C51)*0.0035/C51**
E51: **(C51−$F$42)*0.0035/C51**
F51: **@IF(D51<$F$40,−$F$9,@IF(D51>$F$41, +$F$9,(($F$9*D51)/($F$41))))**
G51: **@IF(E51<$F$40,−$F$9,@IF(E51>$F$41, +$F$9,(($F$9*E51)/($F$41))))**

The nested @IF functions use the appropriate value of the strain, calculated in the same row, and then determine the value of the stress for this value of the strain. Referring to Fig. 1.1, if the predetermined strain is less than the value at B, cell F40, then the stress value is $-0.87f_y$; if the strain is > the value at A then the stress is $+0.87f_y$; if it lies between B and A then the stress can be calculated by simple proportion. For this example the values in row H indicate that the value x/d which will satisfy equation 3.15 lies between 0.4 and 0.5 since the residuals in column H change from negative to positive for these values. However a more accurate value for x/d can be obtained by the following method.

First copy block C50–C58 to I50–I58 and copy block H50–H58 to J50–J58. These additional columns, which are not shown on Sheet 3.2, are listed below:

Sheet 3.2 (Page 2)

```
35                B          C          D            E           F            G            H
36              Doubly Reinforced – Rectangular
37              ------------------------------------------------------
38 Additional Data                     Preliminary Calculations
39 ----------------------              -------------------------------------
40 d'                  60 mm           Strain B           -0.0020
41                                     Strain A            0.0020
42                                     d'/d                0.094
43 Top Steel                           Bottom Steel
44 ----------                          ------------
45 No Bars                4            No Bars                 6
46 Dia                24 mm            Dia                 32 mm
47 As'              1810 mm2           As                4825 mm2
48 Rho'            0.0118              Rho         0.0314
49              x/d Values       es          es'          fs           fs'       Check
50                     0.1    0.032        0.000      400.200       43.750      -10.47
51                     0.2    0.014        0.002      400.200      371.875       -5.03
52                     0.3    0.008        0.002      400.200      400.200       -3.11
53                     0.4    0.005        0.003      400.200      400.200       -1.53
54                     0.5    0.003        0.003      400.200      400.200        0.06
55                     0.6    0.002        0.003      400.200      400.200        1.64
56                     0.7    0.001        0.003      300.000      400.200        6.37
57                     0.8    0.001        0.003      175.000      400.200       11.88
58                     0.9    0.000        0.003       77.778      400.200       16.52
59 Calc. x/d         0.496    0.004        0.003      400.200      400.200        0.00
60 Trial x/d         0.496    0.004        0.003      400.200      400.200       -0.01
61
62 Doubly Reinforced                   Interpolation for x/d
63 Final Calculation                   ---------------------
64 -----------------                          0.40        -1.53
65 Mu/bd2            10.40                    0.5          0.06
66 Mu             1022 kNm                    x/d         0.496
```

	Check H	*x/d* I	*Check* J
50	**−10.47**	**0.1**	**−10.47**
51	**−5.03**	**0.2**	**−5.03**
52	**−3.11**	**0.3**	**−3.11**
53	**−1.53**	**0.4**	**−1.53**
54	**0.06**	**0.5**	**0.06**
55	**1.64**	**0.6**	**1.64**
56	**6.37**	**0.7**	**6.37**
57	**11.88**	**0.8**	**11.88**
58	**16.52**	**0.9**	**16.52**

Note that column J, which is a copy of column H, has been inserted so that two blocks can be defined, H50 to I58 and I50 to J58. The first of these blocks enables the user to enter the table using the out-of-balance as the entry value and the second block can be entered using the value of *x/d*. Both blocks are used in the next section.

Next insert the following formulae in the block E64 to F66:

E64: **@VLOOKUP(0.001,H50..I58,1)**

in which a small positive value is used to locate the value of *x/d* when the residual in column H is the lowest negative value, −1.53 in this case. This results in a value of 0.4 for *x/d*.

Note that if the smallest positive value in column H had been smaller than 0.001 then a small error would result in the value found for *x/d* but the user would be aware of this since one of the residuals in column H would already indicate a zero balance and therefore indicate the correct value of *x/d*. Additionally the user has to enter the final value taken for *x/d* in cell C60 and any required adjustments can be made at this time. Details of this adjustment are given later.

F64: **@VLOOKUP(0.001,H50..I58,0)**

to locate the value of the residual in column H for the *x/d* calculated in E64. This results in a value of −1.53.

E65: **0.1+E64**

in order to output the next larger value of *x/d* to that found in E64.

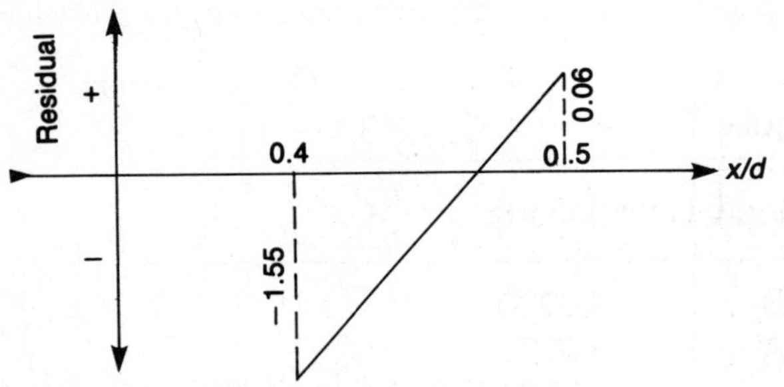

Figure 3.7 Simple proportion diagram

F65: **@VLOOKUP(E65+0.001,I50..J58,1)**

which uses a different block to locate the value of the residual in column H for the x/d calculated in E65. This results in a value of 0.06. The constant 0.001 is included to overcome rounding errors.

As illustrated in Fig. 3.7 there is now sufficient information for the spreadsheet to perform a simple proportion to find the value of x/d which would result in a zero residual for Eqn 3.15.

The required proportion formula is entered in cell F66 in the form:

F66: **((−F64*0.1)/(−F64+F65))+E64**

and then copied to cell C59. This calculated value of x/d is then used to determine the strains ϵ_s and ϵ_s' so that the required stresses f_s and f_s' can be found for insertion into Eqn 3.15 to determine the residual. These calculations which are performed in row 59 use formulae which are similar to those given above for rows 50 and 51.

For the example shown in Sheet 3.2 the residual in cell H59 is shown as zero indicating that the correct value of x/d, 0.4960, has been determined. Even so the user must complete the input by entering a trial value of x/d in cell C60. This input may be identical to the value in cell C59 or a slightly adjusted figure but a value *must* be entered since subsequent calculations use the values in row 60 and not those in row 59.

The final calculation consists of the determination of M_u/bd^2 in cell C65, using a form of Eqn 3.16, and M_u in cell C66. The required formulae are:

C65: **+F48*F60*(1−(C60*F18))+C48*G60 *((F18*C60)−F42)**
C66: **(C65*C13*F8*F8)/1000000**

The value of M_u calculated in cell C66 should be greater than the value of M entered in cell C20 and if not some adjustment to the amount of reinforcement is necessary.

3.4 Use of spreadsheets 3.1 and 3.2

Stage A. Enter initial data in C13 to C20 and stress block selection in G15. A value of M_u will appear in either cell C24 or F26 and it is necessary to ensure that this value should be a little larger than the design value of M entered in C20. Adjust the amount of reinforcement until this is the case (cells C18 and C19). When this adjustment is achieved either block B22 to C27 or block E22 to F29 will appear.

Stage B. If block B22 to C27 appears when the above adjustment of the amount of reinforcement has been completed then the section only requires single reinforcement and no further action is necessary. The design is complete as shown on Sheet 3.1.

Stage C. If block E22 to F29 appears then it is probable that the adjustment referred to in Stage A cannot be achieved with single reinforcement and that the value of x/d in cell F25 is greater than the recommended value of 0.5. This limit of 0.5 is introduced for singly reinforced sections to avoid the possibility of a sudden brittle failure. A note appears on the spreadsheet as a reminder that compression reinforcement is required. Proceed to page 2 of the spreadsheet.

Stage D. Enter values in C40, C45, C46, F45 and F46. A new value of M_u will appear in cell C66 and it is now necessary to adjust the area of reinforcement so that this value of M_u exceeds the design value in cell C20. For each adjustment of the amount of reinforcement the spreadsheet performs a search for the value of x/d and this value appears in cell C59. This value of x/d, or a slightly adjusted value if necessary, *must* be entered in cell C60 each time a change is made in any of the values which define the reinforcement (C45, C46, F45, F46). The value required in cell C60 is the one that results in the 'check' value in cell H60 being as near to zero as possible.

3.5 Alternative approach for rectangular beams

This method uses tables from the Institution of Structural Engineers *Manual for the Design of Reinforced Concrete Building Structures* (1985) and the simplified stress block from BS 8110.

Stage A. Using the simplified rectangular block of BS 8110 and assuming that the value of x/d is at the limiting value of 0.5 it can be shown that the maximum moment M_u, for a singly reinforced section, is given by the equation:

$$M_u = 0.156 f_{cu} b d^2 \qquad [3.17]$$

This formula can be derived by taking moments about the reinforcement of the equivalent rectangular stress block shown in Fig. 3.8 and equating x to $0.5d$.

$$\text{Area of Stress Block} = k_1 f_{cu} b x$$

$$\text{Moment of Area} = (\text{Area}).(d - k_2 x)$$

Taking $k_1 = 0.405$ and $k_2 = 0.45$ results in Eqn 3.17.

Stage B. If the design moment M is less than M_u then only tension steel is required and the amount is given by:

$$A_s = M/(0.87 f_y z) \qquad [3.18]$$

where the value of the lever arm z can be obtained from the following table from the Institution of Structural

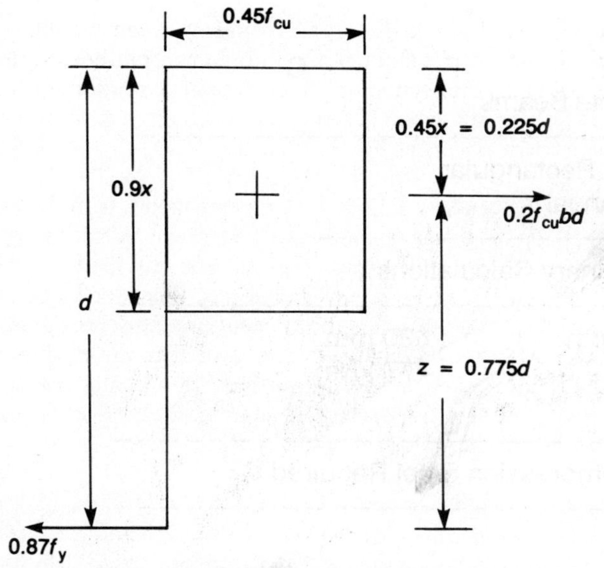

Figure 3.8 Alternative stress block (x/d = 0.5)

Engineers *Manual for the Design of Reinforced Concrete Building Structures* (1985) in which K is defined by:

$$K = M/bd^2f_{cu} \ldots \quad [3.19]$$

Stage C. If the design moment M is > M_u then compression steel required, $A_{s'}$, is given by:

$$A_{s'} = (M - M_u)/0.87f_y(d - d') \quad [3.20]$$

with the proviso that if $d'/x > (1 - f_y/800)$ then the value $700(1 - d'/x)$ must be used in Eqn 3.20 instead of $0.87f_y$.

The amount of tensile reinforcement is given by the equation:

$$A_s = M_u/(0.87f_yz) + A_{s'} \ldots \quad [3.21]$$

where z/d = 0.775, which is the value from Table 3.1 when K = 0.156 and as shown in Fig. 3.8. Note that if $d'/x > (1 - f_y/800)$ then the value of $A_{s'}$ in Eqn 3.21 must be multiplied by $f_{s'}/0.87f_y$, where $f_{s'} = 700(1 - d'/x)$.

3.6 Layout of alternative spreadsheets

In Sheets 3.3 and 3.4, the initial data is entered in cells C9 to C15 and the values from Table 3.1 together with the difference ratios required for interpolation in the block B19 to R23. Only part of this table is visible on Sheets 3.3 and 3.4. The last three column entries for this table are:

	P	Q	R
19	**0.150**	**0.156**	**0.4**
20	**0.79**	**0.78**	**0.78**
21	**0.47**	**0.50**	**0.50**
22	**−1.96**		
23	**5.00**		

Some of the cells of the difference ratio rows are left blank as explained in chapter 1. (See Section 1.13, Reading from Tables.)

The effective depth is calculated in F9, the value of M_u using Eqn 3.17 in F10 and the decision on whether compressive steel is required in E13. The formulae in these cells are:

```
F9:  (C10-C11)
F10: 0.156*C13*C9*(F9^2)/1000000
E13: @IF(F10<=C15,
     "Compression steel required",
     "No compression steel required")
```

A value of K is calculated in C25 using equation 3.19, and z/d and x/d, found by interpolation from the values in Table 3.1, determined in C26 and C27. The required steel areas are calculated in cell C28 for the case of a singly reinforced section, using Eqn 3.18, and in H26 and H27 for a doubly reinforced section, using Eqns 3.20 and 3.21, allowing for the limit imposed by the value of d'/x in H25. The required formulae are:

```
C25: (C15*1000000)/(C9*F9*F9*C13)
C26: @HLOOKUP($C$25,$C$19..$R$23,1)
     +(C$25-(@HLOOKUP($C$25,$C$19..$R$23,0)))
     *(@HLOOKUP($C$25,$C$19..$R$23,3))
C27: @HLOOKUP($C$25,$C$19..$R$23,2)
     +($C$25-(@HLOOKUP($C$25,$C$19..$R$23,0)))
     *(@HLOOKUP($C$25,$C$19..$R$23,4))
C28: (C15*1000000)/(0.87*C14*E26)
H25: @IF(C12/E27>(1-(C14/800)),700
     *(1-(C12/E27)),0.87*C14)
H26: ((C15-F10)*1000000)/(H25*(F9-C12))
H27  ((F10*1000000)/(0.87*C14*0.775*F9))
     +H26*(@IF(C12/E27>(1-(C14/800)),
     (H25/(0.87*C14)),1))
```

At this stage the user has to select the number and diameter of the bars to ensure that the area provided will be > than that required. This selection process could be for a

Table 3.1

K	0.05	0.06	0.07	0.08	0.09	0.10	0.104	0.110	0.119	0.130
z/d	0.94	0.93	0.91	0.90	0.89	0.87	0.87	0.86	0.84	0.82
x/d	0.13	0.16	0.19	0.22	0.25	0.29	0.30	0.32	0.35	0.39

K	0.132	0.140	0.144	0.150	0.156
z/d	0.82	0.81	0.80	0.79	0.775
x/d	0.40	0.43	0.45	0.47	0.50

Sheet 3.3

1/A	B	C	D	E	F	G	H
2		Design of Reinforced Concrete Beams					
3		------------------------------------					
4		Singly or Doubly Reinforced – Rectangular					
5			Using I.Struc.E.Manual				
6			------------------------				
7	Initial Data			Preliminary Calculations			
8	-------------			--------------------------			
9	Breadth	220 mm		Eff.Depth	630 mm		
10	Depth	680 mm		Mu	477 kNm		
11	Cover	50 mm					
12	d'	60 mm		--------------------------			
13	fcu	35 N/mm2		No Compression Steel Required			
14	fy	460 N/mm2		--------------------------			
15	DesignM	350 kNm					
16	------------------						
17	Table of lever–arm and neutral axis depth factors.						
18	--						
19	K	0.050	0.060	0.070	0.080	0.090	0.100
20	z/d	0.94	0.93	0.91	0.90	0.89	0.87
21	x/d	0.13	0.16	0.19	0.22	0.25	0.29
22	Diff_1	–1.28	–1.32	–1.36	–1.41	–1.46	–1.50
23	Diff_2	3.00	3.00	3.00	3.00	4.00	2.50
24	--						
25	K	0.11			d'/xlimit		400
26	z/d	0.85	z	536 mm	As'reqd		–556
27	x/d	0.34	x	211 mm	As reqd		1884
28	As_reqd.	1633 mm2					
29							
30	Case which applies						
31	Bottom Steel Only			Top Steel	&	Bottom Steel	
32	------------------			--------------------------			
33	No Bars	4		No Bars	4	No Bars	6
34	Dia	24 mm		Dia	24	Dia	32
35	As_prov.	1810 mm2		As_prov.	1810	As_prov.	4825
36		O.K.					

singly or doubly reinforced section and information on the case which applies for the current data is output in either cell B30 or E30 using:

B30: **@IF(C15<F10,"Case which applies"," ")**
E30: **@IF(C15>F10,"Case which applies"," ")**

It would be possible to design the spreadsheet so that only information relevant to the case which applies appears on the spreadsheet. This has not been done for the present example although an illustration of how it can be achieved was given for blocks B22 to C27 and E22 to F28 of Sheets 3.1 and 3.2.

Information about the reinforcement is now entered in C33 and C34 for the singly reinforced case or F33, F34, H33 and H34 for the doubly reinforced case. The area of steel provided is then calculated in C35, F35 and H35 and these values compared with the previously calculated areas of steel required.

Finally, the results of the check on the amount of steel provided with that required is output to C36 for the singly reinforced case and to F36 and H36 for the doubly reinforced case. These formulae are written in such a way that output only appears if the case applies:

C36: **@IF(B30< > " ",@IF(C35 >=C28,"O.K.","Check steel")," ")**
F36: **@IF(E30< > " ",@IF(F35 >=H26,"O.K.","Check steel")," ")**
H36: **@IF(E30< > " ",@IF(H35 >=H27,"O.K.","Check steel")," ")**

Sheet 3.4

1/A	B	C	D	E	F	G	H
2	Design of Reinforced Concrete Beams						
3	------------------------------------						
4	Singly or Doubly Reinforced – Rectangular						
5	Using I.Struc.E.Manual						
6	----------------------						
7	Initial Data			Preliminary Calculations			
8	------------------------------------						
9	Breadth	240	mm	Eff.Depth	640 mm		
10	Depth	720	mm	Mu	613 kNm		
11	Cover	80	mm				
12	d'	60	mm	------------------------			
13	fcu	40	N/mm2	Compression Steel Required			
14	fy	460	N/mm2	------------------------			
15	DesignM	1000	kNm				
16	------------------------						
17	Table of lever–arm and neutral axis depth factors.						
18	------------------------------------						
19	K	0.050	0.060	0.070	0.080	0.090	0.100
20	z/d	0.94	0.93	0.91	0.90	0.89	0.87
21	x/d	0.13	0.16	0.19	0.22	0.25	0.29
22	Diff_1	−1.28	−1.32	−1.36	−1.41	−1.46	−1.50
23	Diff_2	3.00	3.00	3.00	3.00	4.00	2.50
24	------------------------------------						
25	K	0.25			d'/xlimit		400
26	z/d	0.78	z	497 mm	As'reqd		1665
27	x/d	0.50	x	320 mm	As reqd		4756
28	As_reqd.	5026	mm2				
29							
30				Case which applies			
31	Bottom Steel Only			Top Steel	&	Bottom Steel	
32	----------------			------------------------			
33	No Bars	4		No Bars	4	No Bars	6
34	Dia	24	mm	Dia	24	Dia	32
35	As_prov.	1810	mm2	As_prov.	1810	As_prov.	4825
36				O.K.		O.K.	

3.7 Use of Spreadsheets 3.3 and 3.4

Stage A. Input the intial data in cells C9 to C15 and information will appear in cell E13 as to whether compression reinforcement is required. If no compression reinforcement is needed then the amount of tensile steel required is shown in cell C28. If compression reinforcement is needed then the amounts of both the compression and tension steel required are shown in cells H26 and H27.

Stage B. Input sufficient steel in rows 33 and 34 until row 35 indicates that it is acceptable using:

(a) C33 and C34 for the singly reinforced case by ensuring that the value in C35 is > the value in C28.
(b) F33, F34, H33 and H34 for the doubly reinforced case by ensuring that the value in F35 is > the value in H26 and that the value in H35 is > the value in H27.

Stage C. Note that the value of M_u in cell F10 is not used in the adjustment. Also the negative value for the area of the reinforcement in cell H26 of Sheet 3.3 has no significance since the 'no compression steel' case applies. It would, however, improve the appearance of the spreadsheet if this value was shown as zero for the singly reinforced case and a method for achieving this is shown on Sheet 3.4R.

3.8 Allowing for redistribution

Although methods for determining the amount of moment redistribution that can occur in continuous beams is beyond the scope of this text an illustration is given of how such a redistribution can be allowed for in design. The BS code permits up to 30 per cent redistribution of moment.

Basic equations

A moment redistribution ratio β_b has been defined in the code as:

$$\beta_b = (100 - \text{Redistribution\%})/100. \qquad [3.22]$$

and a limit placed on the value of x/d so that:

$$x/d \leq (\beta_b - 0.4) \qquad [3.23]$$

It can be shown that the coefficient 0.156 in Eqn 3.17 which defined the maximum moment capacity for a singly reinforced beam applies only for moment redistributions of 0 to 10 per cent. For other percentages of redistribution the coefficient changes in accordance with Table 3.2 according to the *Manual for the Design of Reinforced Concrete Building Structures* (Institution of Structural Engineers, 1985).

Note that values of K′ for other values of redistribution can be obtained from Table 3.2 by interpolation.

Equation 3.17 can now be rewritten in the form:

$$M_u = K' f_{cu} b d^2 \qquad [3.24]$$

and this equation compared with:

$$M = K f_{cu} b d^2 \qquad [3.25]$$

in which M represents the design moment.

Design procedure can now follow the procedure outlined in Part 3 of the British Code:

If $K < K'$ then no compression reinforcement is required and the following equations apply:

$$z = d(0.5 + \sqrt{(0.25 - K/0.9)} \text{ and } < 0.95d \qquad [3.26]$$

$$x = (d - z)/0.45 \qquad [3.27]$$

$$A_s = M/0.87 f_y z \qquad [3.28]$$

If $K > K'$ then compression reinforcement is required and the following equations apply:

$$z = d(0.5 + \sqrt{(0.25 - K'/0.9)} \text{ and } < 0.95d \qquad [3.29]$$

$$x = (d - z)/0.45 \qquad [3.30]$$

$$A_s' = (K - K') f_{cu} b d^2 / 0.87 f_y (d - d') \qquad [3.31]$$

$$A_s = K' f_{cu} b d^2 / 0.87 f_y z + A_s' \qquad [3.32]$$

in which z represents the lever arm from the centroid of the compression block to the level of the tension reinforcement.

Also, if:

Table 3.2

Redist.%	*0–10*	*15*	*20*	*25*	*30*
K′	0.156	0.144	0.132	0.119	0.104

$$d'/x > (1 - f_y/800) \qquad [3.33]$$

then $700(1 - d'/x)$ must be substituted for $0.87 f_y$ in Eqn 3.31 since the compressive stress in the upper reinforcement will not attain the maximum value. Also, the value of A_s' in Eqn 3.32 has to be multiplied by $f_s'/0.87 f_y$ if the stress in the upper reinforcement reaches the yield stress (see Section 3.5 Stage C).

3.9 Spreadsheet layout

Initial values

The initial values of Sheet 3.4R are similar to those used in Sheet 3.3 with the addition of a value for the percentage redistribution entered in cell C16.

Preliminary calculations

The cell entries in this section are as follows:

F9: **(C10−C11)**
F10: **0.87*C4**
F11: **@VLOOKUP(C16,B21..D27,1)**
+(C16−(@VLOOKUP(C16,B21.D27,0)))
***(VLOOKUP(C16,B21..D27,2))**

where F11 uses Table 3.2 plus a difference column in order to perform the necessary interpolation for other values of the percentage redistribution. See the redistribution table in block B21 to D26.

F12: **(C15*1000000)/(C9*C13*(F9^2))**
F13: **+F11*C13*C9*(F9^2)/1000000**
F14: **@IF(F12<=F11,((@SQRT(0.25−(F12/0.9))+0.5)*F9),((@SQRT(0.25−(F11/0.9))+0.5)*F9))**

In F14 the value of z is calculated using either Eqn 3.26 or Eqn 3.29 depending on the values of K and K′.

F15: **@MIN((0.95*F9),F4)**

and F15 is used to ensure that z does not exceed $0.95d$.

F16: **(F9−F15)/0.45**

in order to calculate x from equation 3.27 or 3.30.

Final calculations

The spreadsheet first determines if compression steel is required by checking if $K > K'$ in cell F19:

F19: **@IF(F12<=F11,**
"Compression steel required",
"No compression steel required")

and then proceeds in accordance with the result obtained from this comparison as follows:

(a) If $K < K'$ then the amount of tension steel required is calculated in G23 using Eqn 3.28 and the user must enter sufficient steel in cells F28, F29 and/or G28, G29 until the amount of steel provided (cell G31) is > the amount required (cell G23). Note that

Sheet 3.4R

```
1/A      B              C            D            E              F              G          H
 2          Design of Reinforced Concrete Beams
 3          ------------------------------------
 4             Singly or Doubly Reinforced – Rectangular
 5       Using I.Struc.E.Manual. (Allowing for Redistribution)
 6       ________________________________________________________
 7 Initial Data                                Preliminary Calculations
 8 ____________                                ________________________
 9 Breadth              250 mm                 Eff.Depth        700 mm
10 Depth                730 mm                 0.87fy           400.2 N/mm2
11 Cover                 30 mm                 K'               0.144
12 d'                    60 mm                 K                0.184
13 fcu                   40 N/mm2              Mu                    706 kNm
14 fy                   460 N/mm2              z                     560 mm
15 DesignM              900 kNm                z                     560 mm
16 Redist.               15 %                  x                     311 mm
17 ______________________________________________________________________
18 Redistribution Table
19 ______________________________________Compression Steel Required
20       Redist.          K'          Diff      _____________________________
21             0       0.156             0      d'/xlimit              400
22            10       0.156       -0.0024      As'_reqd.              759 mm2
23            15       0.144       -0.0024      As_reqd.              3907 mm2
24            20       0.132       -0.0026
25            25       0.119        -0.003
26            30       0.104                    Bottom Steel Only
27                                              _________________________
28                                        No Bars        2              2
29                                            Dia       24             24 mm
30                                       As_prov.      905            905 mm2
31                                 Total Provided                    1810 mm2
32
33                    Case which applies
34 Top Steel                                    Bottom Steel
35 ______________________________________________________________________
36       No Bars           3            0                 5              0
37           Dia          20           24 mm             32             24
38      As_prov.         942            0 mm2          4021              0
39 Total Provided                     942 mm2                         4021
40                              O.K.                            O.K.
```

provision is made for two different diameter bars at the same level. If all the bars are of the same diameter then a zero value is entered in either F28 or G28. For K > K′ the information in the rows below row 33 is not used.

(b) If K > K′ then the amount of compression and tension steel required is calculated in G22 and G23 respectively using Eqns 3.31 and 3.32. The limit on d'/x given by Eqn 3.33 is checked in cell G21 so that the correct value of f_s is used for calculating A_s'.

The formulae in cells G21 to G23 are:

```
G21: @IF(C12/F16>(1-(C14/800)),700
     *(1-(C12/F16)),F10)
G22: @IF(F12>=F11,((F12-F11)
     *C13*C9*(F9^2))/(G21*(F9-C12)),0)
G23: @IF(F12<=F11,(C15*1000000)/(F10*F15),
     (F13*1000000)/(F10*F15)+G22
     *(@IF(C12/F16>(1-(C14/800)),
     (G21/(0.87*C14)),1)))
```

Details about the top and bottom steel to be provided for K > K′ are entered in rows 36 and 37 as shown. Note that provision is made for two different diameter bars at the same level for both the top and bottom steels. If the bars at the top or bottom are of the same diameter then a zero value is entered in either D36 and/or G36. For K > K′ the information in the rows 28 to 31 is not used.

Similar formulae for calculating the areas of the reinforcement are entered in F30, G30, C38, D38, F38 and G38 and these are represented by:

F30: **(@PI*F29*F29*F28)/4**

Finally, information is included in cells F25 and C33 to indicate which case applies, singly or doubly reinforced, and in cells G32, D40 and G40 to indicate if there is sufficient steel. Typical formulae to achieve these are:

F25: **@IF(C15<F13,"Case which appies"," ")**
G32: **@IF(F25<>" ",@IF(G31>=G23,"OK", "Check steel")," ")**

The formula in cell G32 first checks if anything has been written in F25, to ascertain if this is the case which applies, and if it finds something there then checks if the amount of steel provided is greater, or equal, to that required. It then outputs either a suitable comment or the null string.

In this example allowance has been made for the inclusion of two different diameter bars at each level. This is not the same as the case of two sets of compression or tension bars at different levels. To obtain an accurate analysis for this would require some changes to the equations. However a close approximation can usually be obtained by using the average effective depth for the case of two layers of tension steel and the average value of d′ for two layers of compression steel as in Kong and Evans (1987).

3.10 Use of spreadsheet 3.4R

Stage A. Input the initial data in cells C9 to C16 and information will appear in cell F19 as to whether compression reinforcement is required. If no compression reinforcement is needed then the amount of tensile steel required is shown in cell G23. If compression reinforcement is needed then the amounts of both the compression and tension steel required are shown in cells G22 and G23 respectively.

Stage B. If no compression steel is required then input sufficient steel in rows 28 and 29 until row 31 indicates that the amount of tension steel provided is acceptable. The output to cell G32 will show if the amount provided is sufficient, although this reminder is hardly required since a visual comparison of cell G31 and cell G23 will make this obvious.

Two bars of different diameter, at the same effective depth, can be input to block F28 to G29. If all bars are of the same diameter then input 0 into either F28 or G28. If there are bars at different depths then an approximate solution can be obtained using the average effective depth.

Stage C. If compression steel is required then input sufficient steel in rows 36 and 37 until row 39 indicates that the amounts of both the compression and tension steel provided are acceptable. Reminders in cells D40 and G40 will state if the amount provided is sufficient.

Two bars of different diameter, at the same effective depth, for both compression and tension reinforcement can be input to block C36 to D37 and block F36 to G37. If all bars are of the same diameter then input 0, as appropriate, into one or more of the cells in row 36. If there are bars at different depths then an approximate solution can be obtained using the average effective depth and/or average depth to the compression reinforcement.

3.11 Comments

Spreadsheets 3.1 and 3.2 give similar results to those given by Sheets 3.3 and 3.4 in spite of the differences in both design and usage. Sheet 3.4R is similar to Sheets 3.3 and 3.4 except that moment redistribution has been included.

For the first set, Sheets 3.1 and 3.2, the design proceeds from trial information about the steel to a calculation of the ultimate moment of resistance M_u. This value of M_u is then compared with the design ultimate moment, M, and the design process continued until an acceptable comparison between M and M_u is found. The true value of x/d is used.

For the second set, Sheets 3.3 and 3.4, the design proceeds from the calculation of the ultimate moment of resistance on the basis of $x/d = 0.5$. The amount of steel required can then be determined and a comparison made between the reinforcement provided and that required.

Design has been for flexure only although there would be no difficulty in extending the spreadsheets to include design for shear, and checks for deflection and stability. These additional checks are included in the next section on the design of reinforced masonry beams.

Design has been for rectangular sections only although similar spreadsheets to the above can be written for T and L beams. Additional information would be required with regard to the end conditions of the beam since the effective width taken for the section may be based on these but, having decided on the effective width, the design is very similar to that outlined above.

REINFORCED MASONRY

3.12 Design for flexure in masonry

The basic equations which have to be satisfied for the design of reinforced masonry beams can be expressed as follows (BS 5628):

$$M_d/bd^2 = (\rho f_y/\gamma_{ms})(1-J_1\rho f_y\gamma_{mm}/f_k\gamma_{ms}) \quad [3.34]$$

$$M_d/bd^2 \leq f_k J_2/\gamma_{mm} \quad [3.35]$$

where the second equation represents a limiting condition which ensures that the beam is underreinforced thus avoiding the possibility of a sudden brittle failure. The constants J_1 and J_2 depend on the initial assumptions regarding the stress/strain relationship for masonry or alternatively the shape assumed for the stress block.

Equations for J_1 and J_2 for some standard stress blocks are given in the next sections.

As far as the spreadsheet solution is concerned the actual values of J_1 and J_2 do not change the logic of the solution. There is therefore no difficulty in using different values from those given in the code.

J_1 and J_2 from BS 5628

In BS 5628 the values of J_1 and J_2 are given as 0.5 and 0.4 respectively. These empirical values are based on the results obtained from a large number of tests and details of some of these tests can be found in Suter and Hendry (1975), Withey (1933), Parsons *et al.* (1932), Scheider (1976) and Thomas and Simms (1981). Useful summaries are provided in Edgell (1981) and Rathbone (1981).

Design aids have been produced which relate the three parameters M_d/bd^2, f_k and ρ so that given any two of these quantities the third can be determined from a graph, see Fig. 3.9.

In Davies and Hendry (1986) these graphs are constructed with a cut-off line for shear. This line is illustrated in Fig. 3.9. Only a nominal allowance for shear would be required for points below the shear limit line whereas for points above the limit line shear becomes a limiting factor.

It will be shown in this chapter that the spreadsheet solution is even more versatile than the design aid graphs. Only the very basic parameters need be entered, for example the bar diameter, and there is no need for the designer to calculate the value of a group of parameters. Additionally it will be shown that design for shear can be incorporated without difficulty.

J_1 and J_2 assuming a parabolic stress block

If the stress/strain relationship for masonry is assumed to be parabolic it can be represented theoretically by the equation:

$$f/\gamma_{mm} = f_k/\gamma_{mm}\,[(2\epsilon/\epsilon_m) - (\epsilon/\epsilon_m)^2] \qquad [3.36]$$

The relationship is illustrated in Fig. 3.10.

There is a great deal of experimental evidence to suppport the above assumption of a parabolic distribution although there has been some discussion on whether the curve has a falling branch as shown in Fig. 3.10. With no falling branch ϵ_u/ϵ_m would equal 1.0, otherwise this ratio would be greater than 1.0.

Having decided on the form of the stress/strain relationship it is possible to draw a diagram showing the variation in the compressive stresses in the masonry from the extreme fibre to the neutral axis. This diagram, which is referred to as the 'stress block', is in fact the same as the stress/strain diagram except that the X axis is related to the distance from the neutral axis instead of the value of the strain. This similarity arises because of the linear variation assumed for the strain throughout the depth.

The stress block resulting from the relationship shown in Fig. 3.10 is as shown in Fig. 3.11.

In this diagram k_3 is related to the stress value at maximum strain and can be determined from the value of the stress when $\epsilon = \epsilon_u$ so that:

$$k_3 = 2\epsilon_u/\epsilon_m - (\epsilon_u/\epsilon_m)^2 \qquad [3.37]$$

and, because of the similarity between the stress/strain diagram and the stress block, Eqn 3.37 can also be expressed as:

$$k_3 = 2d_c/dm_m - (d_c/d_m)^2 \qquad [3.38]$$

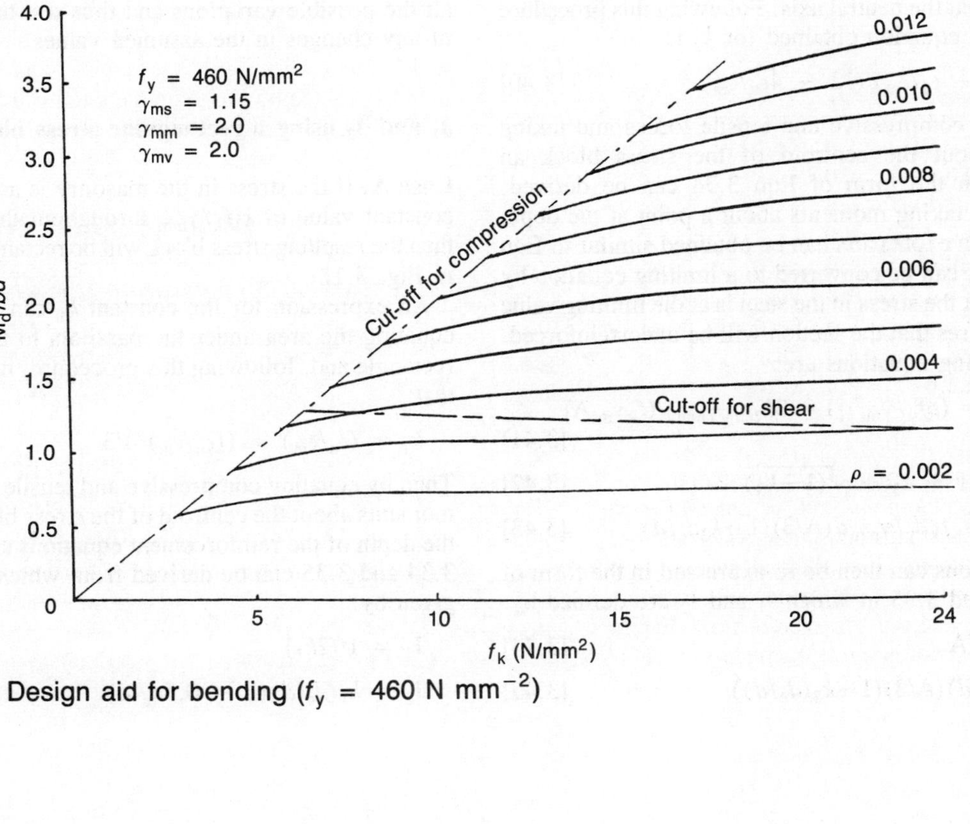

Figure 3.9 Design aid for bending (f_y = 460 N mm^{-2})

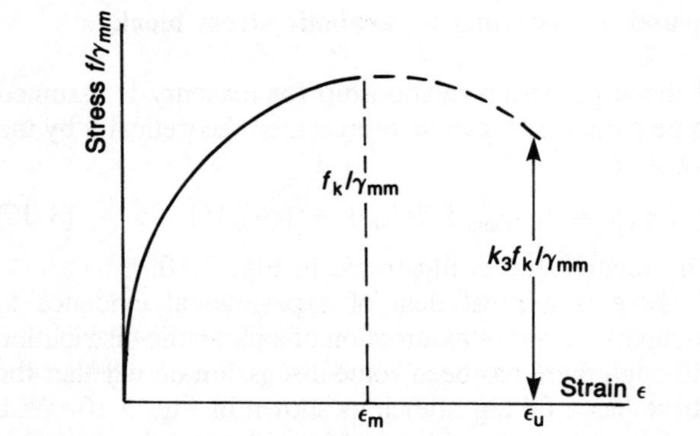

Figure 3.10 Idealised stress/strain diagram for masonry

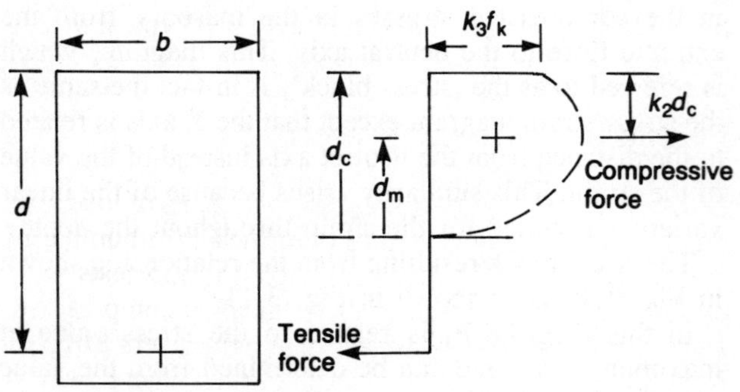

Figure 3.11 Cross-section and stress block

from which:

$$d_c/d_m = 1 + \sqrt{(1-k_3)} \quad [3.39]$$

the negative root having been discarded since it relates to values of $d_c < d_m$.

Since k_2 represents the depth to the centroid of the stress block this can be determined by first calculating the area of the stress block and then taking moments of elements about the neutral axis. Following this procedure the resulting equation obtained for k_2 is:

$$k_2 = (4 - \epsilon_u/\epsilon_m)/(12 - 4\epsilon_u/\epsilon_m) \quad [3.40]$$

By equating compressive and tensile forces and taking moments about the centroid of the stress block an expression in the form of Eqn 3.34 can be derived. Similarly by taking moments about a point at the depth of the steel an expression can be obtained similar to Eqn 3.35 and this can be converted to a limiting equation by assuming that the stress in the steel is at the limiting value f_y. This ensures that the section will be underreinforced.

The resulting equations are:

$$M_d/bd^2 = (\rho f_y/\gamma_{ms})[1-(3k_2\rho f_y\gamma_{mm})/(f_k\gamma_{ms}A)] \quad [3.41]$$

$$\text{where } A = 1 + k_3 + \sqrt{(1-k_3)} \quad [3.42]$$

$$M_d/bd^2 = f_k d_c/\gamma_{mm} d(A/3)(1-k_2 d_c/d) \quad [3.43]$$

These equations can then be re-expressed in the form of Eqns 3.34 and 3.35 in which J_1 and J_2 are defined by:

$$J_1 = 3k_2/A \quad [3.44]$$

$$J_2 = (d_c/d)(A/3)(1-k_2(d_c/d)) \quad [3.45]$$

Table 3.3 J_1 parabolic

ϵ_u/ϵ_m	*1.0*	*1.5*	*1.75*
	0.563	0.556	0.617

Table 3.4 J_2 parabolic

ϵ_u/ϵ_m	*1.0*	*1.5*	*1.75*
f_y			
460	0.250	0.275	0.262
425	0.233	0.257	0.246
410	0.227	0.250	0.239
250	0.174	0.193	0.186

J_1 is a function of the ratio ϵ_u/ϵ_m and J_2 a function of the ratios ϵ_u/ϵ_m and ϵ_u/ϵ_y. As shown above the first ratio equals 1.0 if the parabola representing the stress/strain relationship for masonry is assumed to have no falling branch and a value larger than 1.0 if it has. The second ratio depends on the value of f_y/γ_{ms} since ϵ_y can be determined from f_y/γ_{ms}, using the equation for the strain at B given in section 1.14 of Chapter 1, and ϵ_u can be taken as 0.0035.

Some typical values of J_1 for assumed values of ϵ_u/ϵ_m and J_2 for assumed values of ϵ_u/ϵ_m and f_y are given in Tables 3.3 and 3.4.

Thus in order to design a reinforced masonry beam using a parabolic form of the stress block the designer would first have to decide which value of the ratio ϵ_u/ϵ_m to use. Then a value would have to be assumed for ϵ_u, although this is usually taken as 0.0035, and then the value of ϵ_y determined from f_y/γ_{ms}. This would enable the appropriate values of J_1 and J_2 to be calculated. The design could then proceed using Eqn 3.34, taking care not to exceed the value of M_d given by Eqn 3.35.

This process would normally be far too cumbersome but by using a spreadsheet the designer could incorporate all the possible variations and thus test the significance of any changes in the assumed values.

J_1 and J_2 using a rectangular stress block

Case A. If the stress in the masonry is assumed to be a constant value of $k_1 f_k/\gamma_{mm}$ throughout the strain range then the resulting stress block will be rectangular as shown in Fig. 3.12.

An expression for the constant k_1 can be derived by equating the area under the parabola to the area of the rectangle and, following this procedure, it can be shown that:

$$k_1 = (\epsilon_u/\epsilon_m) - ((\epsilon_u/\epsilon_m)^2)/3 \quad [3.46]$$

Then by equating compressive and tensile forces, taking moments about the centroid of the stress block and about the depth of the reinforcement equations similar to Eqns 3.34 and 3.35 can be derived from which J_1 and J_2 are given by:

$$J_1 = 1/(2k_1) \quad [3.47]$$

$$J_2 = k_1(d_c/d)(1 - 0.5d_c/d) \quad [3.48]$$

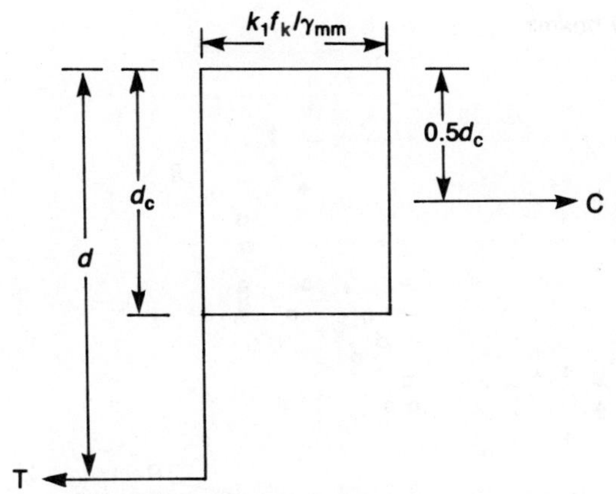

Figure 3.12 Rectangular stress block (Case A)

Table 3.5 J_1 rectangular Case A

ϵ_u/ϵ_m	*1.0*	*1.5*	*1.75*
	0.750	0.667	0.686

Table 3.6 J_2 rectangular Case A

ϵ_u/ϵ_m	*1.0*	*1.5*	*1.75*
f_y			
460	0.233	0.262	0.255
425	0.219	0.246	0.239
410	0.213	0.240	0.233
250	0.167	0.188	0.183

in which $d_c/d = \epsilon_u/(\epsilon_u+\epsilon_y)$. Some typical values of J_1 for assumed values of ϵ_u/ϵ_m and J_2 for assumed values of ϵ_u/ϵ_m and f_y are given in Tables 3.5 and 3.6.

Case B. If the stress in the masonry is assumed to be a constant value of f_k/γ_{mm} for strain values ranging from zero to $k_1\epsilon_u$ then the resulting stress block will be rectangular as shown in Fig. 3.13.

The expression for the constant k_1 will be similar to that given for Case A and is represented by Eqn 3.46.

Equating compressive and tensile forces, taking moments about the centroid of the stress block and about

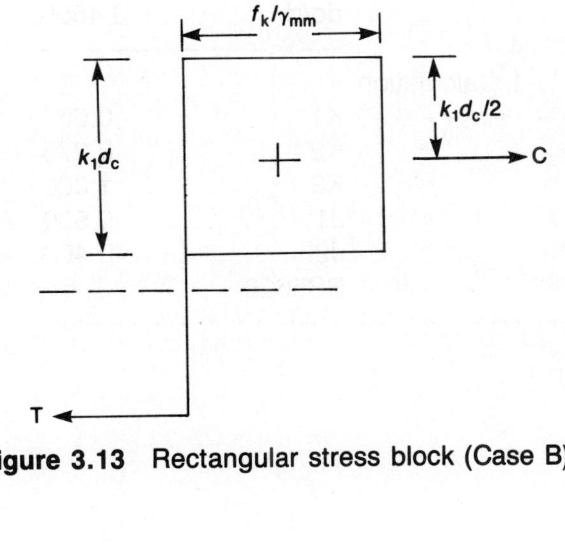

Figure 3.13 Rectangular stress block (Case B)

Table 3.7 J_2 rectangular Case B

f_y	J_2
460	0.349
425	0.328
410	0.320
250	0.251

the depth of the reinforcement equations similar to Eqns 3.34 and 3.35 can be derived for which J_1 and J_2 are given by:

$$J_1 = 0.5 \qquad [3.49]$$

$$J_2 = (d_c/d)(1 - 0.5d_c/d) \qquad [3.50]$$

in which $d_c/d = \epsilon_u/(\epsilon_u+\epsilon_y)$. For this case J_2 is not a function of ϵ_u/ϵ_m but still varies with f_y. Typical values are given in Table 3.7.

Comment on values of J_1 and J_2

Davies and Hendry (1986) discuss the results of load tests on reinforced masonry beams published by a number of research workers. In total the results of 135 tests were analysed and the majority of these failed by compression. Using a plot of the non-dimensional parameters M/bd^2f_k and $\rho f_y/f_k$ the authors illustrated that there was a transition zone above which failure was by compression and below which failure was by tension.

There were 32 test results in the transition zone, seventeen of which failed by tension, 10 by compression and 5 by shear. The range of values in the transition zone arises from statistical variations in the material properties and variations in the test procedures but it is clear that there is a theoretical cut-off for values of M/bd^2f_k below which a tension failure occurs. The values of J_2 calculated in the previous sections represent these cut-off lines for the different stress blocks. The arbitrary value of 0.4 given in BS 5628 can be considered as a safety cut-off figure since all other approaches lead to lower values.

Theoretical plots of M/f_kbd^2 against $\rho f_y/f_k$ for Eqn 3.34 and using different values of J_1 are shown in Fig. 3.14.

In Fig. 3.14 the solid line represents the plot for both BS 5628 and Case B since J_1 equals 0.5 for both these cases. The two plots for the parabolic stress block are for the maximum and minimum values of J_1 calculated in section 3.12. Similarly the two plots for the rectangular stress block represent the range of values calculated in Case A above. It would appear that the value of J_1 given in BS 5628 or calculated for Case B rectangular stress block results in the best plot compared with the test results.

3.13 Spreadsheet layout for stress block coefficients

Case selection. Referring to Sheet 3.5, numbers 1 to 4 are placed in cells E7 to E10 alongside the labels representing the four cases considered. The case to be selected can then be placed in cell E11 by the user as required.

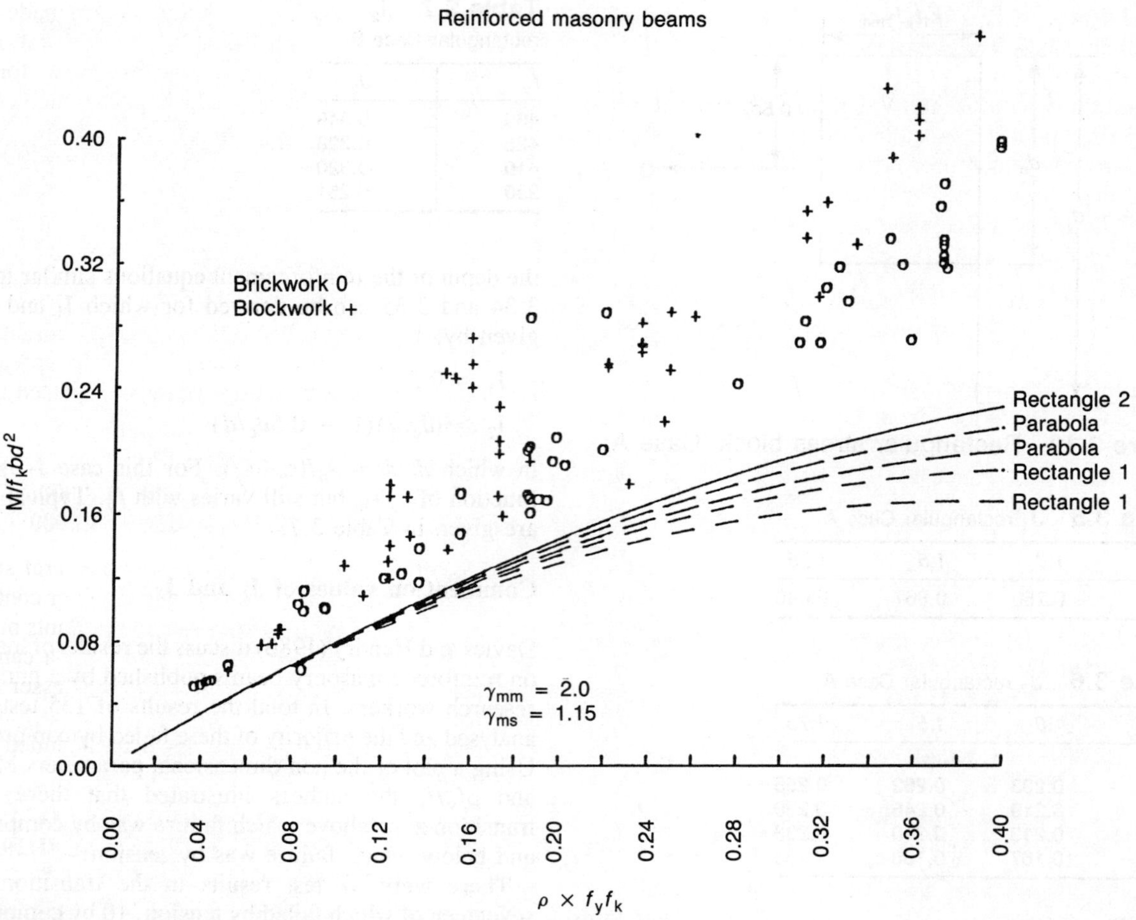

Figure 3.14 Comparison between test results and design curves

Sheet 3.5

1/A	B	C	D	E	F	G	H
2			Reinforced Masonry Beams				
3		Coefficients for Different Stress Blocks					
4	------------	------------	------------	------------	------------	------------	------------
5 Case Selection				Preliminary Calculations			
6	------------	------------		------------	------------	------------	------------
7		BS5628		1		Epsilon_A	0.0016
8		Parabolic		2		Epsilon_y	0.0043
9		Rectangle CaseA		3		dc/d	0.4506
10		Rectangle CaseB		4		------------	------------
11		Select Case ?		1 Calculation			
12		------------	------------	------------	------------	K1	0.667
13 Initial Data						K2	0.375
14		Epsilon_u	**0.0035**			K3	1.000
15		fy	**460**			J1	0.500
16		Gms	**1.15**			J2	0.400
17		eu/em	**1.000**			BS5628	
18		------------	------------	------------	------------	------------	------------

Initial Data. Values of ϵ_u, f_y, γ_{ms} and ϵ_u/ϵ_m together with their labels are placed in the block C14 to D17.

Preliminary calculations. Values of the strains at A and B, see Fig. 1.2, are placed in cells H7 and H8 using the equations previously derived in Chapter 1. Using these values the ratio d_c/d can be determined and placed in cell H9. Cells H7 to H9 contain:

H7: **0.8*D15/(200000*D16)**
H8: **+H7+(0.2*(0.002−H7)/0.03)**
H9: **+D14/(D14+H8)**

Calculation of J_1 and J_2. The values of k_1, k_2, and k_3 can now be determined from Eqns 3.46, 3.40 and 3.37 respectively and placed in cells H12 to H14 and values of J_1 and J_2 determined for the appropriate case, using nested @IF functions, placed in cells H15 and H16. The contents of cells H12 to H16 are:

H12: **+D17−(D17*D17)/3**
H13: **(4−D17)/(12−(4*D17))**
H14: **2*D17−D17*D17**
H15: **@IF(E11=1#OR#E11=4,0.5,**
@IF(E11=2,3*H13/(1+H14+@SQRT(1−H14)),
@IF(E11=3,0.5/H12,"Check entry")))
H16: **@IF(E11=1,0.4,@IF(E11=2,H9*(H13/H15)**
***(1−H13*H9),@IF(E11=3,H12*H9)**
***(1−(0.5*H9)),@IF(E11=4,H9*(1−(0.5*H9)),**
"in cell E11"))))

Although not essential it is convenient to ensure that the number entered by the user in cell E11 is 1, 2, 3 or 4 and to flag any other entry. This can be achieved by making cells H15 and H16 output the label "Check Entry in cell E11" if the user inputs a number outside the required range.

Finally, as a further safeguard, the output in cell G17 is programmed to remind the user of the case selected. The formula in cell G17 is:

G17: **@IF(E11=1,C7,@IF(E11=2,C8,**
@IF(E11=3,C9,@IF(E11=4,C10," "))))

so that for the example shown on Sheet 3.5 the output to cell G17 will be the contents of cell C7, i.e. BS 5628, since the condition E11 = 1 is true.

Use of Sheet 3.5. In this example all cells except E11 and D14 to D17 can be 'protected' since entry is only required for these cells.

The user enters the case in E11 and the initial data in D14 to D17 and these entries result in the required output to cells H15 and H16.

This spreadsheet will be incorporated into Sheets 3.6 to 3.9 as part of the design.

3.14 Spreadsheet layout for flexure

Additional data. Referring to Sheet 3.6, values of f_k, γ_{mm}, breadth, depth, cover to reinforcement, bar diameter, beam span and beam type (simply supported, continuous or cantilever) together with their labels are placed in block C22 to D29.

Preliminary calculations. Calculations are made for the effective depth d, the steel area A_s, the steel ratio ρ and the ratios f_y/γ_{ms} and f_k/γ_{mm}. The formulae for these quantities are placed in cells G22 to G26 which contain:

G22: **(D23−D24−(D27/2))**
G23: **2*@PI*D27*D27/4**
G24: **+G23/(D22*G22)**
G25: **+D16/D17**
G26: **+D25/D26**

Design and safety checks. The design can now be completed using Eqns 3.34 and 3.35. The spreadsheet equivalents of these equations have been placed in cells F32 and F33 which contain:

F32: **+D22*G22*G22*G24*G25**
***(1−(H16*G24*G25/G26))/1000000**
F33: **+D22*G22*G22*G26*H17/1000000**

Code BS 5628 contains a check for the lateral stability of the beam. If the beam is simply supported or continuous then the clear distance between lateral restraints must not exceed the lesser of $60b$ and $250b^2/d$. For a cantilever the corresponding limits are given as the lesser of $25b$ and $100b^2/d$.

A check for lateral stability has been included in cell F34 which contains:

F34: **@IF((D28="SS"#OR#D28="CO")**
#AND#D29*1000<60*D22#AND#D29
***1000<250*D22*D22/G22,"O.K",**
@IF(D28="CA"#AND#D29*1000<25
***D22#AND#D29*1000<100*D22**
***D22/G22,"O.K","Unstable"))**

so that, if the appropriate conditions with regard to beam type and limiting quantities are met, the cell outputs "O.K". Otherwise the output would be "Unstable".

In addition to the lateral stability check the BS code also introduces another limit for the ratio of the span to the effective depth. This avoids the necessity for detailed calculations to check that the limit states of deflection and cracking are not reached. These limiting ratios of 20 for simply supported beams, 26 for continuous beams and 7 for cantilevers are introduced into the spreadsheet in cell F35 which contains the formula:

F35: **@IF(D28="SS"#AND#D29**
***1000/G22<=20,"O.K",@IF(D28="CO"**
#AND#D29*1000/G22<=26,"O.K",
@IF(D28="CA"#AND#D29
***1000/G22<=7,"O.K",**
"Limit exceeded")))

This formula checks the beam type and the corresponding ratio from the table in BS 5628 to check if the recommended ratio is exceeded. If not then the output is "O.K" otherwise the output will be "Limit exceeded".

Use of Sheet 3.6. The first 17 rows are identical with those on Sheet 3.5 so that J_1 and J_2 can be determined for the particular case required.

Apart from the unprotected cells in the first 17 rows (see Sheet 3.5), only cells D22 to D29 and cell F31 require

Sheet 3.6

```
 1/A    B          C                D           E          F           G           H
 2             DESIGN FOR FLEXURE (Rectangular Masonry Beams)
 3             ----------------------------------------------------------------
 4
 5 Case Selection                                  Preliminary Calculations
 6 ---------------                                 ----------------------------
 7             BS5628                              1          Epsilon_A    0.0009
 8             Parabolic                           2          Epsilon_y    0.0084
 9             Rectangle CaseA                     3          dc/d         0.2940
10             Rectangle CaseB                     4          ----------------
11             Select Case ?                       1 Calculation
12             ----------------------------------------------------------
13 Initial Data                                               K1           0.667
14 ---------------                                            K2           0.375
15             Epsilon_u          0.0035                      K3           1.000
16             fy                    250                      J1           0.500
17             Gms                  1.15                      J2           0.400
18             eu/em               1.000                      BS5628
19 ----------------------------------------------------------------------------
20 Additional Data                         Preliminary Calculations
21 ------------------------                -----------------------------------
22             Breadth           215.000              d            322.000
23             Depth             350.000              Area Stl.    402.124
24             Cover              20.000              rho           0.0058
25             fk                 19.200              fy/Gms           217
26             Gmm                     2              Fm/Gmm           9.6
27             Bar diam.              16
28 SS,CO,CA    Beam Type CO
29             Beam Span               4
30             ------------------------              ---------------------
31                 Applied Moment               =        24.000 kNm
32 Design Moment of Resistance                  =        26.297 kNm    O.K
33      Limiting Design Moment                  =        85.602 kNm    O.K
34                 Stability Check              =    O.K
35 Limit. Span/Effective Depth                  =    O.K
36 ----------------------------------------------------------------------------
```

input from the user. The results are shown in cells F32 to F35 and H32 to H33.

3.15 Design for shear

Shear reinforcement must be provided if:

$$v > f_v/\gamma_{mv} \quad [3.51]$$

where v is the shear stress due to the design loads, f_v the characteristic shear strength of the masonry and γ_{mv} the partial safety factor for shear strength of masonry. The value of the shear stress can be calculated from:

$$v = V/bd\ldots \quad [3.52]$$

where V is the shear force due to the design loads.

If shear reinforcement is to be provided then the area of reinforcing steel resisting the shear forces, A_{sv}, must satisfy the equation:

$$A_{sv}/S_v \geq b(v - f_v/\gamma_{mv})\gamma_{ms}/f_y \quad [3.53]$$

where S_v, the spacing of the shear reinforcement along the beam, is not to exceed $0.75d$ and v must not be taken to be greater than $2/\gamma_{mv}$ N mm^{-2}.

The formula given in BS 5628 for the shear strength f_v is:

$$f_v = (0.35 + 17.5\rho)[2.5 - 0.25a_s/d] < 1.75 \text{ N mm}^{-2} \quad [3.54]$$

where the quantity in the first bracket represents the shear strength for ratios of shear span to effective depth greater than 6 and the second bracket is a multiplier to be used for ratios less than 6. The maximum allowable value of f_v is given as 1.75 N mm^{-2}.

3.16 Spreadsheet layout for shear

Additional data. The only additional data required are the values of γ_{mv} and the value of the shear force and these are placed in cells D41 and D42 respectively of Sheet 3.7.

It is convenient to insert the value of the spacing of the shear reinforcement in this section and this is done in cell D44. The reason for this is explained below under Design Check for Shear.

Preliminary calculations. The values of the shear stress calculated from equation 3.51, and a_s/d calculated from M/Vd are placed in cells G41 and G42. The shear strength f_v, calculated from equation 3.54, is placed in cell G43. This value is then checked in cell G44 to ensure that the value of f_v does not exceed the limit of 1.75 Nmm^{-2}. The contents of cells G41 and G44 are:

G41: **+D42*1000/(D22*G22)**
G42: **(F31*1000)/(D42*G22)**
G43: **(0.35+(17.5*G24))**
***@IF(G42<=6,(2.5−(0.25*G42)),1)**
G44: **@IF(G43>=1.75,1.75,G43)**

The value of f_v to be used in subsequent operations is the value in cell G44 since this value satisfies Eqn 3.54 and does not exceed 1.75.

Design check for shear. The labels and formulae for the design check for shear are placed in block B46 and H49. Equation 3.51 is placed in cell E46 to ascertain if any shear reinforcement is required and cell E47 used to calculate the value of A_{sv}/S_v of this reinforcement if necessary. Additionally the maximum spacing allowed for the shear reinforcement is placed in cell E48 and the diameter of each leg of the shear stirrup, based on this maximum spacing, is calculated in cell G48.

The actual spacing of the shear reinforcement has already been entered in cell D44 and this value is copied to cell E49 and used to calculate the diameter required for the shear reinforcement in cell G49. The contents of these cells are:

E46: **@IF(G41>=G44/D41,"Yes","None")**
E47: **@IF(E46="Yes",D22*(G41−(G44/D41))/G25," ")**
E48: **@IF(E46="Yes",0.75*G22," ")**
E49: **@IF(E46="Yes",D44," ")**
G48: **@IF(E46="Yes",@SQRT(E47*E48*2/@PI)," ")**
G49: **@IF(E46="Yes",@SQRT(E47*E49*2/@PI)," ")**

Apart from cell E46 values will only appear in these cells if shear reinforcement is required. Similarly the format of cells C47 to C49, cells D47 to D49 and cells F46 to F49 is such that they will be blank for the case of no shear reinforcement. The contents of these cells are:

C47: **@IF(E46="Yes","Asv/Sv"," ")**
C48: **@IF(E46="Yes","Max Sv"," ")**
C49: **@IF(E46="Yes","Sv"," ")**
D47: **@IF(E46="Yes","="," ")**
D48: **@IF(E46="Yes","="," ")**
D49: **@IF(E46="Yes","="," ")**
F46: **@IF(E46="Yes",**
"Diameter of shear stirrup"," ")
F47: **@IF(E46="Yes",**
"With 2 legs at max spacing"," ")
F48: **@IF(E46="Yes","Dia ="," ")**
F49: **@IF(E46="Yes","Dia ="," ")**

To illustrate the effect of this type of format, if the value of the shear force V on Sheet 3.7 is changed from 40 kN to 20 kN, then there would be no requirement for any shear reinforcement. This is illustrated in Sheet 3.8 which shows that because the value in cell E46 is not "Yes" then there is no output to cell F46 or rows 47 to 49. This approach is not essential but has some cosmetic value. For this to work it was necessary to input the actual stirrup spacing in another cell such as D44 and then to make cell E49 a copy of D44 to ensure that cell E49 would be blank for the case of no reinforcement.

This completes the design and at this stage it is convenient to PROTECT the entire sheet and then UNPROTECT only those cells which can be changed by the user. This safeguards against accidental erasure of cells which contain formulae.

For the present example on Sheet 3.7 the only unprotected cells would be:

E11: (Case);
D16: (f_y);
D17: (γ_{ms});
D18: (ϵ_u/ϵ_m);
D22: (b);
D23: (Depth);
D24: (Cover);
D25: (f_k);
D26: (γ_{mm});
D27: (Dia);
D28: (Beam Type);
D29: (Span);
F31: (M);
D41: (γ_{mv});
D42: (V);
D44: (Stirrup Spacing).

All other cells would be protected. Most of the 16 unprotected quantities are standard and the designer would have no difficulty in entering appropriate values. The exceptions to this are the values of E11 and D18. If E11 is 1, representing BS 5628, then the value in D18 would be immaterial since this quantity does not enter into subsequent operations. If E11 contains 2 to 4 then a value would have to be assumed for D18 and this value would depend on the stress/strain parabola used to represent the masonry if the designer is not prepared to accepted a value of unity.

Use of Sheets 3.7 and 3.8. The first 36 rows of Sheets 3.7 and 3.8 are identical to Sheet 3.6 and the only additional entries required are for cells D41, D42 and D44.

Note that an entry of 40 kN has been entered in cell D42 of Sheet 3.7 and an entry of 20 kN on Sheet 3.8. All other equivalent entries are the same for both sheets but the resulting output in rows 46 to 49 is different.

Sheet 3.7

1/A	B	C	D	E	F	G	H
2		DESIGN FOR FLEXURE (Rectangular Masonry Beams)					
3							
4							
5	Case Selection			Preliminary Calculations			
6							
7		BS5628		1	Epsilon_A		0.0009
8		Parabolic		2	Epsilon_y		0.0084
9		Rectangle CaseA		3	dc/d		0.2940
10		Rectangle CaseB		4			
11		Select Case ?		1 Calculation			
12							
13	Initial Data				K1		0.667
14					K2		0.375
15		Epsilon_u	0.0035		K3		1.000
16		fy	250		J1		0.500
17		Gms	1.15		J2		0.400
18		eu/em	1.000		BS5628		
19							
20	Additional Data			Preliminary Calculations			
21							
22		Breadth	215.000	d		322.000	
23		Depth	350.000	Area Stl.		402.124	
24		Cover	20.000	rho		0.0058	
25		fk	19.200	fy/Gms		217	
26		Gmm	2	Fm/Gmm		9.6	
27		Bar diam.	16				
28	SS,CO,CA	Beam Type CO					
29		Beam Span	4				
30							
31		Applied Moment		=	24.000 kNm		
32	Design Moment of Resistance			=	26.297 kNm	O.K	
33	Limiting Design Moment			=	85.602 kNm	O.K	
34		Stability Check		=	O.K		
35	Limit. Span/Effective Depth			=	O.K		
36							
37		DESIGN FOR SHEAR					
38							
39	Additional Data			Preliminary Calculations			
40							
41		Gmv	2.000	Stress_v		0.578 N/mm2	
42	Shear Force		40.000 kN	a/d		1.863	
43	Stirrup spacing			fv		0.919	
44	If required		200.000	fv		0.919	
45							
46	Shear Reinforcement		Yes	Diameter of Shear Stirrup			
47		Asv/sv	=	0.117 With 2 legs at max spacing			
48		Max Sv	=	241.5 Dia =		4.24 mm	
49		Sv	=	200.0 Dia =		3.86 mm	
50							

Sheet 3.8

1/A		B	C	D	E	F	G	H
2			DESIGN FOR FLEXURE (Rectangular Masonry Beams)					
3	------	------	------	------	------	------	------	------
4								
5	Case Selection					Preliminary Calculations		
6	------	------				------	------	
7			BS5628			1	Epsilon_A	0.0009
8			Parabolic			2	Epsilon_y	0.0084
9			Rectangle CaseA			3	dc/d	0.2940
10			Rectangle CaseB			4	------	------
11			Select Case ?			1 Calculation		
12	------	------	------	------	------	------	------	
13	Initial Data						K1	0.667
14	------	------	------	------			K2	0.375
15			Epsilon_u	0.0035			K3	1.000
16			fy	250			J1	0.500
17			Gms	1.15			J2	0.400
18			eu/em	1.000			BS5628	
19	------	------	------	------	------	------	------	------
20	Additional Data				Preliminary Calculations			
21	------	------	------		------	------	------	
22			Breadth	215.000		d	322.000	
23			Depth	350.000		Area Stl.	402.124	
24			Cover	20.000		rho	0.0058	
25			fk	19.200		fy/Gms	217	
26			Gmm	2		Fm/Gmm	9.6	
27			Bar diam.	16				
28	SS,CO,CA		Beam Type CO					
29			Beam Span	4				
30	------	------	------	------		------	------	
31			Applied Moment		=		**24.000 kNm**	
32	Design Moment of Resistance				=		26.297 kNm	O.K
33		Limiting Design Moment			=		85.602 kNm	O.K
34			Stability Check		=	O.K		
35	Limit. Span/Effective Depth				=	O.K		
36			------	------	------	------	------	------
37			DESIGN FOR SHEAR					
38			------	------				
39	Additional Data				Preliminary Calculations			
40	------	------	------		------	------	------	
41			Gmv	2.000		Stress_v	0.289 N/mm2	
42		Shear Force		20.000 kN		a/d	3.727	
43	Stirrup spacing					fv	0.708	
44	If required			200.000		fv	0.708	
45			------	------	------	------	------	------
46	Shear Reinforcement				None			
47								
48								
49								
50			------	------	------	------	------	

3.17 Flanged sections

Dimensions of section. If the rèinforcement in a section is concentrated locally in such a way that the section can act as a flanged beam then a limit is placed on the permitted value of the design moment. This condition usually occurs in walls and an example is shown in Fig. 3.15.

For this type of wall BS 5628 recommends that the thickness of the flange, t_f, should be taken as the thickness of the masonry but with a limiting value of $0.5d$.

The flange width is taken as the least of:

(a) width of rib or pocket + 12 times flange thickness;
(b) the spacing of the pockets or ribs;
(c) one-third the height of the wall.

Moment of resistance. Having defined the dimensions of the section then these values can be used in a similar manner to those for a rectangular section and the design moment of resistance calculated as shown previously in Sheet 3.7. This design moment of resistance is acceptable as long as it does not exceed the value given by the equation:

$$M_d = f_k b t_f (d - 0.5t_f)/\gamma_{mm} \qquad [3.55]$$

where the term in brackets represents the lever arm, see Fig. 3.16.

Limiting dimensions. In addition to the above limit on the moment of resistance there is also a limit on the ratio of the span to the effective depth. This limit is given as 35 for simply supported walls, 45 for continuous walls and 18 for cantilever walls with $\rho \leq 0.005$.

3.18 Spreadsheet layout for flanged sections

Additional data. In Sheet 3.9, changes are made to the 'Additional Data' section of Sheet 3.7 as follows. The breadth label in cell C22 and its value in cell D22 have been removed since this quantity is no longer required and will be replaced by the value of the width of the flange. The flange width is calculated in the 'Preliminary Calculations' section.

Three new values representing the masonry thickness (t), pocket spacing (S_p) and pocket width (w) have been entered in cells D30, G21 and G22 respectively.

There is also a change in that the number of bars in the pockets has been entered as a variable in cell D27 whereas previously it was assumed that there were 2 bars.

Preliminary calculations. Values of the flange thickness and width are determined from the conditions given in section 3.17. The formulae for calculating these values are entered in cells G24 and G25 which contain:

G24: **@IF(D30> =0.5*G26,0.5*G26,D30)**
G25: **@MIN(G22+(12*G24),G21,D29*1000/3)**

Both of these expressions could have been written either in the @IF or the @MIN forms and it will be found that there are usually a number of alternative ways to achieve the same result.

Limiting moment. The limiting moment, as expressed by Eqn 3.55, has been entered in cell F35 which contains the formula:

F35: **+G30*G25*G24*(G26−(0.5*G24))/1000000**

Limiting dimensions. The check on the ratio of span to effective depth to ensure that it does not exceed the limit specified in the code is entered in cell F36. This entry replaces the equivalent entry in cell F35 of Sheet 3.7 except that the limiting ratios now refer to walls rather than beams. The actual formula in cell F36 is:

F36: **@IF(D28="SS"#AND#D29*1000/G26< =35, "O.K",@IF(D28="CO"#AND#D29 *1000/G26< =45,"O.K", @IF(D28="CA"#AND#D29*1000/G26< =18 #AND#G28< =0.005, "O.K","Limit Exceeded")))**

which should be compared with the formula in cell F35 given in section 3.14.

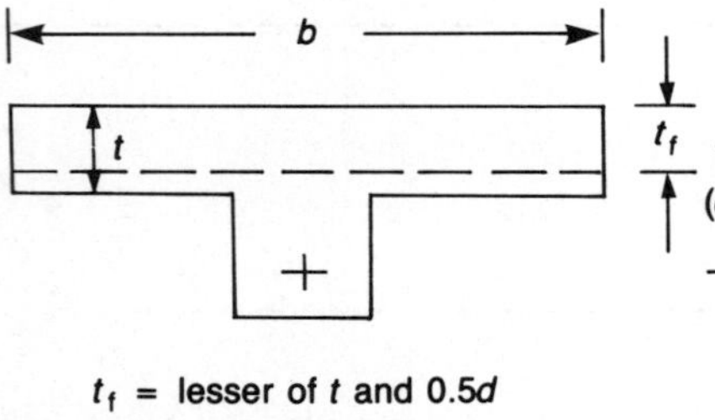

Figure 3.15 Pocket-type wall section

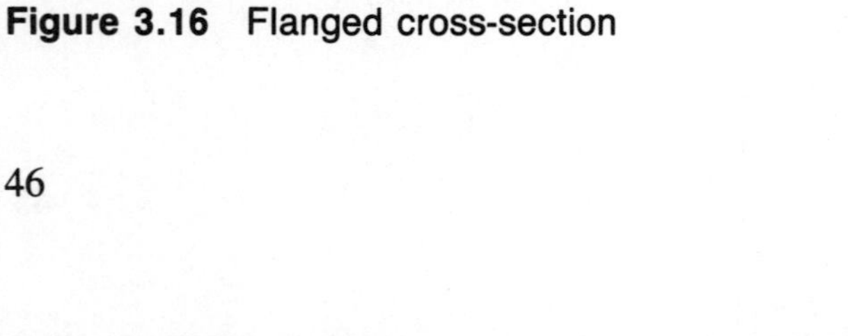

t_f = lesser of t and $0.5d$

Figure 3.16 Flanged cross-section

Example

The dimensions of the 'Pocket' type wall, to which Sheet 3.9 relates, is shown in Fig. 3.17.

Use of Sheet 3.9

Although Sheet 3.9 is very similar to Sheet 3.7 there are some differences in the additional data required for rows 22 to 30. These differences arise due to the requirement for defining the dimensions of the flange.

The additional data cells are D22 to D30 and G21 to G22. The results of the analysis are output in a similar method to that shown on Sheet 3.7.

Sheet 3.9

1/A	B	C	D	E	F	G	H
2		DESIGN FOR FLEXURE (Flanged Masonry Beams)					
3		--					
4	Case Selection				Preliminary Calculations		
5	---------------				----------------------		
6		BS5628		1		Epsilon_A	0.0009
7		Parabolic		2		Epsilon_y	0.0084
8		Rectangle CaseA		3		dc/d	0.2940
9		Rectangle CaseB		4		--------------	
10		Select Case ?		**1**	Calculation		
11		------------------------------------					
12	Initial Data					K1	0.667
13	------------					K2	0.375
14		Epsilon_u	**0.0035**			K3	1.000
15		fy	**250**			J1	0.500
16		Gms	**1.15**			J2	0.400
17		eu/em	**1.000**			BS5628	
18		--					
19							
20	Additional Data						
21	---------------			Pocket Spacing		**400**	
22		Depth	**216.000**	Pocket Width		**150**	
23		Cover	**20.000**	Preliminary Calculations			
24		fk	**19.200**	Flange Thickness		94	
25		Gmm	**2**	Flange Width		400.000	
26		Bar diam.	**16**		d	188.000	
27		No of bars	**2**		Area Stl.	402.124	
28	SS,CO,CA	Beam Type **SS**			rho	0.0053	
29	Height or	Span	**2**		fy/Gms	217	
30	Masonry Thickness		**100**		Fk/Gmm	9.6	
31		------------------			----------------		
32		Applied Moment		=	12.000 kNm		
33	Design Moment of Resistance			=	15.440 kNm		O.K
34	Limiting Design Moment			=	54.288 kNm		O.K
35	Additional Limiting Moment			=	50.895 kNm		O.K
36	Limit. Span/Effective Depth			=	O.K		
37		--					
38		DESIGN FOR SHEAR					
39		-----------------					
40	Additional Data			Preliminary Calculations			
41	---------------			--------------------------			
42		Gmv	**2.000**		Stress_v	0.532 N/mm2	
43		Shear Force	**40.000** kN		a/d	1.596	
44	Stirrup spacing				fv	0.932	
45	If required		**200.000**		fv	0.932	
46		--					
47	Shear Reinforcement			Yes	Diameter of Shear Stirrup		
48		Asv/sv	=	0.121	With 2 legs at max spacing		
49		Max Sv	=	141	Dia =	3.30 mm	
50		Sv	=	200.0	Dia =	3.93 mm	
51		---------------------------------------					

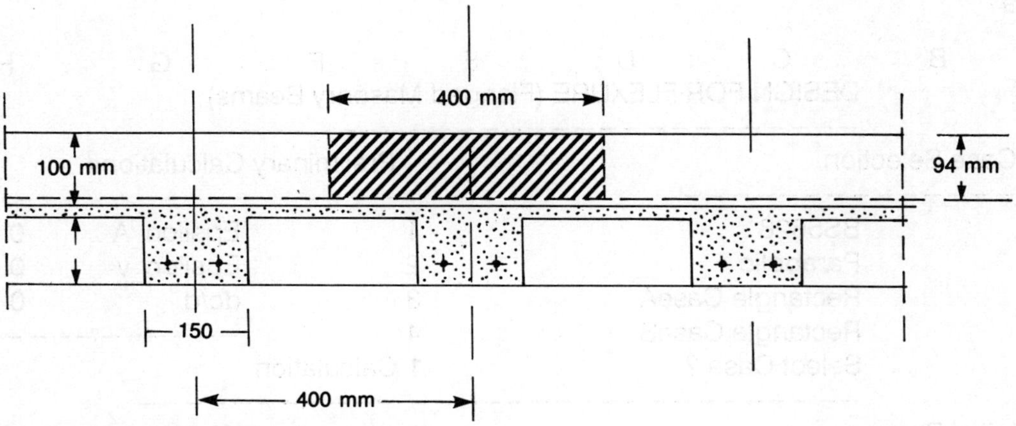

Figure 3.17 Wall dimensions (Sheet 3.9)

4 Reinforced columns

4.1 Introduction

In this chapter examples are given of methods of designing spreadsheets for use in the design of columns of either reinforced masonry or reinforced concrete. The design of such columns can be separated into a number of different categories and in the first part of the chapter a full explanation is given of the development of a series of spreadsheets for use with each category until finally a spreadsheet is developed which can cater for all cases. The categories which apply to columns of reinforced masonry are:

(a) *Short columns (slenderness ratio < 12)*
Uniaxial bending:
(i) For low values of the axial load;
(ii) For higher values of axial load;
(iii) An alternative approach to (ii) above.
Biaxial bending.

(b) *Slender columns (slenderness ratio > 12)*
Uniaxial bending.
Biaxial bending.

The slenderness ratio is defined as the effective height over the effective thickness of the column. The effective height is dependent on the support conditions of the columns. The limiting value of 12 is given in BS 5628, to distinguish between short and slender masonry columns, and the code also refers to an upper limit for slenderness ratios of 18 for cantilever columns and 27 for other columns.

In the second part of the chapter a more direct approach is used to develop spreadsheets for use in the design of columns of reinforced concrete. Design principles for the two materials are very similar and the experience gained with reinforced masonry design, in the first part, can be used to simplify the presentation in the second part. The categories which will be considered for reinforced concrete are:

(a) *Short columns (slenderness ratio < 15)*
Both uniaxial and biaxial bending.

(b) *Slender columns (slenderness ratio > 15)*
Both uniaxial and biaxial bending.

REINFORCED MASONRY

4.2 Short columns — uniaxial bending

For this case the loading consists of a design axial load N and a design bending moment M. The equations that follow assume that the plane of bending is in a direction perpendicular to the XX axis — see Fig. 4.1.

Basic equations

Case A. The basic equation for Case A is given by:

$$N_d = f_k b(t - e_x)/\gamma_{mm} \qquad [4.1]$$

Case A design is recommended when the design axial load N does not exceed the value of N_d given by Eqn 4.1 and for this case only a minimum amount of reinforcement is required. Although the equation refers to a limit on the axial load only, a limit is also implied for the design moment since the ratio M/N which appears in the equation as e_x, is not allowed to exceed 0.5t.

Case B. The basic equations for Case B are:

$$N_d = f_k b d_c/\gamma_{mm} + 0.83 f_y A_{s1}/\gamma_{ms} - f_{s2} A_{s2}/\gamma_{ms} \qquad [4.2]$$

$$M_d = 0.5 f_k b d_c (t - d_c)/\gamma_{mm} + 0.83 f_y A_{s1}(0.5t - d_1)/\gamma_{ms} + f_{s2} A_{s2}(0.5t - d_2)/\gamma_{ms} \qquad [4.3]$$

Most of these symbols have been defined previously and others are illustrated in Fig. 4.1.

To satisfy design criteria the values of N_d and M_d given by Eqn 4.2 and 4.3 must be greater than the applied load (N) and the applied moment (M). In other words:

Applied values must be < Design resistance values.

Case C. The basic design equation for Case C is given by:

$$M_a = M + N(t/2 - d_1) \qquad [4.4]$$

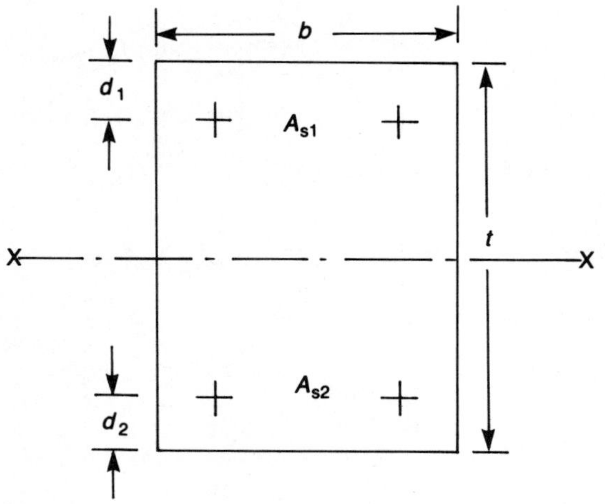

Figure 4.1 Column cross-section

and this case can be used as an alternative method to Case B when the eccentricity is greater than $(t/2 - d_1)$. When design is in accordance with Case C then the area of tension reinforcement may be reduced by:

$$\text{Area reduction} = N\gamma_{ms}/f_y \qquad [4.5]$$

Design strategy

The usual procedure adopted for the design of reinforced masonry columns is as follows:

Step 1. Values of the following parameters are either known or assumed for a trial design: N, M, f_k, f_y, γ_{mm}, γ_{ms}, b, t, d_1, d_2, A_{s1}, A_{s2}.

Step 2. From these values: e_x = M/N, N_d from Eqn 4.1, and $(t/2 - d_1)$ can be calculated.

Step 3. The appropriate design case which applies can now be determined:

(a) If N_d from Eqn 4.1 is > N and e_x is < $0.5t$ then Case A applies and only a minimum amount of reinforcement is required.
(b) If $e_x > (t/2 - d_1)$ then Case C can be applied, as an alternative to Case B design, by designing for M_a from Eqn 4.4 and ignoring the axial load. The area of tension reinforcement may be reduced using Eqn 4.5.
(c) If N_d from Eqn 4.1 is < N then Case B applies and a solution of Eqns 4.2 and 4.3 is required.

Step 4. If Case A or Case C design applies then the solution of the appropriate equations are straightforward. If Case B design applies then additional decisions and assumptions must be made in order to complete the design procedure. Case B design is therefore described in more detail in the next section.

Case B design. The main difficulty in obtaining solutions of the Eqns, 4.2 and 4.3, arises because f_{s2} is a function of the strain at the level of the lower reinforcing steel which is itself a function of d_c the depth of the section in compression. Therefore to obtain a solution, a value for the depth d_c, must be assumed so that the strain ϵ_2 can be calculated and the stress f_{s2} obtained from the stress/strain relationship for steel. Values of N_d and M_d can then be calculated from Eqns 4.2 and 4.3.

If in the above the value assumed for d_c is large (say equal to t) then, since the full depth of section would be carrying the axial load, the solution would result in a high value of N_d and a low value of M_d. If d_c is assumed to be small (say $0.1t$) then, since most of the section would be carrying the bending moment, the solution would result in low values of N_d and high values of M_d. This is illustrated in Fig. 4.2 which shows the typical graphical form of the solution of Eqns 4.2 and 4.3 over the full range of d_c/t values.

If point B in Fig. 4.2 represents the solution for $(d_c/t = 1)$ then for the design to be acceptable the applied values of N and M would have to be within the rectangle OABC. If point E represents the solution for $(d_c/t = 0.2)$ then for the design to be acceptable the applied values of N and M would have to be within the rectangle ODEF and within the curve. There are so many possibilities that to obtain a meaningful solution the designer has to solve the equations for a number of assumed values of d_c. If the values of N and M are outside the curve then new trial values for the section and/or the amount of steel must be introduced and the process repeated.

This process can obviously be tedious and charts have been produced to enable the solution to be obtained more easily for doubly reinforced rectangular sections with a symmetrical arrangement of the reinforcement in Hendry (1991).

It will be shown later that with the spreadsheet approach solutions can be obtained for rectangular sections with symmetrical or unsymmetrical reinforcement.

The relationship between the strain ϵ_2 and d_c is represented by the equation:

$$\epsilon_2 = 0.0035(t - d_c - d_2)/d_c \qquad [4.6]$$

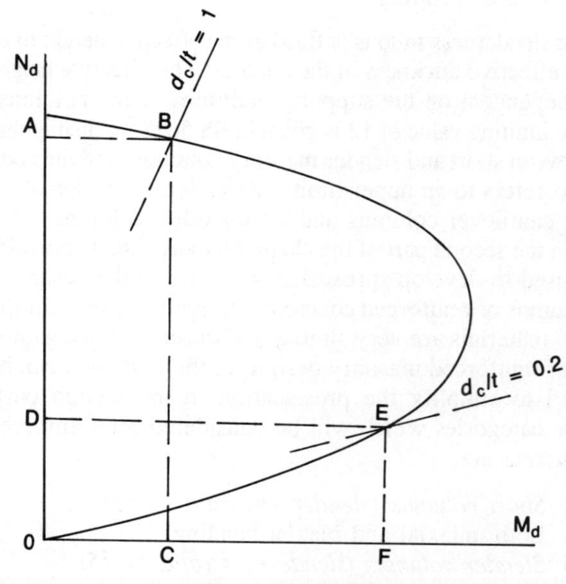

Figure 4.2 Idealised interaction curve

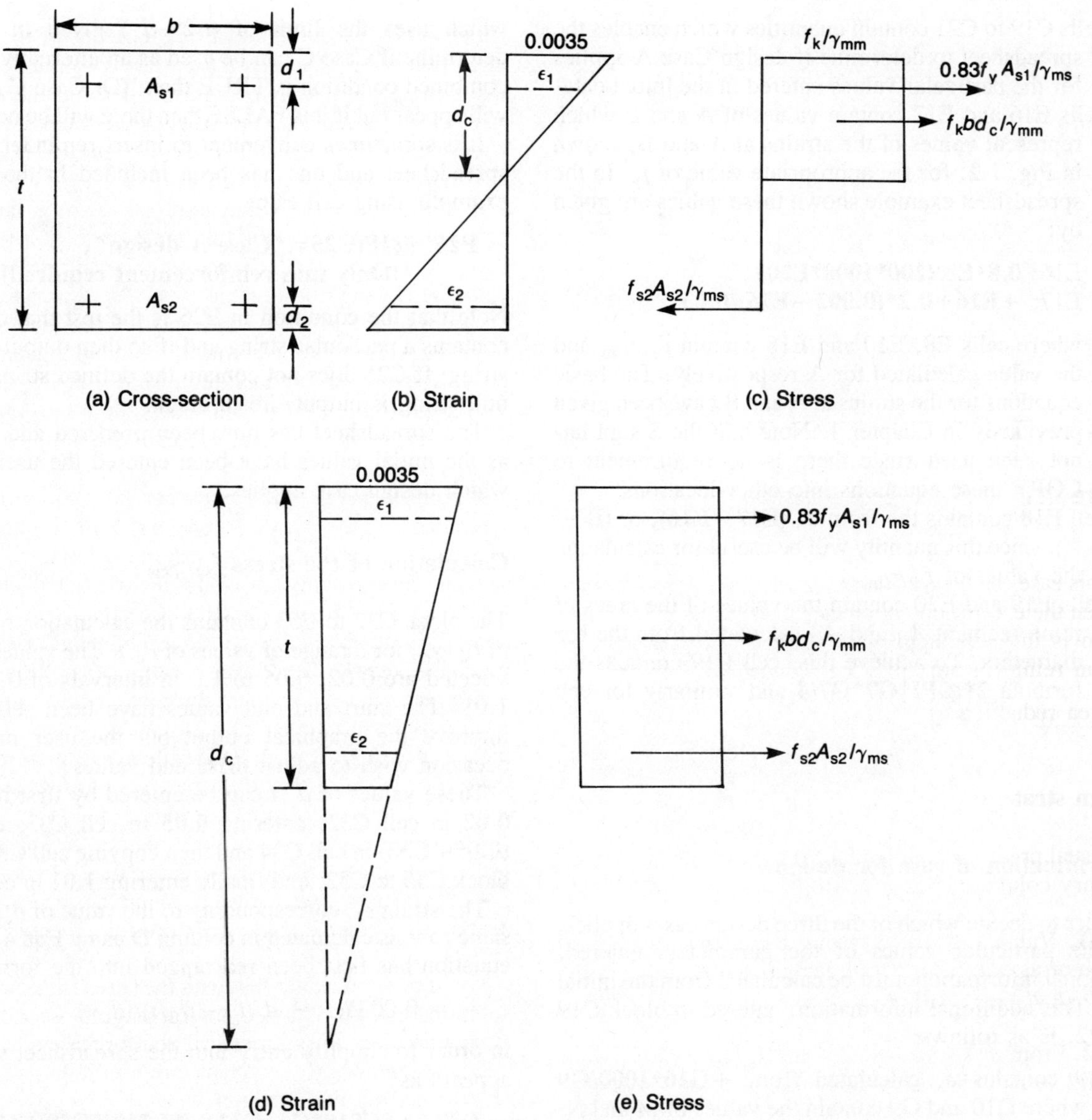

Figure 4.3 Stress and strain variation

which is derived from Fig. 4.3 which shows the strain variation throughout the depth of the section. Two cases are possible but both are represented by Eqn 4.6 if a distinction is made between positive tensile and negative compressive strain.

Note that Eqn 4.6 assumes that the maximum strain in the outermost compression fibre at failure is 0.0035.

Once the strain ϵ_2 has been determined the value of f_{s2} can be obtained from the stress/strain graph for steel. The link between ϵ_2 and f_{s2}, using the graph, is dependent on f_y and, in BS 5628, additional simplifying assumptions are made to make the link between f_2 and d_2 independent of f_y. These simplifying assumptions are not required in the spreadsheet solution and no advantage is gained by their use.

In Chapter 1 it was shown how the stress f_{s2} could be determined from the strain ϵ_2 by use of the @VLOOKUP type of function. The result can also be obtained using a series of nested @IF functions and this will be the method illustrated in this chapter.

4.3 Spreadsheet layout, short columns — uniaxial bending

Initial data. Values of the known or assumed parameters are entered in a convenient block of the spreadsheet. In the example (Sheet 4.1), the block B7 to G10 contains the labels and values of the following: N, M, f_k, f_y, γ_{mm}, γ_{ms}, b, t, d_1, d_2, and the diameters of the upper (Dia1) and lower (Dia2) reinforcing bars.

Preliminary calculations. Using the values of initial data the values of other quantities which are required in subsequent operations can be calculated. Details of these quantities are given below:

Cells C16 and C17 contain f_k/γ_{mm} and f_y/γ_{ms}.

Cell C18 contains the assumed value of the maximum compressive strain in the masonry. The code value of 0.0035 has been entered but this could be changed if necessary.

Cells C19 to C21 contain quantities which enables the spreadsheet to determine if design Case A applies for the particular values entered in the initial data.

Cells E16 and E17 contain values of A and B which represent values of the strains at A and B, shown in Fig. 1.2, for the appropriate value of f_y. In the spreadsheet example shown these values are given by:

E16: **0.8*E8/(200*1000*E10)**
E17: **+E16+0.2*(0.002−E16)/0.03**

where cells E8, E10 and E16 contain f_y, γ_{ms} and the value calculated for A respectively. The basic equations for the strains at A and B have been given previously in Chapter 1. Note that the $ sign has not been used since there is no requirement to COPY these equations into other locations.

Cell E18 contains the formula **(E17−E16)**, or (B − A), since this quantity will be useful for calculating the values of f_{s2}/γ_{ms}.

Cells E19 and E20 contain the values of the areas of reinforcement A_{s1} and A_{s2} calculated from the bar diameters. To achieve this, cell E19 contains the formula **2*@PI*G7*G7/4** and similarly for cell E20.

Determination of case for design

In order to decide which of the three design cases applies, for the particular values of the parameters entered, additional information must be calculated from the initial data. This additional information, entered in block C19 to C22, is as follows:

C19 contains e_x calculated from **+G10*1000/G9** where G10 and G9 contain the values of M and N.
C20 contains Eqn 4.1 in the form:

+E7*C7*(C8−(2*C19))/(E9*1000)

C21 contains $0.5t$ calculated from **0.5*C8.**
C22 contains the value $(t/2-d_1)$ in the form:

(C8/2)−C9

Using a basic @IF construction the spreadsheet can be made to output the design case which applies. For the present example this @IF function is placed in cell C25 in the following form:

C25: **@IF(G9<=C20#AND#C19<=C21,
"Case A design","Case B design")**

which uses the limit implied by Eqn 4.1 and the fact that e_x must not exceed $0.5t$ to determine whether 'Case A design' is TRUE. If it is FALSE the string 'Case B design' will be output.

A similar construction is used in cell E25, of Sheet 4.2, to distinguish between Case B and Case C design:

E25: **@IF(C25="Case B design"
#AND#C19>=C22,
"(Or Case C design)"," ")**

which uses the limit of $(t/2-d_1)$ given in C22 to determine if Case C can be used as an alternative. If the combined condition is TRUE then '(Or Case C design)' will appear but if it is FALSE then there will be no output.

It is sometimes convenient to insert reminders on the spreadsheet and one has been included in the present example using cell E26.

E26: **@IF(C25="Case A design",
"(Only min reinforcement required)"," ")**

Note that the condition in E26 is the test that cell C25 contains a particular string and if so then output another string. If C25 does not contain the defined string then a null string is output, i.e. a blank.

The spreadsheet has now been prepared and as soon as the initial values have been entered the user knows which design case applies.

Calculation of the stress f_{s2}/γ_{ms}

The block C32 to E53 contains the calculation of values of f_{s2}/γ_{ms} for a range of values of d_c/t. The values of d_c/t selected are 0.02, 0.05 to 1.0 in intervals of 0.05, and 1.01. The start and end values have been selected to improve the graphical output but the user might on occasion wish to adjust these end values.

These values of d_c/t can be entered by first inserting **0.02** in cell C32, entering **0.05** in cell C33, entering **(0.05+C33)** in cell C34 and then copying cell C34 to the block C35 to C52, and finally entering **1.01** in cell C53.

The strain ϵ_2, corresponding to the value of d_c/t in the same row, is calculated in column D using Eqn 4.6. This equation has first been rearranged into the form:

$$\epsilon_2 = 0.0035(1 - d_c/t - d_2/t)/(d_c/t).. \qquad [4.7]$$

in order to simplify entry into the spreadsheet where it appears as:

D32: **+$C$18*(1−C32−($C$10/$C$8))/C32**
D33: **+$G$18*(1−C33−($C$10/$C$8))/C33**

and similarly for the rest of the column. Entries in this column are obtained by entering the formula into cell D32 and then copying the contents of cell D32 to the block D33..D53.

Using the strains calculated in column C the corresponding steel stress can now be determined and these, expressed as f_{s2}/γ_{ms}, are shown in column E. In Chapter 1 this calculation was performed using the @VLOOKUP function but in this chapter, in order to illustrate an alternative method, a nested @IF function has been used.

Cells E32 and E33 contain:

E32: **@IF(D32<=−0.002,−0.83*$C$17,
@IF(D32>−0.002#AND#D32<=−$E$16,
−0.8*$C$17−0.03*$C$17
*(D32≠$E$16)/($E$16−0.002),
@IF(D32>−$E$16#AND#D32<$E$16,
0.8*$C$17*D32/$E$16,
@IF(D32>=$E$16#AND#D32<$E$17,
0.8*$C$17+0.2*$C$17*(D32−$E$16)/$E$18,
@IF(D32>=$E$17,$C$17,0)))))**

```
E33: @IF(D33<=-0.002,-0.83*$C$17,
     @IF(D33>-0.002#AND#D33<=-$E$16,
     -0.8*$C$17-0.03*$C$17
     *(D33+$E$16)/($E$16-0.002),
     @IF(D33>-$E$16#AND#D33<$E$16,
     0.8*$C$17*D33/$E$16,
     @IF(D33>=$E$16#AND#D33<$E$17,
     0.8*$C$17+0.2*$C$17*(D33-$E$16)/$E$18,
     @IF(D33>=$E$17,$C$17,0)))))
```

Only E32 need be entered directly since entries to the block E33..E53 can be achieved by copying. Each of these nested @IF functions divides the stress/strain curve into five regions and then determines which region contains the given value of strain. The stress can then be calculated by the use of a simple interpolation formula for the region. The arbitrary 0 value at the end is used only to complete the input requirements of the function.

Note that the maximum number of characters that can be entered in a formula, for most spreadsheets, is 250 and the entries in cells E32 to E53 are at this limit. In fact it was necessary to reduce the number of characters in the original formulae by using $E$18 instead of ($E$17-$E$16) thus saving 8 characters. Cell E18 contains the formula (E17-E16). There are other ways of achieving the same result. For example each of the five nested @IF functions can be placed in five separate columns so that they output the calculated value if the conditions are true and output zero if they are false. Each row of these five columns would then contain zero's in four of the columns and the calculated value in the other column. It would then be necessary to determine the absolute maximum of these five values.

Calculation of N_d and M_d

The equations in BS 5628 for N_d and M_d each contain three terms, the first relating to the brickwork in compression, the second relating to the compression steel and the third to the steel in tension. It is convenient to keep the calculation for each of these terms separate since the influence of any change in the value of the parameters can then be followed more readily.

The three parts of the equation for N_d are placed in columns B, C, and D starting at row 64. Similarly the three parts of the equation for M_d are placed in columns F, G and H. Values of N_d and M_d are placed in columns E and I.

The formulae placed in cells B64 to I64 are:

```
B64: ($C$16*$C$7*C32*$C$8)/1000
C64: (0.83*$C$17*$E$19)/1000
D64: (E32*$E$20)/1000
E64: (B64+C64-D64)
F64: (0.5*$C$16*$C$7*C32*$C$8
     *($C$8-(C32*$C$8)))/1000000
G64: (0.83*$C$17*$E$19
     *((0.5*$C$8)-$C$9))/1000000
H64: (E32*$E$20*((0.5*$C$8)-$C$10))/1000000
I64: (F64+G64+H64)
```

These entries can be copied to the block B65..I85 in the usual way. The numbers 1000 and 1000000 have been used to maintain the correct dimensions since N and M are usually quoted in kN and kNm respectively whilst all other parameters use Newtons and millimetres.

The spreadsheet is now complete and a direct solution can be obtained for any set of the parameters. Values of N_d and M_d are calculated for the complete range of d_c values and the designer can quickly ascertain if any changes are required and also which changes are most effective. These decisions are aided by use of the GRAPHICS command.

Use of graphical output

The graphics command enables the user to obtain an XY plot of N_d against M_d and this can be achieved by defining cells I64 to I85 as coordinates along the X axis and cells E64 to E85 as the corresponding Y coordinates. The location on this graph of the point with coordinates (N,M) indicates to the designer if any changes are necessary, see Fig. 4.4.

For the explanation that follows it is probably best to think of the point (N,M) as being a fixed point on the sheet and to think of the position of the curve as variable so that the curve can move outwards to engulf the point or inwards and thus move away from the point.

If the point (N,M) is in a region outside the curve, for example point A, then the steel areas and/or the section dimensions must be increased in order to move the curve outwards towards the point. If the point is inside the curve, such as point B the steel areas and/or section dimensions can be reduced in order to move the curve inwards. A point on the curve would represent the most economical set of properties.

In conjunction with the graphical output the designer is also able to study the effect of a change in any parameter by noting the changes that result in columns B, C, D, F, G, and H.

It will be found convenient if some additional values are placed in columns E and I as follows. The values of N and M are placed in cells E62 and I62 respectively and

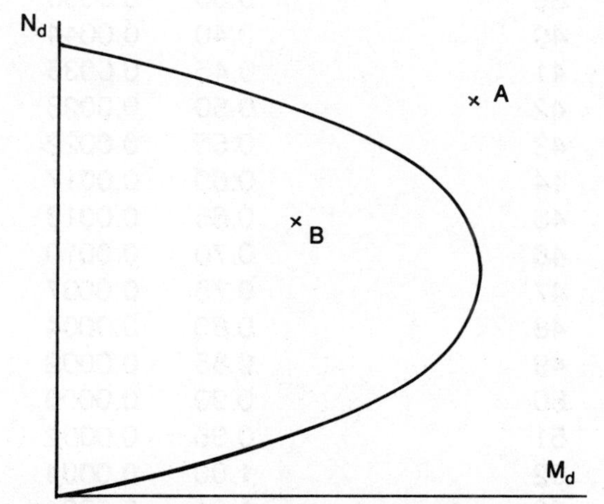

Figure 4.4 Typical interaction curve

Sheet 4.1 (Page 1)

1/A	B	C	D	E	F	G	H	I
2	Design of Short Columns of Reinforced Masonry							
3	--							
4	Uniaxial Bending – Cases A and B							
5	Initial Data							
6	------------							
7	Breadth	500	fk	10	Dia1	14		
8	Depth	500	fy	460	Dia2	14		
9	d1	50	Gmm	2	N	600		
10	d2	50	Gms	1.15	M	119		
11								
12	--							
13								
14	Preliminary Calculations							
15	------------------------							
16	Fk/Gmm	5.00	A	0.0016				
17	Fy/Gms	400.00	B	0.0043				
18	epsilon	0.0035	B–A	0.0027				
19	M/N(Ex)	198.33	As1	307.88				
20	Case_A	258.33	As2	307.88				
21	0.5t	250.00						
22	(t/2–d1)	200.00						
23	--							
24								
25	Case B Design							
26	---------------							
27								
28								
29	Calculation of Fs2/Gms							
30	------------------------							
31		dc/t	epsilon2	Fs2/Gms				
32		0.02	0.1540	400.00				
33		0.05	0.0595	400.00				
34		0.10	0.0280	400.00				
35		0.15	0.0175	400.00				
36		0.20	0.0122	400.00				
37		0.25	0.0091	400.00				
38		0.30	0.0070	400.00				
39		0.35	0.0055	400.00				
40		0.40	0.0044	400.00				
41		0.45	0.0035	377.00				
42		0.50	0.0028	356.00				
43		0.55	0.0022	338.82				
44		0.60	0.0017	324.50				
45		0.65	0.0013	269.23				
46		0.70	0.0010	200.00				
47		0.75	0.0007	140.00				
48		0.80	0.0004	87.50				
49		0.85	0.0002	41.18				
50		0.90	0.0000	0.00				
51		0.95	–0.0002	–36.84				
52		1.00	–0.0004	–70.00				
53		1.01	–0.0004	–76.24				
54								

Sheet 4.1 (Page 2)

58	B	C	D	E	F	G	H	I
59			Calculation of Nd and Md					
60			------------------------------					
61	N(brick)	N(As1)	N(As2)	Nd	Md(brick)	Md(As1)	Md(As2)	Md
62	Design Axial Load =			600		Design Moment=		119
63				0				0
64	25.00	102.21	123.15	4.06	6.13	20.44	24.63	51.20
65	62.50	102.21	123.15	41.56	14.84	20.44	24.63	59.92
66	125.00	102.21	123.15	104.06	28.13	20.44	24.63	73.20
67	187.50	102.21	123.15	166.56	39.84	20.44	24.63	84.92
68	250.00	102.21	123.15	229.06	50.00	20.44	24.63	95.07
69	312.50	102.21	123.15	291.56	58.59	20.44	24.63	103.67
70	375.00	102.21	123.15	354.06	65.63	20.44	24.63	110.70
71	437.50	102.21	123.15	416.56	71.09	20.44	24.63	116.17
72	500.00	102.21	123.15	479.06	75.00	20.44	24.63	120.07
73	562.50	102.21	116.07	548.65	77.34	20.44	23.21	121.00
74	625.00	102.21	109.60	617.61	78.13	20.44	21.92	120.49
75	687.50	102.21	104.31	685.40	77.34	20.44	20.86	118.65
76	750.00	102.21	99.91	752.31	75.00	20.44	19.98	115.42
77	812.50	102.21	82.89	831.83	71.09	20.44	16.58	108.11
78	875.00	102.21	61.58	915.64	65.62	20.44	12.32	98.38
79	937.50	102.21	43.10	996.61	58.59	20.44	8.62	87.66
80	1000.00	102.21	26.94	1075.28	50.00	20.44	5.39	75.83
81	1062.50	102.21	12.68	1152.04	39.84	20.44	2.54	62.82
82	1125.00	102.21	0.00	1227.21	28.12	20.44	0.00	48.57
83	1187.50	102.21	−11.34	1301.06	14.84	20.44	−2.27	33.02
84	1250.00	102.21	−21.55	1373.77	0.00	20.44	−4.31	16.13
85	1262.50	102.21	−23.47	1388.19	−3.16	20.44	−4.69	12.59

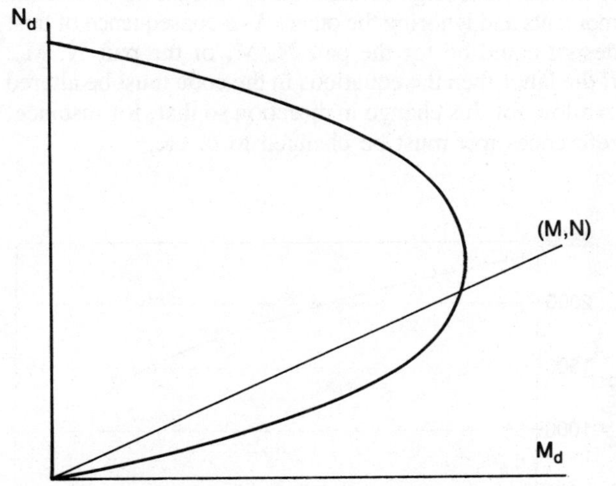

Figure 4.5 Applied M,N relative to curve

the value 0.0 placed in cells E63 and I63. If the X axis is now defined as I62 to I85 and the corresponding Y values as E62 to E85 then the graph will be drawn in the form shown in Fig. 4.5. The design curve for all values of d_c will be accompanied by a straight line from the origin to the point representing the values of N,M. The designer is therefore able to see at a glance what action is required to improve the design.

Example

Consider a design using the following values:

b = 500 mm;
t = 500 mm;
d_1 = 50 mm;
d_2 = 50 mm;
f_k = 10 N mm^{-2};
f_y = 460 N mm^{-2};
γ_{mm} = 2;
γ_{ms} = 1.15;
Dia1 = 14 mm;
Dia2 = 14 mm;
N = 600 kN;
M = 119 kN m

The results for this initial data are shown in Sheet 4.1.

The effect of changes to the bar diameters can be studied by obtaining solutions for a number of cases. For example with diameters of 12 mm, 14 mm and 16 mm the three

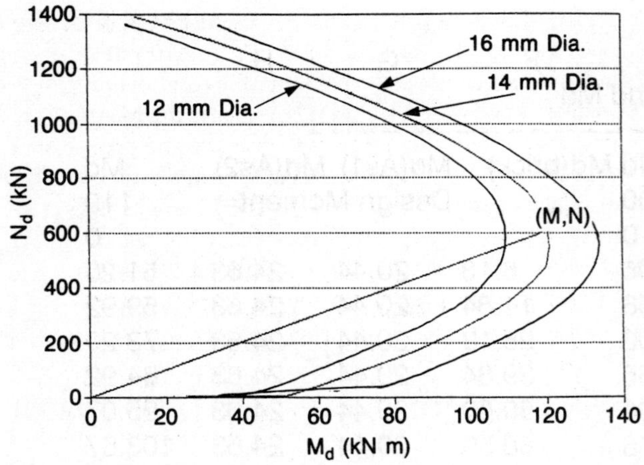

Graph 4.1 Uniaxial bending for short columns of reinforced masonry (A or B)

solutions can be represented by Graph 4.1. Note that only the centre curve was drawn using the spreadsheet, the other two curves were drawn by hand in order to illustrate the comparison. The point N, M is the same for all the diameters but the curves move inwards as the diameters decrease.

With four bars of 16 mm diameter the design is safe since the point representing N,M is within the curve. With four bars of 12 mm diameter the point is outside the curve and therefore the design unacceptable. With four bars of 14 mm diameter the point is just within the curve and therefore the design is safe and economical. In this example if all values except M are kept the same it can be shown that for values of M below 78 kN m Case A design applies and for values of M greater than 78 Case B applies. If the value of M is above 119 then Case C can be used as an alternative and this information will appear on the spreadsheet at cell E25. Additional details about Case C design are given below.

Design for Case C

As stated in section 4.2 the code allows an alternative set of design rules for cases in which $e_x > (t/2 - d_1)$ and the spreadsheet already indicates, in cell E25, when this form of design is permitted. Additional information can be included on the spreadsheet to assist with this form of design as follows.

The increased moment M_a, which has been renamed M_{ax} to indicate that it acts perpendicular to the X axis, is given by:

$$M_{ax} = M + N(t/2 - d_1) \qquad [4.8]$$

and can be entered in cell E21 (Sheet 4.2) in the form:

E21: **@IF(C19>=C22,G10+(G9*C22/1000),0)**

and the allowable reduction in the area of the tensile reinforcement $N\gamma_{ms}/f_y$ entered in cell G22 in the form:

G22: **@IF(C19>=C22,G9*E10*1000/E8,0)**

The @IF functions are used to ensure that the appropriate values are only calculated when Case C is permitted. The design can now be completed using M_{ax} and the principles outlined in Chapter 3 but making allowance for the possible reduction in the amount of tensile reinforcement in accordance with the value in cell G22.

Example

Consider a design using the following values:

b = 500 mm;
t = 600 mm;
d_1 = 50 mm;
d_2 = 60 mm;
f_k = 15 N mm^{-2};
f_y = 460 N mm^{-2};
γ_{mm} = 2;
γ_{ms} = 1.15;
Dia1 = 12 mm;
Dia2 = 14 mm;
N = 600 kN;
M = 150 kN m

The results for this example are shown in Sheet 4.2 and Graph 4.2. Case C can be applied and the column designed as a beam with a moment of 300 kN m, ignoring the axial force, as an alternative to the Case B approach.

4.4 Short columns — biaxial bending

For this case the loading consists of a design axial load N and design moments M_x and M_y where M_x is assumed to act in a plane of bending parallel to t and M_y parallel to b. In this section it is shown that the design philosophy is to treat the design as uniaxial by increasing one of the moments and ignoring the other. As a consequence of this, design could be for the pair N, M_x or the pair N, M_y. If the latter then the equations in the code must be altered to allow for this change in direction so that, for instance, references to t must be changed to b, etc.

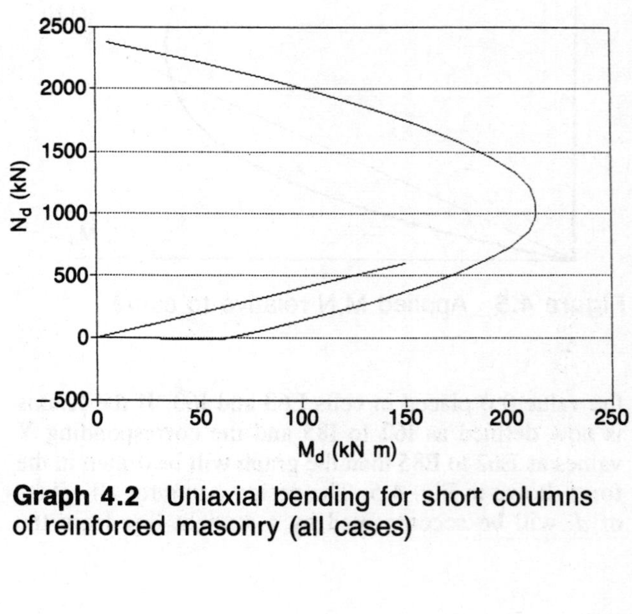

Graph 4.2 Uniaxial bending for short columns of reinforced masonry (all cases)

Basic equations

Where columns are subjected to biaxial bending they can be designed using the method outlined for uniaxial bending but using an increased moment. This increased moment is given by one of the equations:

$$M_{x'} = M_x + \alpha(h'/b')M_y \qquad [4.9]$$

$$M_{y'} = M_y + \alpha(b'/h')M_x \qquad [4.10]$$

depending on whether $M_x/h' \geq M_y/b'$, in which case Eqn 4.9 is used. Otherwise Eqn 4.10 is used.

In these equations h' and b' are dimensions in directions perpendicular to the X or Y axis respectively, Fig. 4.6. M_x and M_y are the design moments about the two axes (corresponding to M for the uniaxial case), and α is defined in the code as a coefficient derived from Table 4.1.

N in Table 4.1 is the design axial load and:

$$N_{dz} = f_k bt \qquad [4.11]$$

Note that Table 4.1 and the equations for $M_{x'}$ and $M_{y'}$ are different from those given in the BS code as originally published.

Design strategy

The strategy for design is similar to that outlined for Design Strategy in section 4.2 except that instead of using

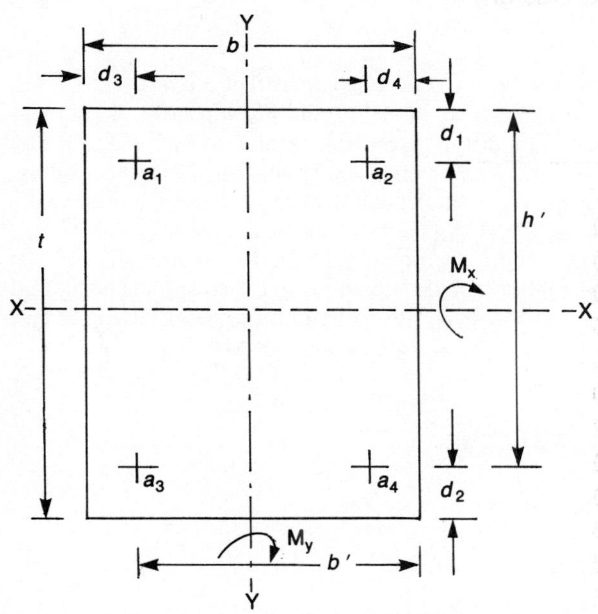

Figure 4.6 Cross-section, biaxial bending

Table 4.1

N/N_{dz}	α
0.0	1.00
0.1	0.88
0.2	0.77
0.3	0.65
0.4	0.53
0.5	0.42
≥ 0.6	0.30

N and M for Step 1 use must be made of either N and $M_{x'}$ or N and $M_{y'}$ so that some preliminary steps must be made to ascertain which case applies and to determine the values of $M_{x'}$ or $M_{y'}$. In other words the biaxial case is first converted into a uniaxial case and the procedures outlined previously for the solution of uniaxial bending applied. A change is required for the Case C solution in that the moment M_a could be either in the X or Y directions and allowance must be made for this.

The additional preliminary steps required are:

Step 1. Calculate N/N_{dz} using:

$$N/N_{dz} = N/(f_k bt) \qquad [4.12]$$

and determine α from the table.

Step 2. Determine which is the larger of M_x/h' and M_y/b'.

Step 3. Calculate either $M_{x'}$ or $M_{y'}$ as appropriate. If this calculation shows that the design parameters are N and $M_{x'}$ then proceed as described in section 4.2. If the design parameters are N and $M_{y'}$ then some of the equations and limiting conditions must be changed to allow for the change in direction. The required changes to Steps 1 to 4 of section 4.2 are listed below as 1a to 4a.

Step 1a. The bending moment M will now become M_x and M_y and it is convenient to represent the area of each bar separately using a_1, a_2, a_3, a_4 and to introduce some additional parameters d_3, d_4. These are shown in Fig. 4.6.

Step 2a. The value e_x is calculated from M_x/N and three additional calculations are required as follows:

$$e_y = M_y/N$$

$$N_d = f_k t(b - e_y)/\gamma_{mm} \qquad [4.1a]$$

and the limit of $(b/2 - d_3)$ on the eccentricity for Case C when N, $M_{y'}$ are used.

Step 3a. Step 3 in section 4.2 applies if the design is for N and $M_{x'}$. If N, $M_{y'}$ apply then changes are required as follows:

(a) If N_d from Eqn 4.1a is > N and e_y is < $0.5b$ then Case A applies.
(b) If $e_y > (b/2 - d_3)$ then Case C can be applied. It is convenient to replace M_a (Eqn 4.4) with M_{ax} and to introduce the term M_{ay} defined as:

$$M_{ay} = M_{y'} + N(b/2 - d_3) \qquad [4.4a]$$

(c) If N_d from Eqn 4.1a is < N then Case B applies and a solution of Eqns 4.2 and 4.3, modified as shown below, is required.

$$N_d = f_k t d_c/\gamma_{mm} + 0.83 f_y A_{s3}/\gamma_{ms} - f_{s2} A_{s4}/\gamma_{ms} \qquad [4.2a]$$

$$M_d = 0.5 f_k t d_c (b - d_c)/\gamma_{mm} + 0.83 f_y A_{s3}(0.5b - d_3)/\gamma_{ms} + f_{s2} A_{s4}(0.5b - d_4)/\gamma_{ms} \qquad [4.3a]$$

Sheet 4.2 (Page 1)

```
 1/A     B         C         D         E         F         G         H         I
 2 Design of Short Columns of Reinforced Masonry
 3 ----------------------------------------------------------------------------
 4          Uniaxial Bending – Cases A, B and C
 5                    Initial Data
 6                    ------------
 7 Breadth           500 fk              15 Dia1                12
 8 Depth             600 fy             460 Dia2                14
 9 d1                 50 Gmm              2 N                  600
10 d2                 60 Gms           1.15 M                  150
11
12 ----------------------------------------------------------------------------
13
14          Preliminary Calculations
15          ---------------------------
16 Fk/Gmm           7.50 A           0.0016
17 Fy/Gms         400.00 B           0.0043
18 epsilon        0.0035 B–A         0.0027
19 M/N(Ex)        250.00 As1         226.19
20 Case_A         375.00 As2         307.88
21 0.5t              300 Max            300
22 (t/2–d1)          250 Reduction in steel area =          1500
23 ----------------------------------------------------------------------------
24
25          Case B Design       (Or Case C Design)
26          --------------
27
28
29            Calculation of Fs2/Gms
30            ------------------------
31                  dc/t epsilon2  Fs2/Gms
32                  0.02   0.1540   400.00
33                  0.05   0.0595   400.00
34                  0.10   0.0280   400.00
35                  0.15   0.0175   400.00
36                  0.20   0.0122   400.00
37                  0.25   0.0091   400.00
38                  0.30   0.0070   400.00
39                  0.35   0.0055   400.00
40                  0.40   0.0044   400.00
41                  0.45   0.0035   377.00
42                  0.50   0.0028   356.00
43                  0.55   0.0022   338.82
44                  0.60   0.0017   324.50
45                  0.65   0.0013   269.23
46                  0.70   0.0010   200.00
47                  0.75   0.0007   140.00
48                  0.80   0.0004    87.50
49                  0.85   0.0002    41.18
50                  0.90   0.0000     0.00
51                  0.95  –0.0002   –36.84
52                  1.00  –0.0004   –70.00
53                  1.01  –0.0004   –76.24
54
```

Sheet 4.2 (Page 2)

57	B	C	D	E	F	G	H	I
58			Calculation of Nd and Md					
59			------------------------------					
60	N(brick)	N(As1)	N(As2)	Nd	Md(brick)	Md(As1)	Md(As2)	Md
61	Design Axial Load =			600		Design Moment=		150
62				0				0
63	45.00	75.10	123.15	−3.05	13.23	18.77	29.56	61.56
64	112.50	75.10	123.15	64.45	32.06	18.77	29.56	80.39
65	225.00	75.10	123.15	176.95	60.75	18.77	29.56	109.08
66	337.50	75.10	123.15	289.45	86.06	18.77	29.56	134.39
67	450.00	75.10	123.15	401.95	108.00	18.77	29.56	156.33
68	562.50	75.10	123.15	514.45	126.56	18.77	29.56	174.89
69	675.00	75.10	123.15	626.95	141.75	18.77	29.56	190.08
70	787.50	75.10	123.15	739.45	153.56	18.77	29.56	201.89
71	900.00	75.10	123.15	851.95	162.00	18.77	29.56	210.33
72	1012.50	75.10	116.07	971.53	167.06	18.77	27.86	213.69
73	1125.00	75.10	109.60	1090.49	168.75	18.77	26.30	213.83
74	1237.50	75.10	104.31	1208.28	167.06	18.77	25.04	210.87
75	1350.00	75.10	99.91	1325.19	162.00	18.77	23.98	204.75
76	1462.50	75.10	82.89	1454.71	153.56	18.77	19.89	192.23
77	1575.00	75.10	61.58	1588.52	141.75	18.77	14.78	175.30
78	1687.50	75.10	43.10	1719.49	126.56	18.77	10.34	155.68
79	1800.00	75.10	26.94	1848.16	108.00	18.77	6.47	133.24
80	1912.50	75.10	12.68	1974.92	86.06	18.77	3.04	107.88
81	2025.00	75.10	0.00	2100.10	60.75	18.77	0.00	79.52
82	2137.50	75.10	−11.34	2223.94	32.06	18.77	−2.72	48.11
83	2250.00	75.10	−21.55	2346.65	0.00	18.77	−5.17	13.60
84	2272.50	75.10	−23.47	2371.07	−6.82	18.77	−5.63	6.32

Step 4a. Equation 4.6 must be changed to:

$$\epsilon_2 = 0.0035(b - d_c - d_4)/d_c \qquad [4.6a]$$

4.5 Spreadsheet layout, short columns — biaxial bending

Initial data. The changes required in the initial data are the insertion of M_x, M_y, d_3, d_4, Dia3 and Dia4 and the removal of M. There are also some changes in the layout and these can be followed by reference to Sheet 4.3.

Preliminary calculations. In order to allow for the possibility that quantities may be required in either the X or Y direction a number of changes to Sheet 4.2 must be introduced both to the labels and the formulae. Most of these changes are based on the relative values of M_x/h' and M_y/b' since these are the quantities which decide which of Eqns 4.9 or 4.10 applies.

The values M_x/h' and M_y/b' are entered in cells G17 and G18 in the forms **+G12/C8** and **+G13/C7** respectively and changes are made to cells C20 to C23 as follows:

C20: **@IF(G17>=G18,G19*1000/G11, G20*1000/G11)**
C21: **E7/(E9*1000)*@IF(G17>=G18#ANDC20 >=C22,C7*(C8−(2*C20)), C8*(C7−(2*C20)))**
C22: **0.5*@IF(G17>=G18,C8,C7)**
C23: **@IF(G17>=G18,(C8/2)−C9,(C7/2)−C11)**

In this way the calculations are related to the X or Y direction as appropriate.

Cell E20 contains the formula **@PI*G7*G7/4** and cells E21 to E23 similar formulae to calculate the areas of the other bars.

Using Eqn 4.12 the value N/N_{dz} can be calculated, and inserted in cell E24, in the form:

E24: **+G11*1000/(E7*C7*C8)**

To calculate α the values given in Table 4.1 are required and these have been included on the spreadsheet in block B35..C42 and the additional column of differences, as described in Chapter 1, added in block D35..D40. For example cell D35 contains:

D35: **(C36−C35)/(B36−B35)**

The value of α can now be determined using @VLOOKUP functions as shown in cell E25 which contains the expression:

Sheet 4.3 (Page 1)

	A	B	C	D	E	F	G	H	I
2		Design of Short Columns of Reinforced Masonry							
3		--							
4		Uniaxial and Biaxial Bending (Cases A,B,C)							
5				Initial Data					
6				------------					
7		Breadth	**500**	fk	**15**	Dia1	**16**		
8		Depth	**600**	fy	**460**	Dia2	**14**		
9		d1	**50**	Gmm	**2**	Dia3	**15**		
10		d2	**60**	Gms	**1.15**	Dia4	**16**		
11		d3	**50**			N	**600**		
12		d4	**60**			Mx	**130**		
13						My	**70**		
14		--							
15			Preliminary Calculations						
16			------------------------						
17		Fk/Gmm	7.50	A	0.0016	Mx/h'	0.217		
18		Fy/Gms	400.00	B	0.0043	My/b'	0.140		
19		epsilon	0.0035	B−A	0.0027	Mx'	200.84		
20		M'/N	334.73	a1	201.06	My'	161.36		
21		Case_A	−260.50	a2	153.94	Mcase_bx	200.84		
22		0.5torb	300.00	a3	176.71	Max	350.84		
23		(t/2−d1)	250.00	a4	201.06	Reduction in			
24				N/Ndz	0.13	Area	1500		
25				Alpha	0.84				
26									
27		--							
28			Case B Design		(Or Case C Design)				
29			------------						
30			(Biaxial Bending)						
31									
32			Table		Calculation of Fs2/Gms				
33			-----		------------------------				
34		N/Ndz	Alpha	Diff.	dc/t	epsilon2	Fs2/Gms		
35		0.0	1.00	−1.20	0.02	0.1540	400.00		
36		0.1	0.88	−1.10	0.05	0.0595	400.00		
37		0.2	0.77	−1.20	0.10	0.0280	400.00		
38		0.3	0.65	−1.20	0.15	0.0175	400.00		
39		0.4	0.53	−1.10	0.20	0.0122	400.00		
40		0.5	0.42	−1.20	0.25	0.0091	400.00		
41		0.6	0.30		0.30	0.0070	400.00		
42		1.0	0.30		0.35	0.0055	400.00		
43					0.40	0.0044	400.00		
44					0.45	0.0035	377.00		
45					0.50	0.0028	356.00		
46					0.55	0.0022	338.82		
47					0.60	0.0017	324.50		
48					0.65	0.0013	269.23		
49					0.70	0.0010	200.00		
50					0.75	0.0007	140.00		
51					0.80	0.0004	87.50		
52					0.85	0.0002	41.18		
53					0.90	0.0000	0.00		
54					0.95	−0.0002	−36.84		
55					1.00	−0.0004	−70.00		
56					1.01	−0.0004	−76.24		
57									

Sheet 4.3 (Page 2)

A	B	C	D	E	F	G	H	I
62			Calculation of Nd and Md					
63			--					
64	N(brick)	N(As1)	N(As2)	Nd	Md(brick)	Md(As1)	Md(As2)	Md
65	Design Axial Load =			600	Moment+Biaxial Effect =			200.8
66	Design Axial Load =			600	Design Moment =			130
67				0				0
68	45.00	117.86	151.11	11.75	13.23	29.46	36.27	78.96
69	112.50	117.86	151.11	79.25	32.06	29.46	36.27	97.79
70	225.00	117.86	151.11	191.75	60.75	29.46	36.27	126.48
71	337.50	117.86	151.11	304.25	86.06	29.46	36.27	151.79
72	450.00	117.86	151.11	416.75	108.00	29.46	36.27	173.73
73	562.50	117.86	151.11	529.25	126.56	29.46	36.27	192.29
74	675.00	117.86	151.11	641.75	141.75	29.46	36.27	207.48
75	787.50	117.86	151.11	754.25	153.56	29.46	36.27	219.29
76	900.00	117.86	151.11	866.75	162.00	29.46	36.27	227.73
77	1012.50	117.86	142.42	987.94	167.06	29.46	34.18	230.71
78	1125.00	117.86	134.49	1108.37	168.75	29.46	32.28	230.49
79	1237.50	117.86	128.00	1227.36	167.06	29.46	30.72	227.25
80	1350.00	117.86	122.59	1345.27	162.00	29.46	29.42	220.89
81	1462.50	117.86	101.71	1478.65	153.56	29.46	24.41	207.44
82	1575.00	117.86	75.56	1617.30	141.75	29.46	18.13	189.35
83	1687.50	117.86	52.89	1752.47	126.56	29.46	12.69	168.72
84	1800.00	117.86	33.06	1884.80	108.00	29.46	7.93	145.40
85	1912.50	117.86	15.56	2014.80	86.06	29.46	3.73	119.26
86	2025.00	117.86	0.00	2142.86	60.75	29.46	0.00	90.21
87	2137.50	117.86	−13.92	2269.28	32.06	29.46	−3.34	58.19
88	2250.00	117.86	−26.44	2394.30	0.00	29.46	−6.35	23.12
89	2272.50	117.86	−28.80	2419.16	−6.82	29.46	−6.91	15.74

E25: @VLOOKUP(E24,B35..D41,1)
+(E24−(@VLOOKUP(E24,B35..D41,0)))
***(@VLOOKUP(E24,B35..D41,2))**

Using this value of α the values of $M_{x'}$ and $M_{y'}$ can be determined and the decision taken as to which of these moments should be used for the design. This is done in cells G19, G20 and G21 which contain:

G19: +G12+(E25*G13*C8)/C7
G20: +G13+(E25*G12+C7)/C8
G21: @IF(G13=0,G12,
@IF(G17>G18,G19,G20))

This last statement first checks if the input is for a uniaxial or biaxial case and then checks if M_x/h' is greater than M_y/b'. Cell G21 will therefore calculate the design value of the moment; for the uniaxial case M_x will be output and for the biaxial case either M_x or M_y as appropriate.

Although not essential it is convenient to allow the spreadsheet to select an appropriate label in cell F21 by using:

F21:
@IF(G17>=G18,"Mcase_bx","Mcase_by")

so that the user knows which direction is critical.

A similar approach is used for the label in cell F22 which contains:

F22: @IF(G17>=G18,"Max","May")

where M_{ax} and M_{ay} are the values of M_a in the X and Y directions respectively when Case C design applies. The appropriate value is calculated in cell G22 using Eqn 4.4 or equivalent.

G22: @IF(G17>=G18#AND#C20
>=C23,G21+G11*C23/1000,0)

With this formula a zero value would be output if Case C design does not apply.

The allowable reduction in area of the tensile reinforcement for Case C is calculated in cell G24 using:

G24: @IF(G22>0,G11*E10*1000/E8,0)

Calculation of f_{s2}/γ_{ms}

Since ϵ_2 is determined from either Eqn 4.6 or 4.6a a change is required to the values in the block F35 to F37. Typically cell F35 contains:

F35: ($C$19/E35)*(1−E35−@IF($G$17
>=$G$18,($C$10/$C$8),($C$12/$C$7)))

in which it is assumed that cell E35 refers to either d_c/t or d_c/b depending on whether or not G17 ≥ G18.

Calculation of N_d and M_d and graphical output

Although the calculation of these quantities is similar to that described in Section 4.3 for the uniaxial case certain changes are required to allow for the possibility that the moment can be in either the X or Y directions. These changes and the cells in which they occur are listed below:

C68: **(0.83*$C$18/1000)*@IF($G$17>=$G$18, ($E$20+$E$21),($E$20+$E$22))**
D68: **(G35/1000)*@IF($G$17>=$G$18, $E$22+$E$23,$E$21+$E$23)**
F68: **((0.5*$C$17*$C$7*E35*$C$8 *(1−E35))/1000000)*@IF($G$17 >=$G$18,$C$8,$C$7)**
G68: **0.83*$C$18/1000000*@IF($G$17>=$G$18, ($E$20+$E$21)*((0.5*$C$8)−$C$9), ($E$20+$E$22)*((0.5*$C$7)−$C$11))**
H68: **(G35/1000000)*@IF($G$17>=$G$18, ($E$22+$E$23)*((0.5*$C$8)−$C$10), ($E$21+$E$23)*((0.5*$C$7)−$C$12))**

Graphical output

Referring to Sheet 4.3 it is convenient to insert an additional row at row 65 and to place the following formulae in cells E65 and I65:

E65: **+G11**
I65: **@IF(G13>0,G21,G12)**

Cells E66 and I66 contain:

E66: +G11
I66: **@IF(G17>=G18,G12,G13)**

So that by including these values in the range of X and Y coordinates the straight line from the origin will show the original moments and also the increase due to the biaxial effect (Graph 4.3).

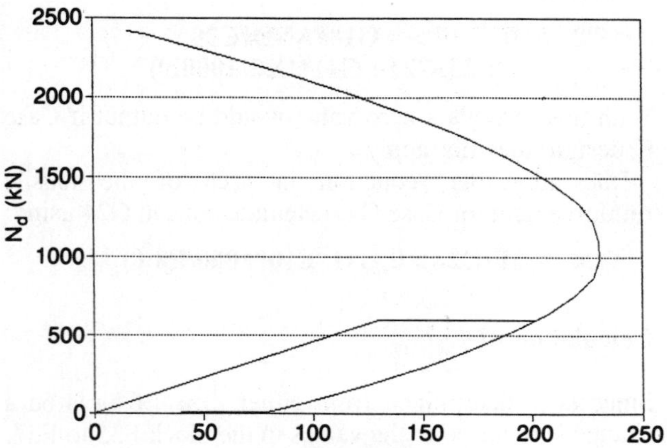

Graph 4.3 Biaxial bending for short columns of reinforced masonry

Example

Consider a design using the following values:

b = 500 mm;
t = 600 mm;
d_1 = 50 mm;
d_2 = 60 mm;
d_3 = 50 mm;
d_4 = 60 mm;
f_k = 15 N mm^{-2};
f_y = 460 N mm^{-2};
γ_{mm} = 2;
γ_{ms} = 1.15;
Dia1 = 16 mm;
Dia2 = 14 mm;
Dia3 = 15 mm;
Dia4 = 16 mm;
N = 600 kN;
M_x = 130 kN m;
M_y = 70 kN m

The results obtained are shown on Sheet 4.3 and Graph 4.3. The horizontal line at N = 600 represents the increase in M_x to $M_{x'}$ due to the biaxial effect. The graph shows that the design is economical.

4.6 Short columns — uniaxial or biaxial bending

Modifications to spreadsheet

With a very simple modification the spreadsheet can be made to determine whether the design being considered is for the case of uniaxial or biaxial bending and to proceed in accordance with this decision. This change makes the spreadsheet very powerful in that it will now deal with all possible cases of uniaxial or biaxial bending and also symmetrical or unsymmetrical sections.

To achieve this, the uniaxial case, where the design is for a particular set of N and M, must be considered as a design for N and M_x with M_y = 0. In other words the sheet must always include values for N, M_x and M_y but for the uniaxial case M_y would be entered as zero.

The formula in cell G21, which was defined in the preliminary calculations to Section 4.5, already allows for either uniaxial or biaxial bending so that the correct moment will be selected.

In order that subsequent printouts record which case has been considered it is convenient to place a reminder in cell C30:

C30: **@IF(G13=0,"(Uniaxial Bending)", "(Biaxial Bending)")**

Example

Consider a design using the following values:

b = 500 mm;
t = 600 mm;
d_1 = 50 mm;

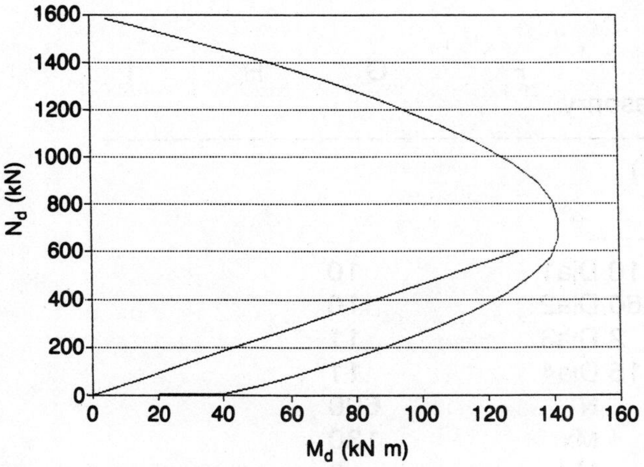

Graph 4.4 Uni/biaxial bending for short columns of reinforced masonry

d_2 = 60 mm;
d_3 = 50 mm;
d_4 = 60 mm;
f_k = 10 N mm^{-2};
f_y = 460 N mm^{-2};
γ_{mm} = 2;
γ_{ms} = 1.15;
Dia1 = 10 mm;
Dia2 = 10 mm;
Dia3 = 11 mm;
Dia4 = 11 mm;
N = 600 kN;
M_x = 130 kN m;
M_y = 0

This example can be considered as the uniaxial equivalent of the biaxial design given in Sheet 4.3 illustrating that the same spreadsheet can be used for either case. The results are shown in Sheet 4.4 and Graph 4.4.

4.7 Slender columns — uniaxial and biaxial bending

Basic design principles. The design principles outlined above for short columns also apply, with a slight modification, to slender columns. The modification arises because, due to the slenderness, there is an increased eccentricity which in turn increases the moment arm of the applied axial load, Fig. 4.7.

This results in an additional moment M_{as} expressed in the BS code as:

$$M_{as} = Nh_{eff}^2/2000t \qquad [4.13]$$

which must be added to the existing moment. In practice the slenderness ratios relative to the XX and YY axis can be different and it is convenient to define M_{asx} and M_{asy} such that:

$$M_{asx} = Nh_{effx}^2/2000t \qquad [4.14]$$

$$M_{asy} = Nh_{effy}^2/2000b \qquad [4.15]$$

where h_{effx} and h_{effy} are the effective heights measured in planes perpendicular to the XX and YY axes respectively.

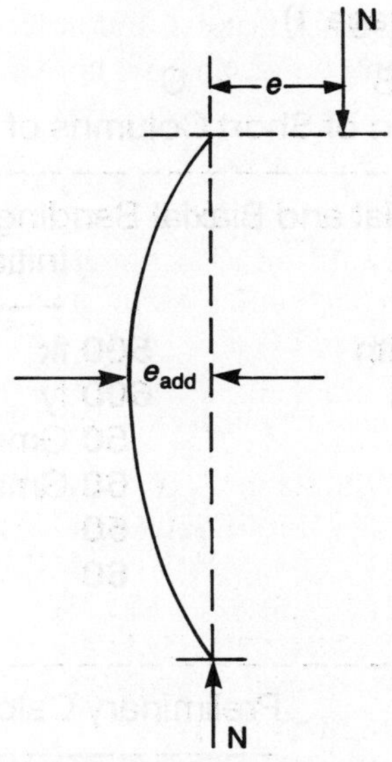

Figure 4.7 Additional eccentricity

Design strategy. The design strategy used above for short columns can also be applied to slender columns if the initial values of M_x and M_y are first increased by the values M_{asx} and M_{asy} respectively. Using these new values for M_x and M_y the procedure for design is identical with that described in the previous sections. Design for uniaxial or biaxial bending and also the preliminary calculation for Case C design can be achieved on one spreadsheet.

In Sheet 4.4 it was shown that, for short columns, both uniaxial and biaxial bending could be solved using the same spreadsheet. The changes required for the solution of slender columns can also be incorporated into the same spreadsheet and this is the approach developed below.

4.8 Spreadsheet layout for slender columns

Initial data. The values of the effective heights in planes perpendicular to the XX and YY axes are entered in cells C11 and C12 respectively (Sheet 4.5).

Preliminary calculations. Values of M_{asx} and M_{asy} are calculated in cells G25 and G26 based on Eqns 4.14 and 4.15.

```
G25: @IF(E11*1000/C8>=12,G11*E11*E11
     *1000/(2000*C8),0)
G26: @IF(E12*1000/C7>=12,G11*E12*E12
     *1000/(2000*C7),0)
```

If the column is slender in either direction then a non-zero value for M_{as} will be calculated for that direction. For short columns a zero value will result. The formulae in cells G17 to G21 are then amended so that they include these calculated values of M_{asx} and M_{asy}:

Sheet 4.4 (Page 1)

1	A	B	C	D	E	F	G	H	I
2	Design of Short Columns of Reinforced Masonry								
3	--								
4	Uniaxial and Biaxial Bending (Cases A,B,C)								
5				Initial Data					
6				------------					
7	Breadth		500	fk	10	Dia1	10		
8	Depth		600	fy	460	Dia2	10		
9	d1		50	Gmm	2	Dia3	11		
10	d2		60	Gms	1.15	Dia4	11		
11	d3		50			N	600		
12	d4		60			Mx	130		
13						My	0		
14	--								
15			Preliminary Calculations						
16			------------------------						
17	Fk/Gmm		5.00	A	0.0016	Mx/h'	0.217		
18	Fy/Gms		400.00	B	0.0043	My/b'	0.000		
19	epsilon		0.0035	B−A	0.0027	Mx'	130.00		
20	M'/N		216.67	a1	78.54	My'	83.42		
21	Case_A		200.00	a2	78.54	Mcase_bx	130.00		
22	0.5torb		300.00	a3	95.03	Max	0		
23	(t/2−d1)		250.00	a4	95.03	Reduction in			
24				N/Ndz	0.20	Area	0		
25				Alpha	0.77				
26									
27	--								
28			Case B Design						
29			-------------						
30			(Uniaxial Bending)						
31									
32			Table		Calculation of Fs2/Gms				
33			-----		----------------------				
34		N/Ndz	Alpha	Diff.	dc/t	epsilon2	Fs2/Gms		
35		0.0	1.00	−1.20	0.02	0.1540	400.00		
36		0.1	0.88	−1.10	0.05	0.0595	400.00		
37		0.2	0.77	−1.20	0.10	0.0280	400.00		
38		0.3	0.65	−1.20	0.15	0.0175	400.00		
39		0.4	0.53	−1.10	0.20	0.0122	400.00		
40		0.5	0.42	−1.20	0.25	0.0091	400.00		
41		0.6	0.30		0.30	0.0070	400.00		
42		1.0	0.30		0.35	0.0055	400.00		
43					0.40	0.0044	400.00		
44					0.45	0.0035	377.00		
45					0.50	0.0028	356.00		
46					0.55	0.0022	338.82		
47					0.60	0.0017	324.50		
48					0.65	0.0013	269.23		
49					0.70	0.0010	200.00		
50					0.75	0.0007	140.00		
51					0.80	0.0004	87.50		
52					0.85	0.0002	41.18		
53					0.90	0.0000	0.00		
54					0.95	−0.0002	−36.84		
55					1.00	−0.0004	−70.00		
56					1.01	−0.0004	−76.24		
57									

Sheet 4.4 (Page 2)

A	B	C	D	E	F	G	H	I
62			Calculation of Nd and Md					
63			------------------------------------					
64	N(brick)	N(As1)	N(As2)	Nd	Md(brick)	Md(As1)	Md(As2)	Md
65	Design Axial Load =			600	Design Moment =			130.0
66	Design Axial Load =			600	Design Moment =			130
67				0				0
68	30.00	52.15	76.03	6.12	8.82	13.04	18.25	40.10
69	75.00	52.15	76.03	51.12	21.38	13.04	18.25	52.66
70	150.00	52.15	76.03	126.12	40.50	13.04	18.25	71.78
71	225.00	52.15	76.03	201.12	57.38	13.04	18.25	88.66
72	300.00	52.15	76.03	276.12	72.00	13.04	18.25	103.28
73	375.00	52.15	76.03	351.12	84.38	13.04	18.25	115.66
74	450.00	52.15	76.03	426.12	94.50	13.04	18.25	125.78
75	525.00	52.15	76.03	501.12	102.38	13.04	18.25	133.66
76	600.00	52.15	76.03	576.12	108.00	13.04	18.25	139.28
77	675.00	52.15	71.66	655.50	111.38	13.04	17.20	141.61
78	750.00	52.15	67.66	734.49	112.50	13.04	16.24	141.78
79	825.00	52.15	64.40	812.75	111.38	13.04	15.46	139.87
80	900.00	52.15	61.68	890.47	108.00	13.04	14.80	135.84
81	975.00	52.15	51.17	975.98	102.38	13.04	12.28	127.69
82	1050.00	52.15	38.01	1064.14	94.50	13.04	9.12	116.66
83	1125.00	52.15	26.61	1150.54	84.37	13.04	6.39	103.80
84	1200.00	52.15	16.63	1235.52	72.00	13.04	3.99	89.03
85	1275.00	52.15	7.83	1319.32	57.37	13.04	1.88	72.29
86	1350.00	52.15	0.00	1402.15	40.50	13.04	0.00	53.54
87	1425.00	52.15	−7.00	1484.15	21.37	13.04	−1.68	32.73
88	1500.00	52.15	−13.30	1565.46	0.00	13.04	−3.19	9.84
89	1515.00	52.15	−14.49	1581.64	−4.55	13.04	−3.48	5.01

G17: **(G12+G25)/C8**
G18: **(G13+G26)/C7**
G19: **(G12+G25)+(E25*G13*C8)/C7**
G20: **(G13+G26)+(E25*G12*C7)/C8**
G21: **@IF(G13=0,G12+G25,@IF(G17 >=G18,G19,G20))**

In G21 the formula first tests if the case is uniaxial. If so it uses (M_x + M_{asx}) and if not it tests if (M_x + M_{asx})/h' is greater than (M_y + M_{asy})/b' in order to calculate the appropriate values of $M_{x'}$ and $M_{y'}$ using $M_{x'}$ + M_{asx} if G17 ≥ G18 and $M_{y'}$ + M_{asy} if it is not.

Calculation of f_{s2}/γ_{ms}, N_d and M_d

The formulae used for the calculation of f_{s2}, N_d and M_d for short columns apply also to slender columns so that no changes are required. It is convenient to enter some additional information in cells F30 and F31 so that the designer is aware of which case is being applied. These cells contain:

F30: **@IF(E11*1000/C8>=12, "Slender Col in XX","Short Col in XX")**
F31: **@IF(E12*1000/C7>=12, "Slender Col in YY","Short Col in YY")**

Use of graphical output

The graphical output is similar to that described above for the biaxial case for short columns except that the horizontal portion of the line starting at the origin represents the sum of the biaxial effect and the slenderness effect. The biaxial effect is defined as the increase applied to one of the biaxial moments to allow for the effect of the other moment. Similarly the slenderness effect is defined as the increase applied to one of the biaxial moments to allow for the effect of the slenderness moment in the other direction. For short columns and biaxial bending there will be a biaxial effect. For slender columns there will be a slenderness effect combined, for the biaxial case, with a biaxial effect.

Example

Consider a design using the following values:

b = 500 mm;
t = 600 mm;
d_1 = 50 mm;
d_2 = 60 mm;
d_3 = 50 mm;

Sheet 4.5 (Page 1)

1A	B	C	D	E	F	G	H	I
2	Design of Slender Columns of Reinforced Masonry							
3	--							
4	Uniaxial and Biaxial Bending (Cases A,B,C)							
5		Initial Data						
6		------------						
7	Breadth	**500**	fk	**15**	Dia1	**16**		
8	Depth	**600**	fy	**460**	Dia2	**14**		
9	d1	**50**	Gmm	**2**	Dia3	**15**		
10	d2	**60**	Gms	**1.15**	Dia4	**16**		
11	d3	**50**	heffx	**6**	N	**600**		
12	d4	**60**	heffy	**8**	Mx	**130**		
13					My	**60**		
14	--							
15		Preliminary Calculations						
16		------------------------						
17	Fk/Gmm	7.50	A	0.0016	Mx/h'	0.2167		
18	Fy/Gms	400.00	B	0.0043	My/b'	0.1968		
19	epsilon	0.0035	B−A	0.0027	Mx'	190.72		
20	M/N(Ex)	317.87	a1	201.06	My'	189.76		
21	Case_A	−134.00	a2	153.94	Mcase_bx	190.72		
22	0.5torb	300.00	a3	176.71	Max	340.72		
23	(t/2−d1)	250.00	a4	201.06	Reduction in			
24			N/Ndz	0.13	Area	1500		
25			Alpha	0.84	Masx	0.00		
26	Treat as beam for crack control				Masy	38.40		
27	--							
28		Case B Design		(Or Case C Design)				
29		------------						
30		(Biaxial Bending)			Short Col in XX			
31					Slender Col in YY			
32		Table		Calculation of Fs2/Gms				
33		-----		----------------------				
34	N/Ndz	Alpha	Diff.	dc/t	epsilon2	Fs2/Gms		
35	0.0	1.00	−1.20	0.02	0.1540	400.00		
36	0.1	0.88	−1.10	0.05	0.0595	400.00		
37	0.2	0.77	−1.20	0.10	0.0280	400.00		
38	0.3	0.65	−1.20	0.15	0.0175	400.00		
39	0.4	0.53	−1.10	0.20	0.0122	400.00		
40	0.5	0.42	−1.20	0.25	0.0091	400.00		
41	0.6	0.30		0.30	0.0070	400.00		
42	1.0	0.30		0.35	0.0055	400.00		
43				0.40	0.0044	400.00		
44				0.45	0.0035	377.00		
45				0.50	0.0028	356.00		
46				0.55	0.0022	338.82		
47				0.60	0.0017	324.50		
48				0.65	0.0013	269.23		
49				0.70	0.0010	200.00		
50				0.75	0.0007	140.00		
51				0.80	0.0004	87.50		
52				0.85	0.0002	41.18		
53				0.90	0.0000	0.00		
54				0.95	−0.0002	−36.84		
55				1.00	−0.0004	−70.00		
56				1.01	−0.0004	−76.24		
57								

Sheet 4.5 (Page 2)

A	B	C	D	E	F	G	H	I
62			Calculation of Nd and Md					
63			------------------------------------					
64	N(brick)	N(As1)	N(As2)	Nd	Md(brick)	Md(As1)	Md(As2)	Md
65	Design Axial Load =			600	Moment+ Biax + Slender =			190.72
66	Design Axial Load =			600	Design Moment =			130
67				0				0
68	45.00	117.86	151.11	11.75	13.23	29.46	36.27	78.96
69	112.50	117.86	151.11	79.25	32.06	29.46	36.27	97.79
70	225.00	117.86	151.11	191.75	60.75	29.46	36.27	126.48
71	337.50	117.86	151.11	304.25	86.06	29.46	36.27	151.79
72	450.00	117.86	151.11	416.75	108.00	29.46	36.27	173.73
73	562.50	117.86	151.11	529.25	126.56	29.46	36.27	192.29
74	675.00	117.86	151.11	641.75	141.75	29.46	36.27	207.48
75	787.50	117.86	151.11	754.25	153.56	29.46	36.27	219.29
76	900.00	117.86	151.11	866.75	162.00	29.46	36.27	227.73
77	1012.50	117.86	142.42	987.94	167.06	29.46	34.18	230.71
78	1125.00	117.86	134.49	1108.37	168.75	29.46	32.28	230.49
79	1237.50	117.86	128.00	1227.36	167.06	29.46	30.72	227.25
80	1350.00	117.86	122.59	1345.27	162.00	29.46	29.42	220.89
81	1462.50	117.86	101.71	1478.65	153.56	29.46	24.41	207.44
82	1575.00	117.86	75.56	1617.30	141.75	29.46	18.13	189.35
83	1687.50	117.86	52.89	1752.47	126.56	29.46	12.69	168.72
84	1800.00	117.86	33.06	1884.80	108.00	29.46	7.93	145.40
85	1912.50	117.86	15.56	2014.80	86.06	29.46	3.73	119.26
86	2025.00	117.86	0.00	2142.86	60.75	29.46	0.00	90.21
87	2137.50	117.86	−13.92	2269.28	32.06	29.46	−3.34	58.19
88	2250.00	117.86	−26.44	2394.30	0.00	29.46	−6.35	23.12
89	2272.50	117.86	−28.80	2419.16	−6.82	29.46	−6.91	15.74

d_4 = 60 mm;
f_k = 15 N mm^{-2};
f_y = 460 N mm^{-2};
γ_{mm} = 2;
γ_{ms} = 1.15;
h_{effx} = 6 m;
h_{effy} = 8 m;
Dia1 = 16 mm;
Dia2 = 14 mm;
Dia3 = 15 mm;
Dia4 = 16 mm;
N = 600 kN;
M_x = 130 kN m;
M_y = 60 kN m

Once these quantities have been entered the designer is informed that the column is short about the XX axis and slender about the YY axis. The column must be designed in accordance with Case B or Case C using: N = 600 kN and M_x = 190.72 kN m for Case B, or N = 0 and M_{ax} = 340.72 kN m for Case C design.

The results are shown on Sheet 4.5 and Graph 4.5.

Note that, even though the column is slender about the YY axis, design for this set of parameters is about the XX axis. For this case M_x is much greater than M_y so that the biaxial effect results in a larger increase about XX than the combined biaxial and slenderness effects about YY. By changing the relative values of the moments and effective heights it is possible to obtain a better understanding of this effect and readers are advised to try this.

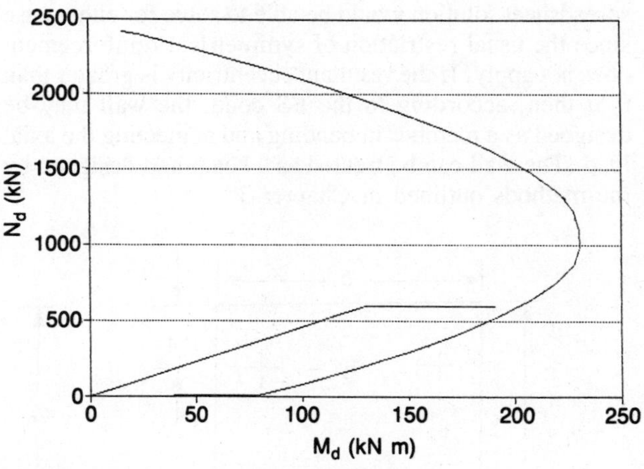

Graph 4.5 Uni/biaxial bending for slender columns of reinforced masonry

4.9 Cracking

Cracking due to bending is unlikely to be a problem if:

$$N > A_m f_k/2 \qquad [4.16]$$

where A_m represents the cross-sectional area of the masonry.

If N is less than this value, then the BS code suggests that the column should be treated as a beam as far as crack control is concerned, see Chapter 3.

This information has been included on Sheet 4.5 in cell B26 in the form:

B26: @IF(G11>=C7*C8*E7/(2*1000),
"Cracking O.K.",
"Treat as beam for crack control")

It will be found that, for the set of parameters included on Sheet 4.5, the output to cell B26 will be 'Treat as beam for crack control'. The column is considered to be lightly loaded for this particular design vertical load of 600 N. If the vertical load had been 2250 N then the output to cell B26 would have been 'Cracking O.K.' and no further action would have been required with regard to cracking. However such an increase in the value of the load would require changes to the values of some of the other parameters to make the design safe.

4.10 Walls

Short walls. The BS code suggests that the wall should be designed from first principles using the basic assumptions given earlier in BS 5628. However, since the design procedures described in this chapter are based on the same assumptions the spreadsheets developed previously can be used for both walls and columns.

Walls may be singly or doubly reinforced but the spreadsheet solution would be able to solve for either case since the usual restriction of symmetrical reinforcement does not apply. If the resultant eccentricity is greater than $0.5t$ then, according to the BS code, the wall may be designed as a member in bending and neglecting the axial load. The wall can be treated as a beam and designed by the methods outlined in Chapter 3.

Slender walls. When the slenderness ratio exceeds 12 the wall is designed following the same procedures as those given for slender columns. The spreadsheet approach described above can therefore be used for the design of both walls and columns.

REINFORCED CONCRETE

4.11 Short columns, rectangular sections

Uniaxial bending

Column design is based on similar assumptions to those used in Chapter 3 for the development of equations for the design of beams in which two alternative types of stress block were used. An idealised parabolic/rectangular stress block could be replaced by a simplified rectangular stress block. This simplification, which was introduced to reduce the amount of computation, is not required with the spreadsheet approach since the amount and complexity of the calculations is not paramount. Consequently the original parabolic/rectangular stress block will be used for the remainder of this chapter.

A sketch of a rectangular cross-section together with the assumed distribution of strain and stress is shown in Fig. 4.8.

In Fig. 4.8(b) the maximum strain has been taken as 0.0035 and the values of k_1 and k_2 can be obtained from equations which were given previously in Chapter 3 (Eqns 3.2 and 3.3).

Equating the forces on the section, as shown in Fig. 4.8(c), results in an equation for the design ultimate axial load:

$$N_d = k_1 f_{cu} b d_c + f_{s1} A_{s1} - f_{s2} A_{s2} \qquad [4.17]$$

and, by taking moments about the mid-depth, an equation for the maximum design moment in the form:

$$M_d = k_1 f_{cu} b d_c (h/2 - k_2 d_c) + f_{s1} A_{s1} (h/2 - d_1) + f_{s2} A_{s2} (h/2 - d_2) \qquad [4.18]$$

Although the partial safety factors are not shown in the above equations they are implied because k_1 already

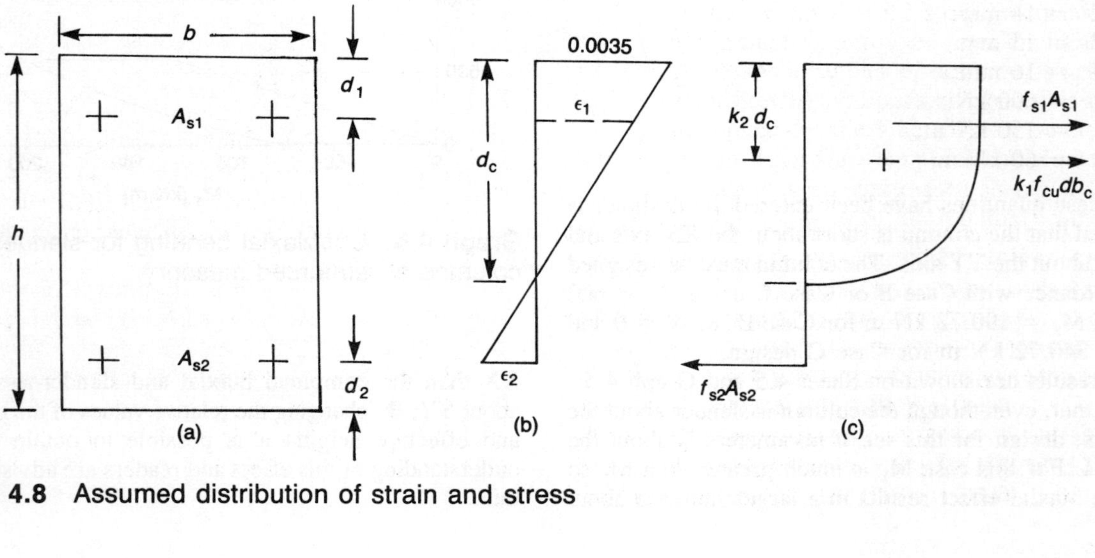

Figure 4.8 Assumed distribution of strain and stress

allows for a partial safety factor of 1.5 for the concrete and a factor of 1.15 will be included when determining f_{s1} and f_{s2} since the maximum steel stress is taken as $0.87f_y$ and not f_y (see Chapter 3).

In the British Standard code there is an additional restriction in that M_d must not be less than $N_d e_{min}$ where e_{min} is the smaller of $0.5h$ or 20 mm. Apart from this restriction the above equations define the problem completely and the only difficulty that arises is that d_c is unknown. In fact d_c can take any value from about $0.01h$ to $1.01h$ and for each assumed value a set of values (N_d,M_d) can be determined using Eqns 4.17 and 4.18. To achieve this, values of ϵ_1 and ϵ_2 must first be obtained by linear proportion from the strain distribution diagram and then values of the stress in the reinforcement obtained for these strains from the stress/strain curve.

This process is quite tedious and interaction diagrams have been produced which enable engineers to avoid a great deal of the calculation. However, only a limited number of curves can be drawn on each interaction diagram and only a limited number of interaction diagrams are available. It may therefore be necessary for the designer to interpolate between curves and/or diagrams in order to obtain a solution for a particular set of parameters.

In contrast the spreadsheet solution is a solution for a selected set of values and by using the graphical function of the spreadsheet the designer is able to obtain a single interaction curve representing the selected values of the variables. Examples of this graphical output were given in relation to the design of columns of reinforced masonry and further examples will be given below. For a design to be acceptable the plot of the applied axial load and applied moment must be within the interaction curve, see Fig. 4.4.

Biaxial bending

Biaxial bending for reinforced concrete is treated in a similar manner to that for reinforced masonry and Eqns 4.9 to 4.11 and the values in Table 4.1 apply to both materials.

Using the notation used in BS 8110 the equations become:

If $M_x/h' \geq M_y/b'$ then use:

$$M_{x'} = M_x + \alpha M_y(h'/b') \qquad [4.19]$$

If $M_x/h' < M_y/b'$ then use:

$$M_{y'} = M_y + \alpha M_x(b'/h') \qquad [4.20]$$

where h' and b' are as illustrated in Fig. 4.9 and α is obtained from Table 4.1 using the parameter $N_d/f_{cu}bh$ for entry into the table.

In Eqns 4.19 and 4.20 M_x and M_y are the design moments and the effect of these equations is to transform the case of biaxial bending into a uniaxial case by using an increased moment. This increased moment is either $M_{x'}$ or $M_{y'}$ depending on the limiting condition stated above the equations. Once this increased moment has been determined then design follows the same procedure as that given for the uniaxial case.

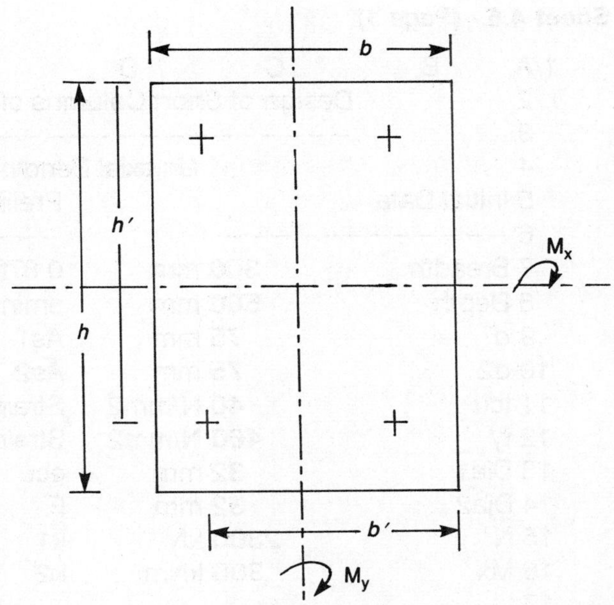

Figure 4.9 Additional notation

4.12 Spreadsheet layout, short columns — uniaxial bending

In this example a rectangular section containing two levels of reinforcement is considered. Refer to Sheet 4.6.

Initial data. The initial data is entered in cells C7 to C16. Values of b, h, d', d_2', f_{cu}, f_y, the diameters of the upper and lower reinforcement, the applied axial load and the bending moment are required. For convenience a value for the maximum strain in the concrete has been entered in cell F13 although there is no strict requirement for this since a constant value of 0.0035, as recommended by the British Standard code, has been used throughout the chapter.

Preliminary calculations. Calculations similar to those described in Chapter 3 are performed in cells F7 to F16. The value of e_{min} in F8 is based on the lesser of $0.5h$ and 20 mm so that this cell contains:

F8: **@MIN(0.05*C8,20).**

The area of the upper and lower reinforcement, assuming two bars at each level, are calculated in F9 and F10 and the strains at the change points, B and A, of the idealised stress strain curve for the reinforcement calculated in F11 and F12 (see Fig. 1.1).

To simplify the cell entries for calculating the values of k_1 and k_2 in F15 and F16 a preliminary calculation for F is made in F14, where F is defined as part of Eqns 3.2 and 3.3 in Chapter 3. The formulae in these cells are:

F14: **(@SQRT(C11))/17.5**
F15: **0.45*(1-(F14/3))**
F16: **(((2-F14)^2)+2)/((4*(3-F14)))**

Sheet 4.6 (Page 1)

1/A	B	C	D	E	F	G	H	I
2		Design of Short Columns of Reinforced Concrete						
3	– –							
4			Uniaxial Bending					
5	Initial Data			Preliminary Calculations				
6	– – – – – – – – – – – – – –			– –				
7	Breadth	300	mm	0.87fy	400.20			
8	Depth	500	mm	emin	20			
9	d'	75	mm	As1	1608.50			
10	d2	75	mm	As2	1608.50			
11	fcu	40	N/mm2	StrainB	−0.0020			
12	fy	460	N/mm2	StrainA	0.0020			
13	Dia1	32	mm	ecu	0.0035			
14	Dia2	32	mm	F	0.361			
15	N	2300	kN	k1	0.396			
16	Mx	300	kNm	k2	0.444			
17	– –							
18	Calculation of Strains & Stresses							
19	– –							
20	dc/h	epsilon1	epsilon2	fs1	fs2			
21	0.184	0.0006	0.0127	129.35	400.20			
22	0.23	0.0013	0.0092	251.28	400.20			
23	0.28	0.0017	0.0070	330.28	400.20			
24	0.33	0.0019	0.0054	385.63	400.20			
25	0.38	0.0021	0.0042	400.20	400.20			
26	0.43	0.0023	0.0034	400.20	400.20			
27	0.48	0.0024	0.0026	400.20	400.20			
28	0.53	0.0025	0.0021	400.20	400.20			
29	0.58	0.0026	0.0016	400.20	318.84			
30	0.63	0.0027	0.0012	400.20	238.49			
31	0.68	0.0027	0.0008	400.20	169.88			
32	0.73	0.0028	0.0006	400.20	110.63			
33	0.78	0.0028	0.0003	400.20	58.93			
34	0.83	0.0029	0.0001	400.20	13.43			
35	0.88	0.0029	−0.0001	400.20	−26.92			
36	0.93	0.0029	−0.0003	400.20	−62.96			
37	0.98	0.0030	−0.0005	400.20	−95.33			
38	1.03	0.0030	−0.0006	400.20	−124.56			
39	1.08	0.0030	−0.0008	400.20	−151.11			
40	1.13	0.0030	−0.0009	400.20	−175.31			
41	1.18	0.0031	−0.0010	400.20	−197.47			

Calculation of strains and stresses

These calculations are placed in block B21 to F41 and each row of the block is based on a 'trial' value of d_c/h. These values of d_c/h have been copied from the range C54 to C74 and at this stage are not changed by the user. Details of the values of d_c/h used will be given later with reference to the calculation of N_d and M_d. For any given value of d_c/h the strains ϵ_1 and ϵ_2 are given by:

$$\epsilon_1 = (1 - d_1/d_c)0.0035 \qquad [4.21]$$

$$\epsilon_2 = (h/d_c - 1 - d_2/d_c)0.0035 \qquad [4.22]$$

and these equations appear in block C21 to D41 in the form:

C21: (1−(($C$9/$C$8)/B21))*$F$13
D21: ((1/B21)−1−(($C$10/$C$8)/B21))*$F$13
C22: (1−(($C$9/$C$8)/B22))*$F$13
D22: ((1/B22)−1−(($C$10/$C$8)/B22))*$F$13

and similar formulae for the remainder.

The stresses can be determined from these strains either

Sheet 4.6 (Page 2)

47								
48	B	C	D	E	F	G	H	I
49		Calculation of N and M				Graphical Co-ords		
50		------------------------------------				------------------		
51							Nd	Md
52							2300	300
53	*-*-*	dc/h	Nd	Md	Md/Nd		0	0
54	Trial->	**0.1840**	1	240	186811		1	240
55	*-*-*	0.23	316	293	928		316	293
56		0.28	562	332	590		562	332
57		0.33	770	361	469		770	361
58		0.38	912	376	412		912	376
59		0.43	1031	384	372		1031	384
60		0.48	1149	389	339		1149	389
61		0.53	1268	392	309		1268	392
62		0.58	1518	369	243		1518	369
63		0.63	1766	344	195		1766	344
64		0.68	1995	320	160		1995	320
65		0.73	2209	296	134		2209	296
66		0.78	2411	271	112		2411	271
67		0.83	2603	245	94		2603	245
68		0.88	2786	218	78		2786	218
69		0.93	2963	190	64		2963	190
70		0.98	3134	160	51		3134	160
71		1.03	3300	128	39		3300	128
72		1.08	3461	94	27		3461	94
73		1.13	3619	59	16		0	0
74		1.18	3773	21	6		0	0

by setting up a stress/strain table of values and using the @VLOOKUP function or by use of an @IF function. Examples of both these methods have been given previously and the latter method is used in block E21 to F21 as illustrated below:

E21: **@IF(C21< =$F$11, – $F$7,**
@IF(C21> =$F$12,$F$7,
(($F$7*C21)/($F$12))))
F21: **@IF(D21< =$F$11, – $F$7,**
@IF(D21> =$F$12,$F$7,
(($F$7*D21)/($F$12))))
E22: **@IF(C22< =$F$11, – $F$7,**
@IF(C22> =$F$12,$F$7,
(($F$7*C22)/($F$12))))
F22: **@IF(D22< =$F$11, – $F$7,**
@IF(D22> =$F$12,$F$7,
(($F$7*D22)/($F$12))))

This fairly standard approach for determining the stresses is illogical with regard to the sign of the stresses since a positive stress for f_{s1} implies compression whilst a positive f_{s2} stress implies tension. This apparent discrepancy is taken care of in Eqn 4.17 by using a negative value of f_{s1} in the equation. The convention is simple when dealing with only two levels of reinforcement but becomes less so when there are many levels since the method of solution requires the use of a range of values of d_c some of which could result in a change of sign in Eqns 4.17 and 4.18. An illustration of how this can be dealt with is given in the next example.

Calculation of N_d and M_d

As stated in the section above, the initial values of d_c/h are entered in the range C54 to C74 and these values are then automatically copied to the range C21 to C41 for calculating the strains and stresses as shown. Additionally, of the range C54 to C74, the user is only required to enter a value in cell C54 since other values of d_c/h are determined automatically by the spreadsheet.

The entry into cell C54 is entered by the user, by trial, until the value of N_d in D54 is both small and positive. Using this trial value as a starting point ensures that the interaction curve is defined near the point where it cuts the M_d axis.

The formula for obtaining the value of d_c/h for the remaining cells of column C is simply the value found in C54 incremented progressively down the column by 0.05. Cells C55 and C56 contain:

C55: **(C54+0.05)**
C56: **(C55+0.05)**

This ensures that the full range of possible N_d values will be used for plotting the ($N_d - M_d$) graph. The range of d_c/h shown on Sheet 4.6 is probably longer than required but this range will be reduced when values are selected later for plotting the interaction curve.

Formulae based on Eqns 4.17 and 4.18 are now used to determine the values of N_d and M_d and typical examples of these are:

D54: **($F$15*$C$11*$C$7*B21*$C$8)/1000 +(E21*$F$9)/1000−(F21*$F$10)/1000**
D55: **($F$15*$C$11*$C$7*B22*$C$8)/1000 +(E22*$F$9)/1000−(F22*$F$10)/1000**
E54: **(($F$15*$C$11*$C$7*B21*$C$8 *(($C$8/2)−($F$16*B21*$C$8)) +(E21*$F$9*(($C$8/2)−$C$9)) +(F21*$F$10*(($C$8/2)−$C$10))))/1000000**
E55: **(($F$15*$C$11*$C$7*B22*$C$8 *(($C$8/2)−($F$16*B22*$C$8)) +(E22*$F$9*(($C$8/2)−$C$9)) +(F22*$F$10*(($C$8/2)−$C$10))))/1000000**

The minimum value of the ratio M_d/N_d was calculated in cell F8 and it is now necessary to determine if the calculated values of N_d and M_d satisfy this limit. As a preliminary, the ratio of M_d/N_d is first calculated in range F54 to F75 using:

F54: **(E54*1000/D54)**

and these values compared with the minimum value from F8 in order to determine the final acceptable values of N_d and M_d. The comparison is made in the block H54 to I74 using:

H54: **@IF(F54>$F$8,D54,0)**
I54: **@IF(F54>$F$8,E54,0)**
H55: **@IF(F55>$F$8,D55,0)**
I55: **@IF(F55>$F$8,E55,0)**

so that zero values will be output if the ratio is below the minimum.

Interaction diagram

The graphical facility of the spreadsheet can now be used to produce an interaction diagram which is specific to the initial data. In order to improve the presentation some additional information is entered in the block H52 to I53.

Values of the design axial load and design moment are transferred to H52 and I52 respectively and zero values are entered in H53 and I53. The effect of this is to introduce an additional line on the interaction diagram drawn from the origin to the point with coordinates (M,N) of the design loads.

The X axis is now defined as the range I52 to I74 and the Y axis as H52 to H74 and the resulting graph is as illustrated in Graph 4.6.

Use of Sheet 4.6

To avoid possible errors it is now convenient to enable the protection facility of the spreadsheet so that initially the whole sheet is protected. Then the cells which require changes for dealing with different designs can be unprotected so that the user would be warned if an illegal entry was attempted. Following this procedure the only unprotected cells would be: C7 to C16 and C54.

To use the spreadsheet the user would enter the required values in C7 to C16 and then adjust the trial value in C54 until the value of N_d in D4 becomes small and positive. For example, in Sheet 4.6, trial values were placed in C54 until a value of 0.184 resulted in a value of 1 in D54.

The interaction diagram can now be called by pressing the appropriate function key related to the graphical facility. If this diagram shows that the end point of the straight line from the origin is within the curve then the design is acceptable although if the point is too far away from the curve then the amount of reinforcement could be reduced. If the point is outside the curve then the amount of reinforcement must be increased.

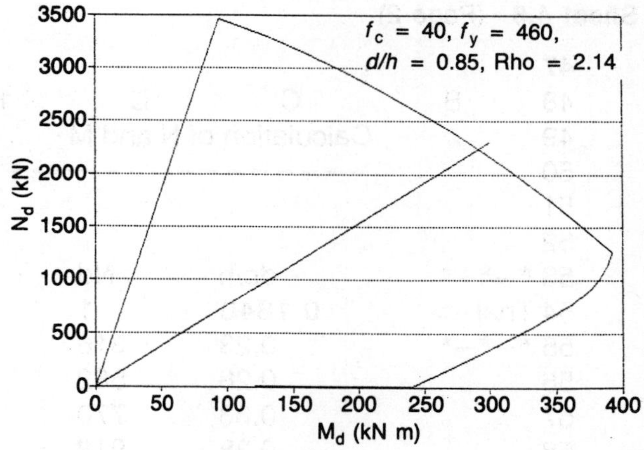

Graph 4.6 Uniaxial bending for short columns of reinforced concrete

4.13 Alternative design for short columns — uniaxial bending

In this example a rectangular section containing many levels of reinforcement is considered and, in order to ensure that the forces and moments are taken in the correct direction, the sign convention used must be defined. The convention adopted is illustrated in Fig. 4.10 in which the depth to the reinforcement is defined as h_{x1}, h_{x2}, h_{x3} . . . etc. so that in general the strain at any level is defined by:

$$\epsilon_x = (1 - h_x/d_c)\epsilon_{cu} \quad [4.23]$$

If $h_x < d_c$ then the strain at the level of the reinforcement determined using Eqn 4.23 will be positive. If $h_x > d_c$ then the strain will be negative. Stresses determined from these strains will retain the same sign. This convention will take care of the force balance equation, Eqn 4.17, except that all terms must initially be considered as positive and allowance for any negative forces determined automatically using the above sign convention. Equation 4.17 can therefore be rewritten as:

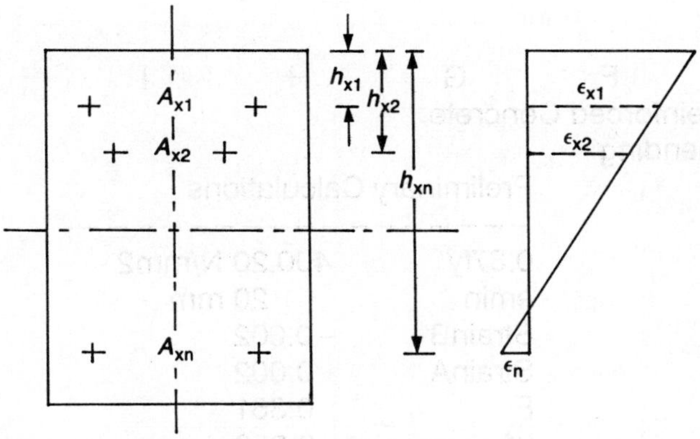

Figure 4.10 Notation for multi-levels of reinforcement

$$N_d = k_1 f_{cu} b d_c + \sum_1^n f_{sx} A_x \qquad [4.24]$$

where n represents the number of levels at which there is reinforcement and f_{sx} the stress at level h_x resulting from the strain ϵ_x.

The sign convention must also be extended to take account of the sign of the lever arm for each force and this can be achieved by defining the lever arm about the mid section as:

$$\text{Lever arm} = h/2 - h_x \qquad [4.25]$$

so that for values of $h_x > h/2$ the level arm is negative. Using this convention the moment equation of 4.18 can be expressed as:

$$M_d = k_1 f_{cu} b d_c (h/2 - k_2 d_c) + \sum_1^n f_{sx} A_x (h/2 - h_x) \qquad [4.26]$$

A spreadsheet for the design of rectangular sections with many levels of reinforcement can now be constructed using Eqns 4.23 to 4.26 and this is illustrated in the next section.

4.14 Spreadsheet layout for alternative design

In this example the layout used differs from that used previously in that details about the reinforcement are kept together in a separate part of the spreadsheet. Additionally the diameters of the reinforcement are placed on the spreadsheet in the same pattern as the intended design so that the designer has a visual picture of the layout. For the example it has been assumed that bars are placed along six sections perpendicular to the height with four bars along the top and bottom sections and two bars at the other four sections. There would be no difficulty in increasing the number of bars if required.

Any of the 16 bars could be of zero diameter so that the spreadsheet could be used for sections with less bars. Additionally the location of the bars in both the X and Y directions has still to be defined so that there is great flexibililty with regard to layout. Examples of different possible layouts are illustrated in Fig. 4.11.

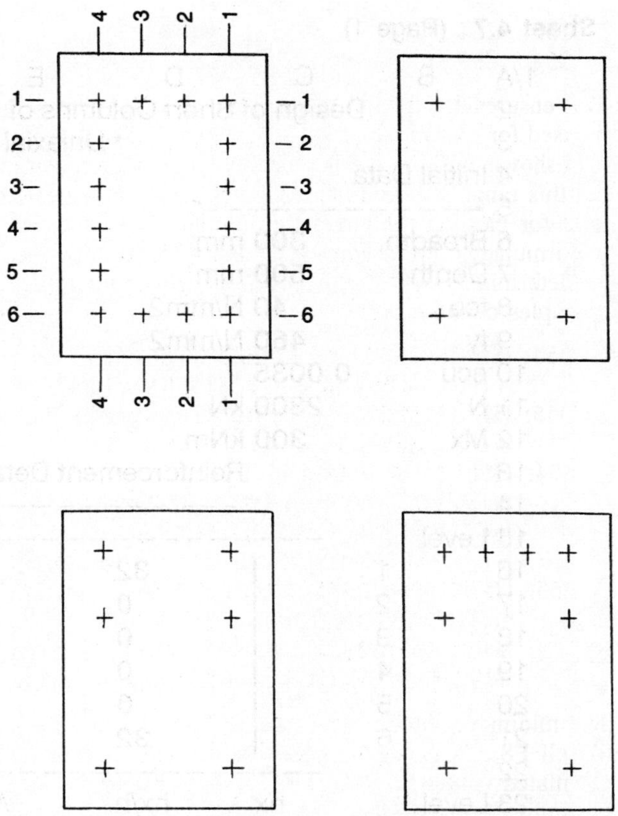

Figure 4.11 Examples of bar layouts

Initial data. Refer to Sheet 4.7. Values of b, h, f_{cu}, f_y, ϵ_{cu}, N and M_x are placed in the range C6 to C12.

As stated above the additional data required to define the amount and position of the reinforcement is placed in a separate part of the spreadsheet. Rows 15 to 29 are used to house this information and details on the form of this input are given under Reinforcement Details below.

Preliminary calculations. These calculations are performed in range H6 to H12 and are similar to those described in section 4.12 with reference to Sheet 4.6. The formulae in these cells might differ slightly because of a change in the location of some of the data but no new principles are involved.

Reinforcement details

As illustrated in Fig. 4.11 the diameter of the bars can be entered in 16 locations and on the spreadsheet these are entered in cells D16 to D21, E16, F16, E21, F21 and G16 to G21. Lines have been drawn around these entries to represent the section. The example chosen on Sheet 4.7 is the same as that solved previously on Sheet 4.6 and only four diameter bars are entered, the remaining entries being zero.

In range C24 to C29 the depth h_x, measured from the top of the section, for each level of reinforcement is entered using a zero value if no reinforcement exists at that level.

Sheet 4.7 (Page 1)

1/A	B	C	D	E	F	G	H	I
2		Design of Short Columns of Reinforced Concrete						
3				Uniaxial Bending				
4 Initial Data						Preliminary Calculations		
5 -------------						-------------------		
6 Breadth		300	mm			0.87fy	400.20	N/mm2
7 Depth		500	mm			emin	20	mm
8 fcu		40	N/mm2			StrainB	−0.002	
9 fy		460	N/mm2			StrainA	0.002	
10 ecu		0.0035				F	0.361	
11 N		2300	kN			k1	0.396	
12 Mx		300	kNm			k2	0.444	
13			Reinforcement Details					
14			----------------------					
15 Level		--						
16	1	|	32	0	0	32 |		
17	2	|	0			0 |		
18	3	|	0			0 |		
19	4	|	0			0 |		
20	5	|	0			0 |		
21	6	|	32	0	0	32 |		
22		--						
23 Level		hx	hx/h	Ax	h/2−hx			
24	1	75	0.15	1608	175			
25	2	0	0.00	0	250			
26	3	0	0.00	0	250			
27	4	0	0.00	0	250			
28	5	0	0.00	0	250			
29	6	425	0.85	1608	−175			
30	---							
31			Calculation of Strains					
32	dc/h	dc	eps_x1	eps_x2	eps_x3	eps_x4	eps_x5	eps_x6
33	0.18	92	0.0006	0.0035	0.0035	0.0035	0.0035	−0.0127
34	0.23	117	0.0013	0.0035	0.0035	0.0035	0.0035	−0.0092
35	0.28	142	0.0017	0.0035	0.0035	0.0035	0.0035	−0.0070
36	0.33	167	0.0019	0.0035	0.0035	0.0035	0.0035	−0.0054
37	0.38	192	0.0021	0.0035	0.0035	0.0035	0.0035	−0.0042
38	0.43	217	0.0023	0.0035	0.0035	0.0035	0.0035	−0.0034
39	0.48	242	0.0024	0.0035	0.0035	0.0035	0.0035	−0.0026
40	0.53	267	0.0025	0.0035	0.0035	0.0035	0.0035	−0.0021
41	0.58	292	0.0026	0.0035	0.0035	0.0035	0.0035	−0.0016
42	0.63	317	0.0027	0.0035	0.0035	0.0035	0.0035	−0.0012
43	0.68	342	0.0027	0.0035	0.0035	0.0035	0.0035	−0.0008
44	0.73	367	0.0028	0.0035	0.0035	0.0035	0.0035	−0.0006
45	0.78	392	0.0028	0.0035	0.0035	0.0035	0.0035	−0.0003
46	0.83	417	0.0029	0.0035	0.0035	0.0035	0.0035	−0.0001
47	0.88	442	0.0029	0.0035	0.0035	0.0035	0.0035	0.0001
48	0.93	467	0.0029	0.0035	0.0035	0.0035	0.0035	0.0003
49	0.98	492	0.0030	0.0035	0.0035	0.0035	0.0035	0.0005
50	1.03	517	0.0030	0.0035	0.0035	0.0035	0.0035	0.0006
51	1.08	542	0.0030	0.0035	0.0035	0.0035	0.0035	0.0008
52	1.13	567	0.0030	0.0035	0.0035	0.0035	0.0035	0.0009
53	1.18	592	0.0031	0.0035	0.0035	0.0035	0.0035	0.0010

Sheet 4.7 (Page 2)

	B	C	D	E	F	G	H	I
58			Calculation of Stresses X Areas					
59								
60		fxAx1	fxAx2	fxAx3	fxAx4	fxAx5	fxAx6	
61								
62		208	0	0	0	0	−644	
63		404	0	0	0	0	−644	
64		531	0	0	0	0	−644	
65		620	0	0	0	0	−644	
66		644	0	0	0	0	−644	
67		644	0	0	0	0	−644	
68		644	0	0	0	0	−644	
69		644	0	0	0	0	−644	
70		644	0	0	0	0	−513	
71		644	0	0	0	0	−384	
72		644	0	0	0	0	−273	
73		644	0	0	0	0	−178	
74		644	0	0	0	0	−95	
75		644	0	0	0	0	−22	
76		644	0	0	0	0	43	
77		644	0	0	0	0	101	
78		644	0	0	0	0	153	
79		644	0	0	0	0	200	
80		644	0	0	0	0	243	
81		644	0	0	0	0	282	
82		644	0	0	0	0	318	
83								
84								
85							Graphical Co−ords	
86								
87		Calculation of Nd and Md					Nd	Md
88							2300	300
89	*−*−*	dc/h	Nd	Md	Md/Nd		0	0
90	Trial−>	**0.1840**	1	240	186811		1	240
91	*−*−*	0.23	316	293	928		316	293
92		0.28	562	332	590		562	332
93		0.33	770	361	469		770	361
94		0.38	912	376	412		912	376
95		0.43	1031	384	372		1031	384
96		0.48	1149	389	339		1149	389
97		0.53	1268	392	309		1268	392
98		0.58	1518	369	243		1518	369
99		0.63	1766	344	195		1766	344
100		0.68	1995	320	160		1995	320
101		0.73	2209	296	134		2209	296
102		0.78	2411	271	112		2411	271
103		0.83	2603	245	94		2603	245
104		0.88	2786	218	78		2786	218
105		0.93	2963	190	64		2963	190
106		0.98	3134	160	51		3134	160
107		1.03	3300	128	39		3300	128
108		1.08	3461	94	27		3461	94
109		1.13	3619	59	16		0	0
110		1.18	3773	21	6		0	0

Calculations are then performed by the spreadsheet for h_x/h, A_x (the area of reinforcement at each level) and $(h/2 - h_x)$. Examples of the formulae required are:

D24: **(C24/$C$7)**
E24: **(@PI/4)*((D16^2)+(E16^2)+(F16^2)+(G16^2))**
F24: **($C$7/2)−C24**
D25: **(C25/$C$7)**
E25: **(@PI/4)*((D17^2)+(G17^2))**
F25: **($C$7/2)−C25**

Calculation of the strains

These calculations are placed in block B33 to I53 and, as for Sheet 4.6, each row of the block is based on a 'trial' value of d_c/h. These values of d_c/h have been copied automatically from the range C90 to C110 and are not entered by the user. For convenience the value d_c has been calculated in column C of the block.

Details of the values of d_c/h used will be given later with reference to the calculation of N_d and M_d. For any given value of d_c/h the strain at any level is given by Eqn 4.23 and this equation can now be used to determine the strain at each of the six levels. Note that if there is no reinforcement at a particular level and h_x has been entered as zero then the strain calculated from Eqn 4.23 will be equal to 0.0035 for all values of d_c/h.

Typically the formulae in some of the cells of row 33 and 34 are:

B33: **+C90**
C33: **(B33*$C$7)**
D33: **(1−($C$24/C33))*$C$10**
E33: **(1−($C$25/C33))*$C$10**
B34: **+C91**
C34: **(B34*$C$7)**
D34: **(1−($C$24/C34))*$C$10**
E34: **(1−($C$25/C34))*$C$10**

and similar formulae for the remainder.

Calculation of the product — (Stress × Area)

Using the value of the strain calculated at any level the stress can be determined from the idealised stress/strain curve and examples of how this is achieved have been given previously. For convenience this stress value is now multiplied by the area of steel at the particular level since this product simplifies subsequent entries. These calculations are performed in block C62 to H82 and typical examples of the required formulae are:

C62: **(@IF(D33< =$H$8,−$H$6*$E$24, @IF(D33> =$H$9,$H$6*$E$24,(($H$6*D33*$E$24)/($H$9)))))/1000**
D62: **(@IF(E33< =$H$8,−$H$6*$E$25, @IF(E33> =$H$9,$H$6*$E$25,(($H$6*E33*$E$25)/($H$9)))))/1000**
C63: **(@IF(D34< =$H$8,−$H$6*$E$24, @IF(D34> =$H$9,$H$6*$E$24,(($H$6*D34*$E$24)/($H$9)))))/1000**
D63: **(@IF(E34< =$H$8,−$H$6*$E$25, @IF(E34> =$H$9,$H$6*$E$25,(($H$6*E34*$E$25)/($H$9)))))/1000**

Calculation of N_d and M_d

As for Sheet 4.6 the initial values of d_c/h are entered in the range C90 to C110 and these values are then automatically copied to the range C33 to C53 for calculating the strains and stresses. Additionally, of the range C90 to C110 the user is only required to enter a value in cell C90 since other values of d_c/h are determined automatically by the spreadsheet.

The entry into cell C90 is entered by the user, by trial, until the value of N_d in D90 is both small and positive. Using this trial value as a starting point ensures that the interaction curve is defined near the point where it cuts the M_d axis.

The formulae for obtaining the value of d_c/h for the remaining cells of column C is simply the value found in C90 incremented progressively down the column by 0.05. Cells C91 and C92 contain:

C91: **(C90+0.05)**
C92: **(C91+0.05)**

This ensures that the full range of possible N_d values will be used for plotting the N_d-M_d graph. The range of d_c/h shown on Sheet 4.6 is probably longer than required but this range will be reduced when values are determined later for plotting the interaction curve.

Formulae based on equations 4.24 and 4.26 are now used to determine the values of N_d and M_d and typical examples of these are:

D90: **(($H$11*$C$8*$C$6*C33)/1000 +(C62)+(D62)+(E62)+(F62)+(G62)+(H62))**
D91: **(($H$11*$C$8*$C$6*C34)/1000 +(C63)+(D63)+(E63)+(F63)+(G63)+(H63))**
E90: **(($H$11*$C$8*$C$6*C33*(($C$7/2)−($H$12*C33)))/1000+(C62*$F$24)+(D62*$F$25)+(E62*$F$26)+(F62*$F$27)+(G62*$F$28)+(H62*$F$29)))/1000**
E91: **(($H$11*$C$8*$C$6*C34*(($C$7/2)−($H$12*C34)))/1000+(C63*$F$24)+(D63*$F$25)+(E63*$F$26)+(F63*$F$27)+(G63*$F$28)+(H63*$F$29)))/1000**

The minimum value of the ratio M_d/N_d was calculated in cell H7 and it is now necessary to determine if the calculated values of N_d and M_d satisfy this limit. As a preliminary the ratio of M_d/N_d is first calculated in range F90 to F110 using formulae of the type:

F90: **(E90*1000/D90)**

and these values compared with the minimum value from H7 in order to determine the final acceptable values of N_d and M_d. The comparison is made in the block H90 to I110 using:

H90: **@IF(F90>$H$7,D90,0)**
I90: **@IF(F90>$H$7,E90,0)**

H91: **@IF(F91>$H$7,D91,0)**
I91: **@IF(F91>$H$7,E91,0)**

so that zero values will be output if the ratio is below the minimum.

Interaction diagram

The graphical facility of the spreadsheet can now be used to produce an interaction diagram which is specific to the initial data. In order to improve the presentation some additional information is entered in the block H88 to I89.

Values of the design axial load and design moment are transferred to H88 and I88 respectively and zero values are entered in H89 and I89. The effect of this addition is to introduce an additional line on the interaction diagram from the origin to the point with the coordinates (M,N) of the design loads.

The X axis is now defined as the range I88 to I110 and the Y axis as H88 to H110. The resulting graph is as illustrated in Graph 4.7b.

In the British Standard code the interaction diagrams represent plots of N_d/bd against M_d/bd^2 and, if required, there would be no difficulty in converting the calculated values of N_d and M_d to the code parameters. No advantage would be gained by such a conversion but the results obtained are illustrated in Graph 4.7a. This graph should be compared with the chart published in the BS code for the same parameters.

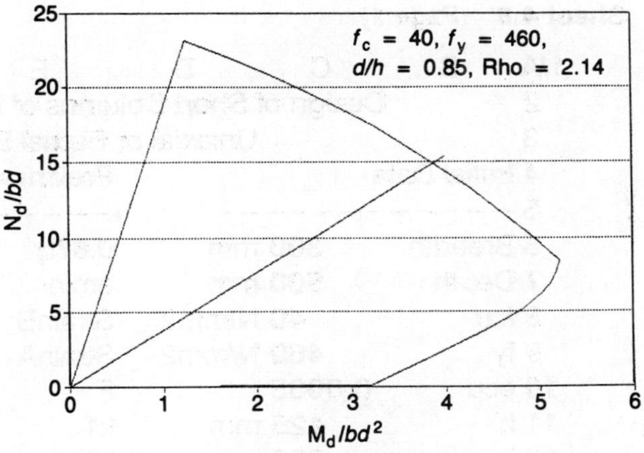

Graph 4.7a Uniaxial bending for short columns of reinforced concrete (alternative method and plotting N_d/bd against M_d/bd^2)

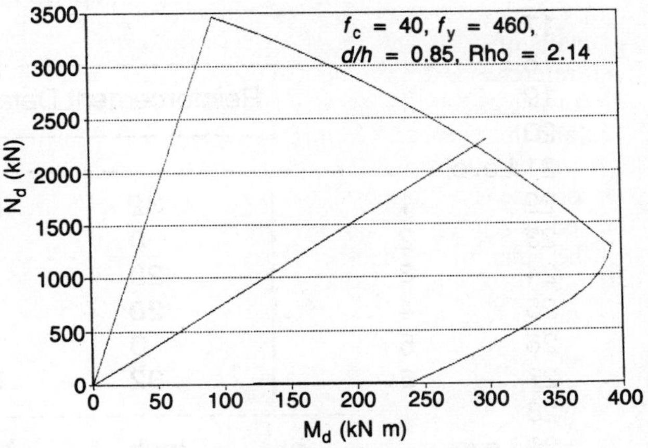

Graph 4.7b Uniaxial bending for short columns of reinforced concrete (alternative method)

Use of Sheet 4.7

To avoid possible errors it is now convenient to enable the protection facility of the spreadsheet so that initially the whole sheet is protected. Then the cells which require changes for dealing with different designs can be unprotected so that the user would be warned if an illegal entry was attempted. Following this procedure the only unprotected cells would be: C6 to C12, D16 to D21, G16 to G21, E16, F16, E21, F21, C24 to C29, and C90.

To use the spreadsheet the user would enter the required values in the above cells and then adjust the trial value in C90 until the value of N_d in D90 becomes small and positive.

The interaction diagram can now be called by pressing the appropriate function key related to the graphical facility. If this diagram shows that the end point of the straight line from the origin is within the curve then the design is acceptable although if the point is too far away from the curve then the amount of reinforcement could be reduced. If the point is outside the curve then the amount of reinforcement must be increased.

4.15 Spreadsheet layout, short columns — biaxial bending

An example of the design of a short column with six possible levels of reinforcement in one direction and four possible levels in the perpendicular direction is given on Sheet 4.8.

It was shown in section 4.11 that the biaxial case can be reduced to a uniaxial case by means of Eqns 4.19 and 4.20. The process involves increasing one of the biaxial moments by a defined amount and then designing for the axial load and the increased moment. In fact once the decision is made as to which moment should be increased and the required increase calculated the design process becomes almost identical with that described for Sheet 4.7. The description given above for the design of Sheet 4.7 applies also to Sheet 4.8 except that additional information and extra calculations will be required to allow for the biaxial effect.

Additional data on Sheet 4.8 (cf. Sheet 4.7)

Values of h′ and b′ and M_y are added to the initial data column C6 to C15. These quantities are illustrated in Fig. 4.9. For this example the reinforcement is symmetrical with regard to position and amount and therefore the direction of the moments, clockwise or anticlockwise, does not affect the results. If there was a lack of symmetry then the direction of moment would be important since the value of h′ and/or b′ might change.

Sheet 4.8 (Page 1)

1/A	B	C	D	E	F	G	H	I
2		Design of Short Columns of Reinforced Concrete						
3			Uniaxial or Biaxial Bending					
4 Initial Data				Preliminary Calculations				
5 -------------				------------------------------				
6 Breadth		300 mm		0.87fy	400.20 N/fbh		Alpha	Diff
7 Depth		500 mm		emin	15	0.00	1.00	−1.20
8 fcu		40 N/mm2		StrainB	−0.002	0.10	0.88	−1.10
9 fy		460 N/mm2		StrainA	0.002	0.20	0.77	−1.20
10 ecu		0.0035		F	0.361	0.30	0.65	−1.20
11 h'		425 mm		k1	0.396	0.40	0.53	−1.10
12 b'		250 mm		k2	0.444	0.50	0.42	−1.20
13 N		1500 kN		N/fcubh	0.25	0.60	0.30	
14 Mx		200 kNm		Mx/h'	0.47	1.00	0.30	
15 My		150 kNm		My/b'	0.60 Design for:−			
16				Alpha	0.71 N		1500 kN	
17						My'	234 kNm	
18 ------------------------------								
19			Reinforcement Details					
20			-------------------------					
21 Level		------------------------------						
22	1	|	32	0	0	32		
23	2	|	0			0 |		
24	3	|	20			20 |		
25	4	|	20			20 |		
26	5	|	0			0 |		
27	6	|	32	0	0	32 |		
28		------------------------------						
29 Level		hx	hx/h	Ax	h/2−hx		Area Selection	
30	1	75	0.15	1608	175			2237
31	2	0	0.00	0	250			0
32	3	175	0.35	628	75			0
33	4	325	0.65	628	−75	Max6%	9000	2237
34	5	0	0.00	0	250	Min0.4%	600	0
35	6	425	0.85	1608	−175	SumArea	4474	0
36							Depth Selection	
37	Level	by	by/b	Ay	b/2−by			100
38	1	50	0.17	2237	100			150
39	2	0	0.00	0	150			150
40	3	0	0.00	0	150			−100
41	4	250	0.54	2237	−100			0
42 REMEMBER ENTRY TO CELL C115								0
43 D115 =		1 O.K.						

Sheet 4.8 (Page 2)

48								
49								
50	B	C	D	E	F	G	H	I
51			Calculation of Strains					
52	bc/b	bc	eps_1	eps_2	eps_3	eps_4	eps_5	eps_6
53	0.22	66	0.0008	0.0035	0.0035	−0.0098	0.0000	0.0000
54	0.27	80.76	0.0013	0.0035	0.0035	−0.0073	0.0000	0.0000
55	0.32	95.76	0.0017	0.0035	0.0035	−0.0056	0.0000	0.0000
56	0.37	110.76	0.0019	0.0035	0.0035	−0.0044	0.0000	0.0000
57	0.42	125.76	0.0021	0.0035	0.0035	−0.0035	0.0000	0.0000
58	0.47	140.76	0.0023	0.0035	0.0035	−0.0027	0.0000	0.0000
59	0.52	155.76	0.0024	0.0035	0.0035	−0.0021	0.0000	0.0000
60	0.57	170.76	0.0025	0.0035	0.0035	−0.0016	0.0000	0.0000
61	0.62	185.76	0.0026	0.0035	0.0035	−0.0012	0.0000	0.0000
62	0.67	200.76	0.0026	0.0035	0.0035	−0.0009	0.0000	0.0000
63	0.72	215.76	0.0027	0.0035	0.0035	−0.0006	0.0000	0.0000
64	0.77	230.76	0.0027	0.0035	0.0035	−0.0003	0.0000	0.0000
65	0.82	245.76	0.0028	0.0035	0.0035	−0.0001	0.0000	0.0000
66	0.87	260.76	0.0028	0.0035	0.0035	0.0001	0.0000	0.0000
67	0.92	275.76	0.0029	0.0035	0.0035	0.0003	0.0000	0.0000
68	0.97	290.76	0.0029	0.0035	0.0035	0.0005	0.0000	0.0000
69	1.02	305.76	0.0029	0.0035	0.0035	0.0006	0.0000	0.0000
70	1.07	320.76	0.0030	0.0035	0.0035	0.0008	0.0000	0.0000
71	1.12	335.76	0.0030	0.0035	0.0035	0.0009	0.0000	0.0000
72	1.17	350.76	0.0030	0.0035	0.0035	0.0010	0.0000	0.0000
73	1.22	365.76	0.0030	0.0035	0.0035	0.0011	0.0000	0.0000
74								
75			Calculation of Stresses X Areas					
76								
77		f1A1	f2A2	f3A3	f4A4	f5A5	f6A6	
78								
79		375	0	0	−895	0	0	
80		596	0	0	−895	0	0	
81		748	0	0	−895	0	0	
82		859	0	0	−895	0	0	
83		895	0	0	−895	0	0	
84		895	0	0	−895	0	0	
85		895	0	0	−895	0	0	
86		895	0	0	−727	0	0	
87		895	0	0	−541	0	0	
88		895	0	0	−384	0	0	
89		895	0	0	−248	0	0	
90		895	0	0	−131	0	0	
91		895	0	0	−27	0	0	
92		895	0	0	65	0	0	
93		895	0	0	146	0	0	
94		895	0	0	219	0	0	
95		895	0	0	286	0	0	
96		895	0	0	345	0	0	
97		895	0	0	400	0	0	
98		895	0	0	450	0	0	
99		895	0	0	496	0	0	

Sheet 4.8 (Page 3)

108	B	C	D	E	F	G	H	I
109	b or h	500						
110	h or b	300				Graphical Co–ords		
111						–––––––––	–––––––––	
112		Calculation of Nd and Md				Nd	Md	
113		––––––––––	––––––––––	––––––––––	––––––––––	1500	234	
114	*–*–*	bc/b	Nd	Md	Md/Nd	1500	150	
115	Trial–>	**0.2192**	1	190	305928	0	0	
116	*–*–*	0.27	340	222	652	1	190	
117		0.32	611	246	402	340	222	
118		0.37	841	264	314	611	246	
119		0.42	995	273	274	841	264	
120		0.47	1114	277	248	995	273	
121		0.52	1233	279	226	1114	277	
122		0.57	1520	262	173	1233	279	
123		0.62	1824	243	133	1520	262	
124		0.67	2100	225	107	1824	243	
125		0.72	2355	207	88	2100	225	
126		0.77	2591	189	73	2355	207	
127		0.82	2814	172	61	2591	189	
128		0.87	3024	154	51	2814	172	
129		0.92	3224	135	42	3024	154	
130		0.97	3416	116	34	3224	135	
131		1.02	3601	96	27	3416	116	
132		1.07	3780	74	20	3601	96	
133		1.12	3953	52	13	3780	74	
134		1.17	4122	29	7	0	0	
135		1.22	4286	4	1	0	0	
						0	0	

Additional information is also included with regard to the level of reinforcement in the perpendicular direction to *h*. This is required since design will be for either N and $M_{x'}$ or N and $M_{y'}$. If the former then the levels of reinforcement are measured in the *h* direction whilst for the latter reinforcement levels would be taken in the *b* direction, see Fig. 4.12.

To anticipate this possible change in direction additional information about the reinforcement is placed in rows 37 to 41. The formulae required for this additional information are similar to those given on Sheet 4.7 when dealing reinforcement levels in the *h* direction. Note that a value of h_x should be included in each row for which there is a non-zero value of A_x. If the value of A_x is zero at any level then the value of h_x should be zero at that level.

Additional preliminary calculations

The values given in Table 4.1 together with the difference column for interpolation are entered in block G7 to I14 and the value of $N_d/f_{cu}bh$ required for entry into the table is calculated in cell F13. The value of α can now be determined and this calculation is performed in F16. The formulae used in these cells are:

F13: **((C13*1000)/(C8*C6*C7))**
F16: **@VLOOKUP(F13,G7.I14,1)**
+(F13–(@VLOOKUP(F13,G7.I14,0)))
***(@VLOOKUP(F13,G7.I14.2))**

The ratios M_x/h' and M_y/b' used to determine which of the moments is to be increased are calculated in F14 and F15 and the actual increase in moment in cell H17. For convenience the label which appears in cell G17 is either $M_{x'}$ or $M_{y'}$ depending on values found in F14 and F15 and the limiting conditions of Eqns 4.19 and 4.20. The formulae used are:

F14: **(C14/C11)**
F15: **(C15/C12)**
G17: **@IF(F14>=F15,"Mx' ","My' ")**
H17: **@IF(F14>=F15,C14+(F16*C11*C15/C12),**
C15+(F16*C12*C14/C11))

The format of the formulae given above for cells G17 and H17 is used often in the rest of the spreadsheet since the design can proceed in either the *h* direction or the *b* direction depending on whether design is for N and $M_{x'}$ or N and $M_{y'}$. The format in G17 is an example of how the appropriate label can be obtained and the format H17 an example of how a choice is made between alternative formulae.

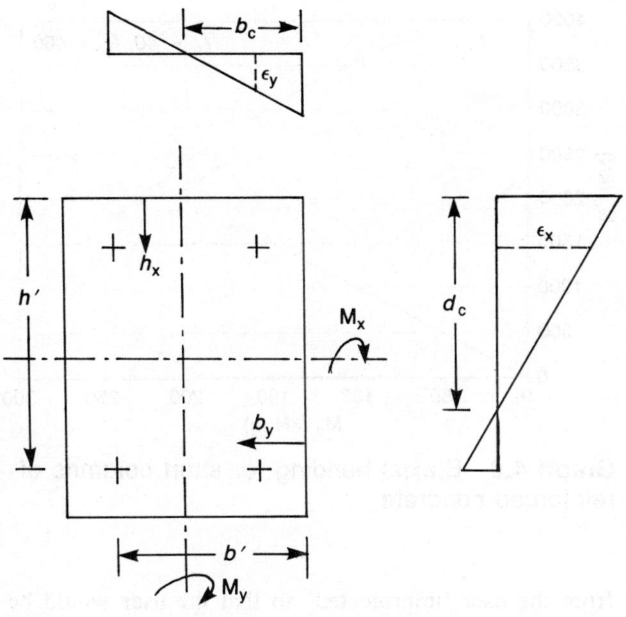

Figure 4.12 Cross-section and strains for biaxial bending

As an example of this the values of the areas of reinforcement to be used at each level and the distance from the datum to that level depend on the values previously determined in F14 and F15. If the value in F14 is greater than the value in F15 then six values of the area are required since design relates to $M_{x'}$. If F14 < F15 then only four levels are defined and the values of these four areas must be determined. Calculation of the areas and depths which apply are made in I30 to I35 and I37 to I42 respectively. Examples of the formulae used are:

I30: **@IF($F$14>=$F$15,$E$30,$E$38)**
I31: **@IF($F$14>=$F$15,$E$31,$E$39)**
I37: **@IF($F$14>=$F$15,$F$30,$F$38)**
I38: **@IF($F$14>=$F$15,$F$31,$F$39)**

A check that the amount of reinforcement provided is within certain limits has to be performed and it is convenient to enter these checks in the block G33 to H35. Assuming that the amount of reinforcement is not to exceed 6 per cent of the sectional area and that the minimum amount is 0.4 per cent, the appropriate formulae can be entered:

H33: **6*C6*C7/100**
H34: **0.4*C6*C7/100**
H35: **@SUM(I30..I35)**

The user can now see at a glance if the area of reinforcement in H35 is between the maximum and minimum values. A similar type of check is required for the minimum spacing between bars but this requires a knowledge of the size of the aggregate.

In row 42 a reminder has been placed to inform the user that the entry for d_c/h (or b_c/b) is required in cell C115. Additionally the value in cell D115 is copied to cell C43 so that the user can tell at a glance if any change is required. Information about this change is output to D43 using:

D43: **@IF(C43<0,"Increase cell C115 value",**
@IF(C43>5,"Decrease cell C115 value",
"O.K."))

where 5 is introduced an an arbitrary small positive limit.

Changes to calculation of strains in sheet 4.8

The only changes that can arise are due entirely to the possible change in direction resulting from the use of $M_{y'}$ instead of $M_{x'}$. Cell entries for the labels of row 52 are:

B52: **@IF(F14>=F15,"dc/h","bc/b")**
C52: **@IF(F14>=F15,"dc","bc")**
D52: **"eps_1**
E52: **"eps_2**

and similarly for F52 to I52 for the other four strains.

Calculation of Stress × Area

If design is for N and $M_{x'}$ then six columns of values are to be calculated in block C79 to H99. If design is for N and $M_{y'}$ then the values output to the fifth and sixth columns would be zero. Examples of the formulae required are:

C79: **(@IF(D53<=$F$8,−$F$6*$I$30,**
@IF(D53>=$F$9,$F$6*$I$30,
(($F$6*D53*$I$30)/($F$9)))))/1000
G79: **@IF($F$14>=$F$15,(@IF(H53<=$F$8,**
−$F$6*$I$34,@IF(H53>=$F$9,$F$6*$I$34,
(($F$6*H53*$I$34)/($F$9)))))/1000,0)
H79: **@IF($F$14>=$F$15,(@IF(I53<=$F$8,**
−$F$6*$I$35,@IF(I53>=$F$9,$F$6*$I$35,
(($F$6*I53*$I$35)/($F$9)))))/1000,0)

Calculation of N_d and M_d

To reduce the input required for the calculation of N_d and M_d it is convenient to make the decision as to whether h or b is to be used on a separate part of the spreadsheet and with this in mind the following entries have been made in cells C109 to C110:

C109: **@IF(F14>=F15,C6,C7)**
C110: **@IF(F14>=F15,C7,C6)**

Statements such as the above will be required later for inclusion into other formulae in order to reduce repetitive input.

The values placed in cells C115 to C135 are either values of d_c/d or b_c/b depending on the moment axis of the design moment. To indicate which of d_c/d or b_c/b applies the formula entered in C114 is:

C114: **@IF(F14>=F15,"dc/d","bc/b")**

As for Sheet 4.6 the initial values of d_c/h are entered in the range C115 to C135 and these values are then automatically copied to the range B53 to B73 for calculating the strains and stresses. Additionally, of the range C115 to C135 the user is only required to enter a value in cell C115 since other values of d_c/h are determined automatically by the spreadsheet.

The entry into cell C115 is entered by the user, by trial, until the value of N_d in D115 is both small and positive. Using this trial value as a starting point ensures that the interaction curve is defined near the point where it cuts the M_d axis.

The formulae for obtaining the value of d_c/h for the remaining cells of column C is simply the value found in C115 incremented progressively down the column by 0.05. Cells C116 and C117 contain:

C116: **(C115+0.05)**
C117: **(C116+0.05)**

This ensures that the full range of possible N_d values will be used for plotting the N_d-M_d graph. The range of d_c/h shown on Sheet 4.8 is probably longer than required but this range will be reduced when values are selected later for plotting the interaction curve.

A check is now made if the ratio of M_d/N_d calculated in column F exceeds the minimum value previously calculated in F7 and, if so, the values of N_d and M_d in columns D and E are transferred as graphical coordinates to columns G and H. If the condition is not satisfied then a zero value is obtained as explained for Sheet 4.7.

Interaction diagram

The graphical facility of the spreadsheet can now be used to produce an interaction diagram which is specific to the initial data. In order to improve the presentation some additional information is entered in the block G113 to H115. The entries to this block are similar to those for Sheet 4.7 except that there is now an additional set of N_d and M_d values in the first row of the block. Cells G113 and G114 contain the value of N copied from H16 and cell H113 either $M_{x'}$ or $M_{y'}$ copied from H17.

The block G113 to H115 thus contains the quantities:

	G	H
113	**N**	**($M_{x'}$ or $M_{y'}$)**
114	**N**	**(M_x or M_y)**
115	**0**	**0**

so that the plot of these three points produces a line from the origin to the point with coordinates N,M and then a horizontal line to the point N,M′, where M and M′ refer to either the X or Y value as appropriate, see Graph 4.8. The increase in the design moment resulting from the biaxial moments is therefore visible on the graph.

The X axis is now defined as the range H113 to H135 and the Y axis as G113 to G135 and the resulting graph is as illustrated in Graph 4.8.

Use of Sheet 4.8

As explained for Sheet 4.7 the spreadsheet can now be 'protected' and only the cells which require information

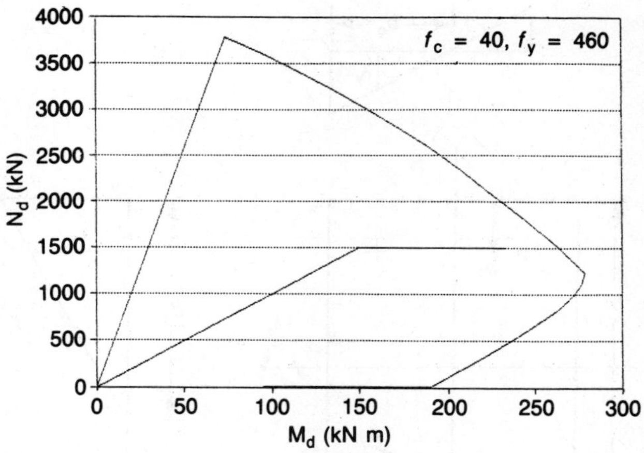

Graph 4.8 Biaxial bending for short columns of reinforced concrete

from the user 'unprotected' so that the user would be warned if an illegal entry were attempted. Following this procedure the only unprotected cells would be: C6 to C15, D22 to D27, G22 to G27, E22, F22, E27, F27, C30 to C35, C38 to C41 and C115.

To use the spreadsheet the user would enter the required values in the above cells and then adjust the trial value in C115 until the value of N_d in D115 becomes small and positive.

The interaction diagram can be called by pressing the appropriate function key related to the graphical facility. If this diagram shows that the end point of the line from the origin is within the curve then the design is acceptable although if the point is too far away from the curve then the amount of reinforcement could be reduced. If the point is outside the curve then the amount of reinforcement must be increased.

4.16 Non-rectangular sections

Spreadsheet designs for non-rectangular sections follow similar principles to those outlined for rectangular sections and the method can easily be extended to include other shapes.

The strain distribution diagram and the stress block are unchanged and the only difference arises from the determination of the area of the concrete in compression.

Consider a column of circular cross-section as shown in Fig. 4.13.

The term in the force balance equation representing the compressive force in the concrete changes from $k_1 f_{cu} b d_c$ to $k_1 f_{cu} A_x$, where A_x represents the area of concrete in compression and there would be no difficulty in calculating this area for each trial value of d_c.

Similarly the term in the moment equation representing the moment due to the compressive stresses in the concrete would be changed from:

$$k_1 f_{cu} b d_c (h/2 - k_2 d_c) \quad \text{to} \quad k_1 f_{cu} A_x (h/2 - k_2 d_c)$$

Applying these changes to Sheet 4.8 would require:

1. The calculation of A_x for each value of d_c.

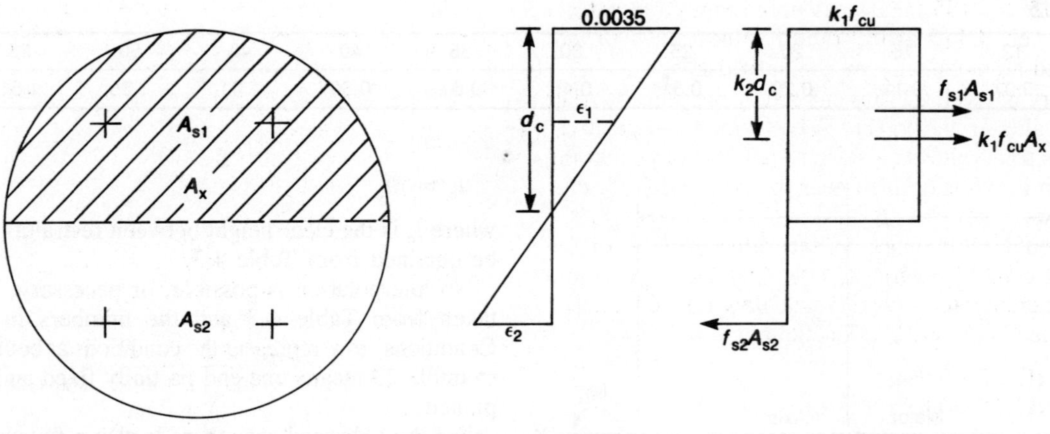

Figure 4.13 Strains and stresses — circular cross-section

2. Changes to the first terms of the equations for calculating N_d and M_d as indicated above.

There would be no difficulty in introducing these changes to Sheet 4.8 and the same principles would apply to other non-rectangular sections such as hexagonal or octagonal cross-sections.

4.17 Design of slender columns

Columns are considered to be slender when either of the slenderness ratios, about the X and Y axes, is greater than 15 for braced columns or 10 for unbraced columns. Braced columns are those for which the resistance to lateral loading has been supplied by other parts of the structure such as shear walls or cross bracing. A column may be braced in one plane and unbraced in the other and this influences how the effective length is evaluated and also the determination of the design moments. However, the spreadsheets developed in this text assume that columns are braced in both directions although an indication of how the user could introduce the necessary changes, to allow for the unbraced case, is given.

The spreadsheet design process outlined in Sheet 4.8 for the biaxial bending of short columns began with knowledge of the applied loading N, M_x, and M_y.

If design is required for a slender column then the applied loading changes to N, M_{tx}, and M_{ty} where M_{tx} and M_{ty} are the total moments arising from the applied moment plus any additional moment due to the effect of slenderness. Once the values of M_{tx} and M_{ty} have been determined then the design process becomes identical with that described in the previous sections. Uniaxial bending can be treated as the biaxial case for which one of the moments is zero.

For calculating M_{tx} and M_{ty} the two cases of bending considered are as illustrated in Fig. 4.14.

In Fig. 4.14 the end moment M_{2x} is considered as positive and the smaller moment M_{1x} as either positive or negative as shown. If M_{1x} is positive then the column is bent in single curvature and if M_{1x} is negative then the column is bent in double curvature. The diagrams in Fig. 4.14 could be duplicated for bending about the minor axis using M_{1y} and M_{2y} and in what follows although the

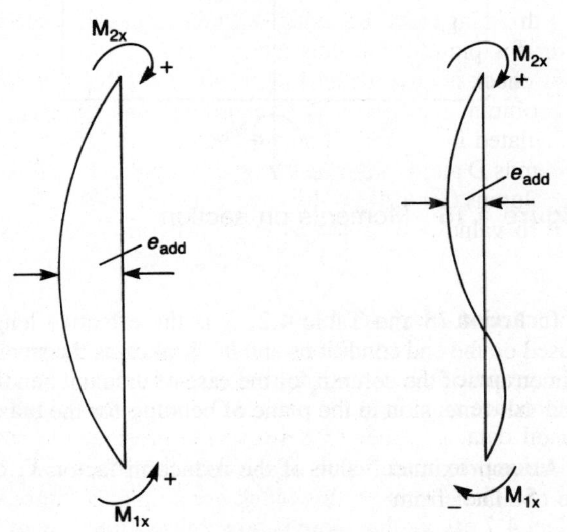

Figure 4.14 Types of end moments

equations are expressed in terms of bending about the major axis similar equations can be written with reference to the minor axis. In this section the total moment M_{tx} is considered as bending about the major axis as shown in Fig. 4.15.

The value of M_{tx} is taken as the greatest of the values calculated from:

$$M_{tx} = M_{ix} + M_{addx}$$
$$M_{tx} = M_{2x}$$
$$M_{tx} = M_{1x} + 0.5M_{addx}$$
$$M_{tx} = Ne_{min}$$

where $M_{ix} = 0.4M_{1x} + 0.6M_{2x} \geq 0.4M_{2x}$. Note that M_{1x} is taken as negative for double curvature. M_{addx} is defined by:

$$M_{addx} = N\beta_a K_r h \quad [4.27]$$

where β_a can be obtained from the equation:

$$\beta_a = (l_e/b')^2/2000 \quad [4.28]$$

or alternatively from Table 4.2.

Table 4.2

l_e/b'	12	15	20	25	30	35	40	45	50	55	60
β_a	0.07	0.11	0.20	0.31	0.45	0.61	0.80	1.01	1.25	1.51	1.80

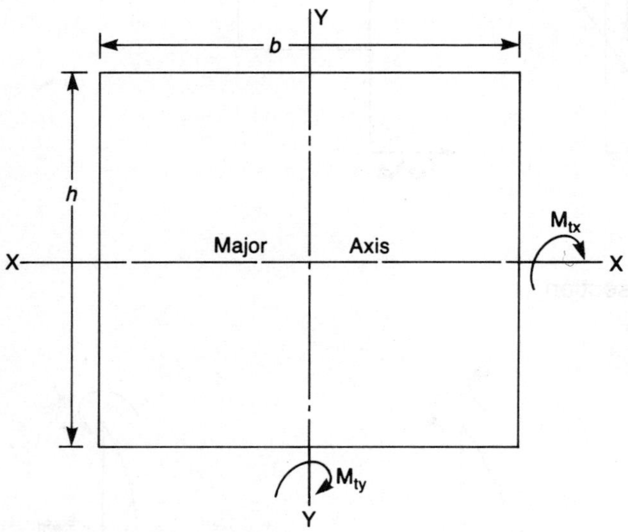

Figure 4.15 Moments on section

In Eqn 4.28 and Table 4.2, l_e is the effective length based on the end conditions and b' is taken as the smaller dimension of the column for the case of uniaxial bending and the dimension in the plane of bending for the biaxial case.

An approximate value of the reduction factor K_p can be obtained from:

$$K_r = (0.45 f_{cu} A_c + 0.87 f_y A_{sc} - N)/(0.2 f_{cu} A_c + 0.87 f_y A_{sc}) \quad [4.29]$$

in which A_c is the area of the concrete and A_{sc} the area of the reinforcement. Normally the value of A_{sc} is unknown and it is usual to assume that K_r equals 1 in order to determine an initial value of A_{sc} which can then be corrected if required by using the value of K_r determined from Eqn 4.29. Some designers ignore the possible reduction in the design moment that results from the use of this optional reduction factor by taking K_p as being equal to 1 and this procedure has been followed in this example. To amend the spreadsheet so as to allow for K_p it would be necessary to calculate K_r from Eqn 4.29 and amend the value of M_{add} in cells F16 and F17 of Sheet 4.9.

End conditions can be considered as 'Fixed' (1), 'Partially Restrained' (2), or 'Pinned' (3) and it is possible to express these conditions in tabular form so that the effective length l_e can be obtained from:

$$l_e = \beta l_o \quad [4.30]$$

where l_o is the clear height between restraints and β can be obtained from Table 4.3.

No interpolation is possible, or necessary, for values taken from Table 4.3 and the numbers in the 'End Conditions' row represent the conditions at both ends. For example 23 means one end partially fixed and the other pinned.

For the unbraced case there is also a fourth condition when the end of the column is unrestrained against both lateral movement and rotation and the coefficient for this case (14) is taken as 2.2. This case is represented by the free end of a cantilever in an unbraced structure.

4.18 Spreadsheet layout, slender columns

Note that this spreadsheet can also be used for short columns.

Additional initial data compared with Sheet 4.8

To allow for the effect of bending in single or double curvature values of M_{1x}, M_{2x}, M_{1y} and M_{2y} have been entered in the range C14 to C17. When entering these moments M_{1x} is the smaller moment about the major axis and M_{1y} the smaller moment about the minor axis. On the spreadsheet these labels are shown as M1x(S) and M1y(S) to show that they refer to the smaller moment. Also the values of M_{1x} and/or M_{1y} are negative for the case of double curvature and a reminder is placed in cells E25 and E26 concerning this.

The height of the column between supports has been entered in C18 and the end conditions for bending about the X and Y axes entered in H17 and I17 respectively. These entries are in the form 'ij', where both i and j = (1,2,3), so that an entry such as 12 means that one end of the column is fixed and the other end is partially fixed. A note about the end conditions has been placed in cells E15 to I15.

To distinguish between the different possible cases of design it is convenient to define:

Case 1. Bending about the minor axis only. In which case, design is for N and the greatest value of M_{tx} and using M_{1x} and M_{2x} about the minor axis.

Case 2. Bending about the major axis and satisfying the following conditions:

Table 4.3

End conditions	*11*	*12*	*13*	*21*	*22*	*23*	*31*	*32*	*33*
Braced	0.75	0.80	0.90	0.80	0.85	0.95	0.90	0.95	1.0
Unbraced	1.2	1.3	1.6	1.3	1.5	1.8	1.6	1.8	—

Sheet 4.9 (Page 1)

1/A	B	C	D	E	F	G	H	I	J
2		Design of Short or Slender Columns of Reinforced Concrete							
3			Uniaxial or Biaxial Bending						
4 Initial Data				Preliminary Calculations					
5 -------------				-------------------------					
6 Breadth		**400**	mm	0.87fy	348.00	N/fbh	Alpha	Diff	
7 Depth		**400**	mm	emin	20	0.00	1.00	−1.20	
8 fcu		**30**	N/mm2	StrainB	−0.002	0.10	0.88	−1.10	
9 fy		**400**	N/mm2	StrainA	0.002	0.20	0.77	−1.20	
10 ecu		**0.0035**		F	0.313	0.30	0.65	−1.20	
11 h'		**350**	mm	k1	0.403	0.40	0.53	−1.10	
12 b'		**350**	mm	k2	0.451	0.50	0.42	−1.20	
13 N		**1800**	kN	N/fcubh	0.38	0.60	0.30		
14 M1x(S)		**100**	kNm	Alpha	0.56	1.00	0.30		
15 M2x(L)		**200**	kNm	End Conditions		1−Fixed,2−Partially,3−Pinned			
16 M1y(S)		**0**	kNm	Madd−x	61		MajorX	MinorY	
17 M2y(L)		**0**	kNm	Madd−y	86	End Conds	**12**	**23**	
18 Height		**6.5**	m	Mix	160	Mi+Madd	221	86	
19 Case		**2**		Miy	0	M2	200	0	
20 1/MinorY, 2/MajorX + Conds(Yes)						M1+0.5Ma	130	43	
21 3/MajorX + (Conds(No), 4/Biaxial						Nemin	36	36	
22 Cond =	Yes			Mx/h'	0.6310	Mtx/y	221	86	
23 lex		5.20	m	My/b'	0.2451				
24 ley		6.18	m	Note:−		Design for:−			
25 SRx		13.0		M1 negative for		N		1800	kN
26 SRy		15.4		Double Curvature		Mtx'		221	kNm
27 --									
28			Reinforcement Details						
29 Level			4	3	2	1			
30		----------------------------------							
31	1	|	**24**	**24**	**0**	**24** |			
32	2	|	**24**			**24** |			
33	3	|	**24**			**24** |			
34	4	|	**0**			**0** |			
35	5	|	**0**			**0** |			
36	6	|	**24**	**24**	**0**	**24** |			
37		----------------------------------							
38 Level		hx	hx/h	Ax	h/2−hx		Area Selection		
39	1	**50**	0.13	1357	150			1357	
40	2	**150**	0.38	905	50			905	
41	3	**250**	0.63	905	−50			905	
42	4	**0**	0.00	0	200	Max6%	9600	0	
43	5	**0**	0.00	0	200	Min0.4%	640	0	
44	6	**350**	0.88	1357	−150	SumArea	4524	1357	
45							Depth Selection		
46	Level	by	by/b	Ay	b/2−by			150	
47	1	**50**	0.13	1810	150			50	
48	2	**200**	0.50	905	0			−50	
49	3	**0**	0.00	0	200			200	
50	4	**350**	0.88	1810	−150			200	
51 REMEMBER ENTRY TO CELL C122								−150	
52 D122 =		1 O.K.							
53									

Sheet 4.9 (Page 2)

59	B	C	D	E	F	G	H	I
60			Calculation of Strains					
61	bc/b	bc	eps_1	eps_2	eps_3	eps_4	eps_5	eps_6
62	0.21	73.7	0.0006	0.0035	0.0035	−0.0103	0.0000	0.0000
63	0.26	91.2	0.0012	0.0035	0.0035	−0.0076	0.0000	0.0000
64	0.31	108.7	0.0016	0.0035	0.0035	−0.0058	0.0000	0.0000
65	0.36	126.2	0.0018	0.0035	0.0035	−0.0045	0.0000	0.0000
66	0.41	143.7	0.0020	0.0035	0.0035	−0.0036	0.0000	0.0000
67	0.46	161.2	0.0022	0.0035	0.0035	−0.0028	0.0000	0.0000
68	0.51	178.7	0.0023	0.0035	0.0035	−0.0022	0.0000	0.0000
69	0.56	196.2	0.0024	0.0035	0.0035	−0.0017	0.0000	0.0000
70	0.61	213.7	0.0025	0.0035	0.0035	−0.0013	0.0000	0.0000
71	0.66	231.2	0.0026	0.0035	0.0035	−0.0009	0.0000	0.0000
72	0.71	248.7	0.0027	0.0035	0.0035	−0.0006	0.0000	0.0000
73	0.76	266.2	0.0027	0.0035	0.0035	−0.0003	0.0000	0.0000
74	0.81	283.7	0.0028	0.0035	0.0035	−0.0001	0.0000	0.0000
75	0.86	301.2	0.0028	0.0035	0.0035	0.0001	0.0000	0.0000
76	0.91	318.7	0.0028	0.0035	0.0035	0.0003	0.0000	0.0000
77	0.96	336.2	0.0029	0.0035	0.0035	0.0005	0.0000	0.0000
78	1.01	353.7	0.0029	0.0035	0.0035	0.0006	0.0000	0.0000
79	1.06	371.2	0.0029	0.0035	0.0035	0.0008	0.0000	0.0000
80	1.11	388.7	0.0030	0.0035	0.0035	0.0009	0.0000	0.0000
81	1.16	406.2	0.0030	0.0035	0.0035	0.0010	0.0000	0.0000
82	1.21	423.7	0.0030	0.0035	0.0035	0.0011	0.0000	0.0000
83								
84			Calculation of Stresses X Areas					
85								
86		f1A1	f2A2	f3A3	f4A4	f5A5	f6A6	
87								
88		313	0	0	−966	0	0	
89		577	0	0	−966	0	0	
90		756	0	0	−966	0	0	
91		886	0	0	−966	0	0	
92		966	0	0	−966	0	0	
93		966	0	0	−966	0	0	
94		966	0	0	−966	0	0	
95		966	0	0	−808	0	0	
96		966	0	0	−603	0	0	
97		966	0	0	−430	0	0	
98		966	0	0	−281	0	0	
99		966	0	0	−151	0	0	
100		966	0	0	−38	0	0	
101		966	0	0	63	0	0	
102		966	0	0	152	0	0	
103		966	0	0	232	0	0	
104		966	0	0	304	0	0	
105		966	0	0	369	0	0	
106		966	0	0	429	0	0	
107		966	0	0	483	0	0	
108		966	0	0	533	0	0	

Sheet 4.9 (Page 3)

	B	C	D	E	F	G	H
115	B	C	D	E	F	G	H I
116	b or h	560					
117	h or b	350				Graphical Co-ords	
118						----------------	
119		Calculation of Nd and Md				Nd	Md
120		----------------------------				2000	342
121	*-*-*	bc/b	Nd	Md	Md/Nd	2000	233
122	Trial->	0.2105	1	240	221388	0	0
123	*-*-*	0.26	420	286	681	1	240
124		0.31	754	320	424	420	286
125		0.36	1039	346	333	754	320
126		0.41	1274	364	286	1039	346
127		0.46	1429	370	259	1274	364
128		0.51	1584	374	236	1429	370
129		0.56	1897	357	188	1584	374
130		0.61	2257	332	147	1897	357
131		0.66	2585	309	119	2257	332
132		0.71	2890	286	99	2585	309
133		0.76	3174	263	83	2890	286
134		0.81	3443	239	69	3174	263
135		0.86	3698	214	58	3443	239
136		0.91	3943	188	48	3698	214
137		0.96	4178	161	39	3943	188
138		1.01	4405	133	30	4178	161
139		1.06	4626	102	22	4405	133
140		1.11	4840	70	15	4626	102
141		1.16	5050	36	7	0	0
142		1.21	5255	1	0	0	0
						0	0

(a) the ratio of longer side to that of shorter side is less than 3;
(b) $l_e/h \leq 20$.

In which case, design is for N and the greatest value of M_{tx} and using M_{1x} and M_{2x} about the major axis.

Case 3. Bending about the major axis and not satisfying both of the above conditions. In which case, design is considered to be biaxial with zero initial moments about the minor axis.

Case 4. Biaxial bending. Design for N, M_{tx} and M_{ty} using Eqns 4.19 and 4.20 to transform to uniaxial design.

Design for Cases 1 and 2 is similar except for the change in direction. Case 3 is treated as a design for biaxial bending even though there are no applied moments about the minor axis. A memo is included in rows 20 and 21 as a reminder of the above case classification and the user is required to input the case which applies in cell C19. A check on the conditions of Case 2 is made in cell C22 as part of the preliminary calculations since this information is required for deciding which case applies.

Additional preliminary calculations

Cells C23 to C26 contain information about the effective lengths and slenderness ratios in directions perpendicular to the X and Y axes assuming the column is braced. For the unbraced case the formula in cell C23 would have to be changed in accordance with the coefficients given in the bottom row of Table 4.3. The contents of cells C22 to C26 are:

```
C22: @IF((@MAX(C6,C7))/@MIN(C6,C7))
     <3#AND#C25<=20,"Yes","No")
C23: (@IF(H17=11,0.75,@IF(H17=22,0.85,
     @IF(H17=33,1,@IF(H17=12#OR#H17
     =21,0.8,@IF(H17=13#OR#H17
     =31,0.9,0.95))))))*C18
C24: (@IF(I17=11,0.75,@IF(I17=22,0.85,
     @IF(I17=33,1,@IF(I17=12#OR#I17
     =21,0.8,@IF(I17=13#OR#I17
     =31,0.9,0.95))))))*C18
C25: +C23*1000/C7
C26: +C24*1000/C6
```

The additional moment due to the eccentricity, as given by Eqn 4.27, is calculated in cells F16 and F17 and the values of M_i, as defined in section 4.17, calculated in cells F18 and F19. These cells contain:

F16: **@IF(C25>15#OR#C26>15,C13 *((((C23*1000)/(@IF(C19=2, (@MIN(C6,C7)),C7)))^2)*C7/2000)/1000,0)**
F17: **@IF(C25>15#OR#C26>15,C13 *((((C24*1000)/(@MIN(C6,C7)))^2) *C6/2000)/1000,0)**
F18: **@MAX(0.4*C14+(0.6*C15),0.4*C15)**
F19: **@MAX(0.4*C16+(0.6*C17),0.4*C17)**

The initial @IF function in F16 and F17 is used to test if the column is slender or short and assumes that the columns are braced. If it is short then M_{add} would be zero. There is also a difference between the form of the formula in F16 compared with that in F17 resulting from the different way that b′ is defined for uniaxial and biaxial bending. In effect this difference results in b′ being always the shorter side for determining M_{addy} but could be either side for determining M_{addx} depending on the type of bending and the second @IF function in F16 allows for this. The value of h in Eqn 4.27 would be taken from C6 for M_{addy} and from C7 for M_{addx}.

In order to determine the maximum values of M_{tx} and M_{ty} the four equations given previously for each of these quantities are used in the block H18 to I21 and the maxima calculated in H22 and I22. Examples of the formulae in some of these cells are:

H18: **(F18+F16)**
H19: **(C15)**
H20: **(C14)+0.5*F16**
H21: **+C13*(@MIN(0.05*C7,20))/1000**
H22: **@MAX(H18..H21)**

Note that the label M_{tx}/y, written in G22 means either M_{tx}, the value in H22, or M_{ty}, the value in I22. These values of M_{tx} and M_{ty} are now used in cells F22 and F23 to determine the ratios M_x/h' and M_y/b' so that the increase due to the biaxial effect can be determined. The formula in F22 is:

F22: **(H22/C11)**

Finally the quantities to be used for the design are gathered together in cells I25, I26 and H26 where the latter is used to produce a variable label. These cells contain:

H26: **@IF((F22>=F23),"Mtx' ","Mty' ")**
I25: **+C13**
I26: **@IF(C19=4#OR#C19=3,@IF(F22>=F23, H22+(F14*C11*I22/C12), I22+(F14*C12*H22/C11), @IF(C19=1,I22,H22))**

where the first @IF function checks if biaxial bending applies and the second @IF calculates the appropriate value if it does. The third @IF function checks whether the uniaxial bending is about the minor or major axis and outputs the moment that applies.

The remainder of Sheet 4.9 is similar to Sheet 4.8. Note that although the arrangement of bars shown in rows 31 to 36 appear as unsymmetrical on the spreadsheet they are in fact symmetrical for this example, since the value of h_x at level 3 is equal to $h/2$. Symmetry or the lack of it depends on the values of h_x and b_x and not on the sketch of the reinforcement.

If the column had been unbraced then the formulae in cells H18 to I21 would have to be replaced. Unbraced columns are generally in double curvature and the maximum moments, both initial and additional, occur at the ends. The governing equation for a column AB becomes the greater of design moment ($M_A + M_{add}$) and design moment:

$$M_B + M_{add}(k_B/k_A)$$

where k_A and k_B are the total joint stiffnesses of all members connected monolithically at the joint, including the columns, and $k_A > k_B$, Martin *et al.* (1989).

Use of Sheet 4.9

As explained for Sheet 4.8 the spreadsheet can now be 'protected' and only the cells which require information from the user 'unprotected' so that the user would be warned if an illegal entry were attempted. Following this procedure the only unprotected cells would be: C6 to C19, H17, I17, D31 to D36, G31 to G36, E31, F31, E36, F36, C39 ro C44, C47 to C50 and C122.

To use the spreadsheet the user would enter the required values in the above cells and then adjust the trial value in C122 until the value of N_d in D122 becomes small and positive.

The interaction diagram can then be called by pressing the appropriate function key related to the graphical facility. If this diagram shows that the end point of the line from the origin is within the curve then the design is acceptable although if the point is too far away from the curve then the amount of reinforcement could be reduced. If the point is outside the curve then the amount of reinforcement must be increased.

Note that the values entered in cells C44 and C50 are the same values respectively as those for h′ and b′ in C11 and C12 as one would expect.

Example

A section 400 mm by 400 mm is subjected to an axial load of 1800 kN and moments at the ends of the column of 100 kN m and 200 kN m about the major axis. The moment direction is such that the column bends in single curvature and the ends of the column are one end fixed and the other partially fixed with regard to the major axis and with one end partially fixed and the other end pinned with regard to the minor axis. It is assumed that f_{cu} is 30 N mm^{-2} and f_y is 400 N mm^{-2}. The column is of height 6.5 m.

Assuming that a cover of 50 mm is to be provided for the top and bottom bars in both directions gives $b' = 350$ mm and $h' = 350$ mm. Initially it will be assumed that 10 bars of diameter 20 mm are to be placed as shown in Fig. 4.16 (these are not the diameters shown in the spreadsheet).

A number of the required cell entries can now be made as follows:

Sheet 4.9 (Example) (Page 1)

1/A	B	C	D	E	F	G	H	I	J
2		Design of Short or Slender Columns of Reinforced Concrete							
3			Uniaxial or Biaxial Bending						
4 Initial Data				Preliminary Calculations					
5 ------------				----------------------------					
6 Breadth		400	mm	0.87fy	348.00	N/fbh	Alpha	Diff	
7 Depth		400	mm	emin	20	0.00	1.00	−1.20	
8 fcu		30	N/mm2	StrainB	−0.002	0.10	0.88	−1.10	
9 fy		400	N/mm2	StrainA	0.002	0.20	0.77	−1.20	
10 ecu		0.0035		F	0.313	0.30	0.65	−1.20	
11 h'		350	mm	k1	0.403	0.40	0.53	−1.10	
12 b'		350	mm	k2	0.451	0.50	0.42	−1.20	
13 N		1800	kN	N/fcubh	0.38	0.60	0.30		
14 M1x(S)		100	kNm	Alpha	0.56	1.00	0.30		
15 M2x(L)		200	kNm	End Conditions		1−Fixed,2−Partially,3−Pinned			
16 M1y(S)		0	kNm	Madd−x	61		MajorX	MinorY	
17 M2y(L)		0	kNm	Madd−y	86	End Conds	12	23	
18 Height		6.5	m	Mix	160	Mi+Madd	221	86	
19 Case		2		Miy	0	M2	200	0	
20 1/MinorY, 2/MajorX + Conds(Yes)						M1+0.5Ma	130	43	
21 3/MajorX + (Conds(No), 4/Biaxial						Nemin	36	36	
22 Cond = Yes				Mx/h'	0.6310	Mtx/y	221	86	
23 lex		5.20	m	My/b'	0.2451				
24 ley		6.18	m	Note:−			Design for:−		
25 SRx		13.0		M1 negative for			N	1800 kN	
26 SRy		15.4		Double Curvature			Mtx'	221 kNm	
27 --									
28				Reinforcement Details					
29 Level			4	3	2	1			
30		---------------------------------------							
31	1	|	24	24	0	24 |			
32	2	|	24			24 |			
33	3	|	24			24 |			
34	4	|	0			0 |			
35	5	|	0			0 |			
36	6	|	24	24	0	24 |			
37		---------------------------------------							
38 Level		hx	hx/h	Ax	h/2−hx		Area Selection		
39	1	50	0.13	1357	150			1357	
40	2	150	0.38	905	50			905	
41	3	250	0.63	905	−50			905	
42	4	0	0.00	0	200	Max6%	9600	0	
43	5	0	0.00	0	200	Min0.4%	640	0	
44	6	350	0.88	1357	−150	SumArea	4524	1357	
45							Depth Selection		
46	Level	by	by/b	Ay	b/2−by			150	
47	1	50	0.13	1810	150			50	
48	2	200	0.50	905	0			−50	
49	3	0	0.00	0	200			200	
50	4	350	0.88	1810	−150			200	
51 REMEMBER ENTRY TO CELL C122								−150	
52 D122 =		1 O.K.							
53									

Sheet 4.9 (Example) (Page 2)

59	B	C	D	E	F	G	H	I
60			Calculation of Strains					
61	dc/h	dc	eps_1	eps_2	eps_3	eps_4	eps_5	eps_6
62	0.28	111.1	0.0019	−0.0012	−0.0044	0.0035	0.0035	−0.0075
63	0.33	131.1	0.0022	−0.0005	−0.0032	0.0035	0.0035	−0.0058
64	0.38	151.1	0.0023	0.0000	−0.0023	0.0035	0.0035	−0.0046
65	0.43	171.1	0.0025	0.0004	−0.0016	0.0035	0.0035	−0.0037
66	0.48	191.1	0.0026	0.0008	−0.0011	0.0035	0.0035	−0.0029
67	0.53	211.1	0.0027	0.0010	−0.0006	0.0035	0.0035	−0.0023
68	0.58	231.1	0.0027	0.0012	−0.0003	0.0035	0.0035	−0.0018
69	0.63	251.1	0.0028	0.0014	0.0000	0.0035	0.0035	−0.0014
70	0.68	271.1	0.0029	0.0016	0.0003	0.0035	0.0035	−0.0010
71	0.73	291.1	0.0029	0.0017	0.0005	0.0035	0.0035	−0.0007
72	0.78	311.1	0.0029	0.0018	0.0007	0.0035	0.0035	−0.0004
73	0.83	331.1	0.0030	0.0019	0.0009	0.0035	0.0035	−0.0002
74	0.88	351.1	0.0030	0.0020	0.0010	0.0035	0.0035	0.0000
75	0.93	371.1	0.0030	0.0021	0.0011	0.0035	0.0035	0.0002
76	0.98	391.1	0.0031	0.0022	0.0013	0.0035	0.0035	0.0004
77	1.03	411.1	0.0031	0.0022	0.0014	0.0035	0.0035	0.0005
78	1.08	431.1	0.0031	0.0023	0.0015	0.0035	0.0035	0.0007
79	1.13	451.1	0.0031	0.0023	0.0016	0.0035	0.0035	0.0008
80	1.18	471.1	0.0031	0.0024	0.0016	0.0035	0.0035	0.0009
81	1.23	491.1	0.0031	0.0024	0.0017	0.0035	0.0035	0.0010
82	1.28	511.1	0.0032	0.0025	0.0018	0.0035	0.0035	0.0011
83								

84		Calculation of Stresses X Areas				
85						
86	f1A1	f2A2	f3A3	f4A4	f5A5	f6A6
87						
88	472	−222	−315	0	0	−472
89	472	−91	−315	0	0	−472
90	472	5	−315	0	0	−472
91	472	78	−292	0	0	−472
92	472	136	−195	0	0	−472
93	472	183	−117	0	0	−472
94	472	222	−52	0	0	−472
95	472	255	3	0	0	−374
96	472	283	49	0	0	−276
97	472	307	89	0	0	−192
98	472	315	124	0	0	−119
99	472	315	155	0	0	−54
100	472	315	182	0	0	3
101	472	315	207	0	0	54
102	472	315	229	0	0	100
103	472	315	248	0	0	141
104	472	315	266	0	0	179
105	472	315	282	0	0	213
106	472	315	297	0	0	244
107	472	315	311	0	0	273
108	472	315	315	0	0	299

Sheet 4.9 (Example) (Page 3)

	B	C	D	E	F	G	H	I
115	B	C	D	E	F	G	H	I
116	b or h	400						
117	h or b	400				Graphical Co-ords		
118								
119		Calculation of Nd and Md				Nd	Md	
120		------------------------				1800	221	
121	*-*-*	dc/h	Nd	Md	Md/Nd	1800	221	
122	Trial*	**0.2778**	1	227	231364	0	0	
123	*-*-*	0.33	228	242	1062	1	227	
124		0.38	421	254	604	228	242	
125		0.43	614	262	427	421	254	
126		0.48	866	263	304	614	262	
127		0.53	1088	264	242	866	263	
128		0.58	1288	262	204	1088	264	
129		0.63	1571	245	156	1288	262	
130		0.68	1839	226	123	1571	245	
131		0.73	2085	207	99	1839	226	
132		0.78	2298	188	82	2085	207	
133		0.83	2490	168	68	2298	188	
134		0.88	2671	148	55	2490	168	
135		0.93	2843	127	45	2671	148	
136		0.98	3007	105	35	2843	127	
137		1.03	3165	82	26	3007	105	
138		1.08	3317	58	18	3165	82	
139		1.13	3464	33	10	0	0	
140		1.18	3607	7	2	0	0	
141		1.23	3746	−21	−6	0	0	
142		1.28	3874	−49	−13	0	0	
						0	0	

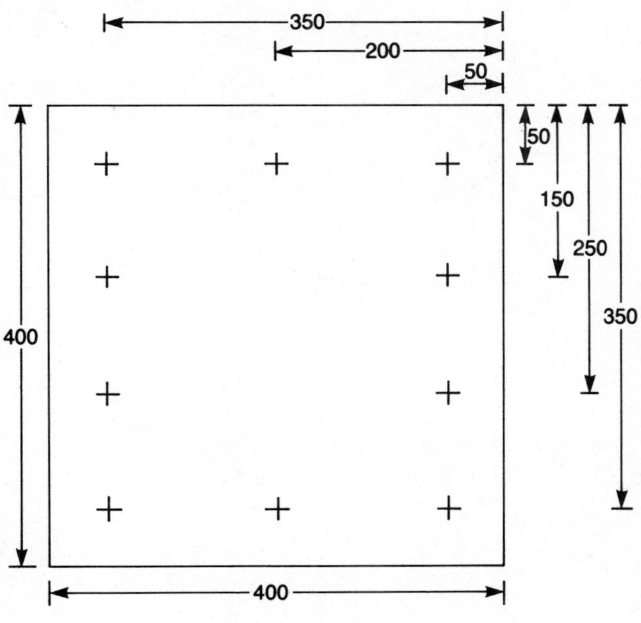

Figure 4.16 Dimensions of Example 4.9 in mm

C6 = 400; C7 = 400; C8 = 30; C9 = 400; C10 = 0.0035; C11 = 350; C12 = 350; C13 = 1800; C14 = 100; C15 = 200; C16 = 0; C17 = 0; C18 = 6.5; C19 = ?; H17 = 12; I17 = 23.

At this stage the value to be entered in C19 is unknown. Since at least one of the values in C25 or C26 is > 15, this is a slender column with bending about the major axis. It is therefore either Case 2 or Case 3 depending on the output in cell C22. If this output is "Yes" then use Case 2, otherwise use Case 3. All other known values, including details about the reinforcement, should be entered first so that the output of cell C22 will be known and the correct entry value for C19 determined.

The value 20 is entered in D31; E31; G31; D32; D33; D36; E36; G36; G32; G33; i.e. 10 bars of diameter 20 mm. The value 0 is entered in the other bar locations D34, D35, G34, G35, F31 and F36.

The levels, shown in Fig. 4.16, are now entered:

C39 = 50; C40 = 150; C41 = 250; C42 = 0; C43 = 0; C44 = 350; C47 = 50; C48 = 200; C49 = 0; C50 = 350.

The value in cell C122 is now adjusted by trial until the value in D122 is small and positive. It will be found that a value of 0.2356 in C122 produces a value of 0 in D122. Using the graphical function key (usually F10), the graph produced will be shown as the inner curve of Graph 4.9. The end of the straight line is outside this curve and therefore the amount of reinforcement must be increased.

Try 10 bars of diameter 24 mm in the same locations. The trial value in C122 will now be found to be 0.2778 when the value in D122 is 1. The graph produced will be shown as the outer curve of Graph 4.9. The end of the straight line is now within the curve and the design is acceptable. Cells H42 to H44 show that the area of reinforcement provided is within the limits.

Other arrangements and bar diameters are also possible and the designer can easily find a suitable combination before proceeding with the rest of the detailing and additional checks.

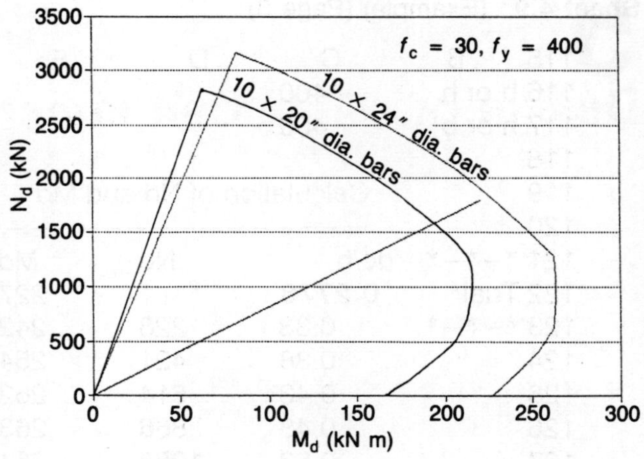

Graph 4.9 (Example) Uni/biaxial bending for short or slender columns of reinforced concrete

5
Design of reinforced concrete slabs

The spreadsheets developed in Chapter 3 for the design of beams can be readily adapted so that they apply to slabs. The breadth of the beam has to be changed into a selected width of slab, usually taken as one metre, and some of the limits that have to be satisfied may be different but there is no difficulty in allowing for these changes.

5.1 Types of slab

The main problem that arises is the determination of the value of the bending moment that acts on the strip. Various methods have been proposed for calculating this moment and the application of these methods depends on the type of slab being considered.

Slabs spanning one way

Single span slabs which span one way present no problem since the moment on the strip can be easily determined.

For continuous slabs, moment and shear coefficients are available which enable the determination of the shear at the supports and the moments at the support and mid spans. These coefficients are based on an elastic analysis but plastic collapse methods are also used.

Slabs spanning two way

For single slabs, if the slab is simply supported then equations are available for calculating approximate values of the moment.

If the slab is restrained then the moments can be determined by the use of a yield line approach although for rectangular slabs moment coefficients are given in BS 8110.

Examples of how spreadsheets can assist in the design for some of the above cases are given below. The methods could be applied to other cases but these are so numerous that not all could be covered in this text. Generally design is for rectangular slabs but design may also be required for irregular shapes and the reader should refer to texts which deal exclusively with the various applications of yield line theory for more information on the methods used for the derivation of the appropriate equations in Johansen (1972), Hognestad (1953), Jones and Wood (1967), Armer (1968), Wood and Armer (1968), and Hillerborg (1982).

5.2 Spreadsheet layout for slab design

The first page of spreadsheet 5.1 is similar to Sheet 3.4R developed in Chapter 3 for the design of rectangular reinforced concrete beams. There are some additional calculations on page 1 of Sheet 5.1 compared with page 1 of Sheet 3.4R and only details of the changes are given below.

Initial data (changes to Sheet 3.4R)

The effective depth has been inserted in cell C12 and the cover to the reinforcement has been removed (from C11 of Sheet 3.4R). Using an assumed effective depth, instead of calculating this value from the cover, at the initial stage, is more convenient and enables a check to be carried out using values of the cover which are related to the diameter of the reinforcement. Details of this check are given later.

The unit breadth of slab used for the design is normally one metre and although there is no real necessity for including this constant on the spreadsheet it has been included in cell C9. It could be changed to some other value if required.

The values of the characteristic strength of the links, f_{yv}, and the shear force (V) have also been included.

Preliminary calculations

The three additional entries at the foot of this section are not strictly preliminary calculations, since they require input from the user into cells G16 to G18, and they are only placed in this location for convenience.

Exposure conditions are classified as:

mild, if protected against the weather;
moderate, if sheltered from severe rain or freezing when wet and/or in contact with non-aggresive soil;
severe, if exposed to severe rain or occasional freezing;
very severe, if exposed to sea water spray, de-icing salts or freezing conditions.

In the British Standard code there is also an 'extreme' exposure condition which occurs when the surface is subject to abrasive action or machinery but this condition has not been included on the present spreadsheet.

The user must select the exposure condition which applies and insert this choice in cell G16. If the spreadsheet package used is case-sensitive then care would have to be taken with the method of entry of this label. The entry is converted into a numerical value in cell F16 since this is more convenient for use with an @VLOOKUP function. The formula in cell F16 is:

F16: **@IF(G16="Mild",1,**
@IF(G16="Moderate",2,
@IF(G16="Severe",3,4)))

In cell G17 the value of the fire rating for the slab is entered in the form of an integer of value 1 to 4 hours. This integer is used later to determine if the slab thickness and the cover to the reinforcement satisfy the conditions pertaining to the selected fire rating.

The minimum area of reinforcement required in each direction is normally taken as 0.13 per cent of the area of the section for f_y = 460 N mm^{-2} and 0.24 per cent for f_y = 250 N mm^{-2}. However, if the control of shrinkage and temperature cracking is considered important then these percentages are increased to 0.25 and 0.3 respectively. In order to allow for this on the spreadsheet a shrinkage control constant, defined as 1 for normal and 2 for the increased values, is entered in cell G18.

Reinforcement details

The method used for calculating the amounts of reinforcement required and the comparison of these values with amounts provided is identical with that described for Sheet 3.4R. The only additional calculations occur in rows 42 and 43 where a check is carried out to ascertain if the amount of steel used satisfies the minimum requirements given in the previous section.

This check uses cells F42, F43 and H43 which contain:

F42: **(@IF(G18=2#AND#C14=460,0.0025,**
@IF(G18=2#AND#C14=250,0.003,
@IF(G18=1#AND#C14=460,
0.0013,0.0024))))*C9*C10
F43: **@IF(F11<=F10,G32,G40)**
H43: **@IF(F43>=F42,"O.K.","Check steel")**

Selection of cover to reinforcement

In the British Standard the cover to the reinforcement is selected on the basis of the strength of the concrete and the exposure conditions. Relevant values taken from the code have been entered in rows 55 to 59. Values of f_{cu} have been placed in the range B55 to B59 and the numerical equivalent of the exposure conditions in range C54 to F54. Zero values have been inserted in the locations where no code value was specified and, for convenience, the table extended using an initial value of f_{cu} = 10 N mm^{-2}.

The significance of the zero values for cover in the table is that it flags a warning that either the concrete strength is too low for the exposure conditions specified or the exposure condition has been entered incorrectly. Automatic interpolation has not been allowed for but the user would be able to make a visual interpolation quite quickly for other values of f_{cu}.

For the exposure condition specified the value obtained from the table (without interpolation) is calculated in cell C60 which contains:

C60: **@VLOOKUP(C13,B55..F59,G16)**

If the value output in C60 is zero then a warning appears in Cell E60 using:

E60: **@IF(C60=0#OR#C13<30,"Check fcu"," ")**

The user may not wish to use this calculated value and is therefore requested to select a value for the cover in cell E61. This selected value must be equal to or greater than the value shown in cell E60.

Check on the values of *d'* and *d*

In order to calculate suitable values for the reinforcement it is first necessary to assume values for d' and d but these assumed values may not be suitable and may require to be adjusted so that the required area of reinforcement can be recalculated.

This has been done in rows 67 and 68. First the values of d' and d are calculated in cells C67 and C68 using:

C67: **@IF(C16<F12,"Not required",**
E61+(0.5*D39))
C68: **@IF(C16<F12,C10−(E61+(0.5*G31)),**
C10−(E61+(0.5*G39))

The actual values of d' and d are output to H67 and H68 so that the user can decide if any correction is needed. If C67 shows that d' is not required then any value that is shown in H67 will not be used and can therefore be ignored.

Check for shear

Using the equation:

$$v = V/bd \qquad [5.1]$$

the value of the shear stress v is calculated in D74 using:

D74: **(C17*1000000)/C9*C12)**

This value of v should be less than the lesser value of $0.8\sqrt{f_{cu}}$ or 5 N mm^{-2}, otherwise the slab thickness should be increased. The condition is checked in F74 using:

Sheet 5.1 (Page 1)

1/A	B	C	D	E	F	G	H
2		Design of Reinforced Concrete Slabs					
3		-----------------------------------					
4			Singly or Doubly Reinforced				
5	Using I.Struc.E.Manual. (Allowing for Redistribution)						
6	---						
7	Initial Data			Preliminary Calculations			
8	-------------			-------------------------			
9	Breadth	**1000**	mm	0.87fy	400.2	N/mm2	
10	Depth	**220**	mm	K'	0.156		
11	d'	**28**	mm	K	0.159		
12	d	**183**	mm	Mu	157	kNm	
13	fcu	**30**	N/mm2	z	142	mm	
14	fy	**460**	N/mm2	z	142	mm	
15	fyv	**250**	N/mm2	x	91	mm	
16	DesignM	**160**	kNm/m	Exposure	1	**Mild**	
17	ShearV	**200**	kN	Fire Rating		**1**	h
18	Redist.	**10**	%	Shrinkage Control		**1**	
19	-------------------			-------------------------			
20	Redistribution Table						
21	------------------------			!! Compression Steel Required !!			
22	Redist.	K'	Diff				
23	0	0.156	0		d'/xlimit	400	
24	10	0.156	−0.0024		As'_reqd.	53	mm2
25	15	0.144	−0.0024		As_reqd.	2807	mm2
26	20	0.132	−0.0026				
27	25	0.119	−0.003				
28	30	0.104			Bottom Steel Only		
29					-----------------		
30					Spacing	**100**	mm
31					Dia	**10**	mm
32					As_prov.	785	mm2
33							
34							
35		!!! Case which applies !!!					
36	Top Steel				Bottom Steel		
37	---						
38		Spacing	**300**	mm	Spacing	**160**	mm
39		Dia	**6**	mm	Dia	**24**	mm
40		As_prov.	94	mm2	As_prov.	2827	mm2
41			O.K.			O.K.	
42	Minimum tension steel required				286	mm2	
43			As provided		2827	mm2	O.K.
44	---						

F74: @IF(D74>(@MIN((0.8*@SQRT(C13)),5)), "Slab thickness O.K.", "Increase slab thickness")

Since it is considered to be impractical to use shear links for slabs less than 200 mm depth a check is also made in D75 using:

D75: @IF(C10<200, "Impractical to use shear links"," ")

If the slab is deeper than 200 mm then there would be no output to cell D75.

To proceed with the remainder of the calculations it is necessary to determine the limiting value of the shear

Sheet 5.1 (Page 2)

```
50     B            C           D           E           F           G           H
51 Cover
52 -----                   Nominal Cover mm
53                         Exposure Condition
54        Fcu          1          2          3          4
55         10          0          0          0          0
56         30         25          0          0          0
57         35         20         35          0          0
58         40         20         30         40         50
59         50         20         20         25         30
60   Cover            25 mm
61 Cover selected by user                   25 mm
62 ---------------------------------------------------------------------
63                         Note on values used for d' and d
64                         -----------------------------------------
65 Required values based on Cover + 0.5dia              Values used above
66 ---------------------------------------------------------------------
67 d'                 28                                d'           28
68 d                 183 mm                             d           183
69               User may wish to change values of d' and d in
70                CELLS C11 and C12 if discrepancy too large
71 ---------------------------------------------------------------------
72                         Check for Shear
73                         ---------------
74 Shear Stress Limit          1.09 N/mm2      Slab thickness O.K.
75
76
77               Effective depth  =           183 100As/bd        1.55
78                           vc =             0.94 N/mm2
79     Shear Reinforcement =        Minimum Links         Constant          1
80               Max Link spacing =           137
81 User spacing if required  =                120 mm
82  Min. Link Area (2 Legs)  =                221               Area required
83     Link dia. if required =                   6 mm
84                         Area    =          236               Area provided
85 ---------------------------------------------------------------------
```

stress v_c. This limiting value can be obtained from a table of values given in BS 8110 or calculated from the equation:

$$v_c = 0.79(100A_s/bd)^{1/3}(400/d)^{1/4}(f_{cu}/25)^{1/3}/\gamma_m \quad [5.2]$$

On the spreadsheet the calculation is performed in E78 having first calculated **$100A_s/bd$** in cell G77 and taking γ_m as 1.25:

G77: **100*@IF(C16>F12,G40,G32)/(C9*C12)**
E78: **(0.79*((G77^0.333)*((400/C12)^0.25)/1.25))*((C13/25)^0.333))**

This value of v_c is now compared with the value of the shear stress calculated in cell D74 in order to decide how much shear reinforcement is required. Three limiting conditions are defined.

Condition (i) $v < v_c$ for which no shear reinforcement is required.

Condition (ii) $v_c \leq v \leq (v_c+0.4)$ for which only a minimum amount of shear reinforcement is required. The minimum area A_{sv}, of the two legs of the shear links, is given by:

$$A_{sv} \geq 0.4bs_v/0.87f_{yv} \quad [5.3]$$

where f_{yv} is the characteristic strength of the link and s_v the link spacing.

Sheet 5.1 (Page 3)

```
92        B            C          D            E            F            G        H
93                                Deflection
94                                ----------
95 Additional Data                           Preliminary Calculations
96 ----------------                          ----------------------------
97 Span               5.5 m                  fs                317
98 BeamType  SS       SS,CO,CA               Factor            0.78
99                                           Sp/Depth        15.69
100                               Limiting Depth               351 Depth O.K
101 ------------------------------------------------------------------------
102 Secondary Reinforcement
103 ---------------------------------
104                  Minimum required =                        286 mm2
105                           Trial dia =                       10 mm
106                       Trial spacing =                      260 mm
107                       Area provided =                      302 mm2      O.K.
108 ------------------------------------------------------------------------
109 Cracking
110 ---------
111 Clear spacing between main bars =                          136 mm
112      Spacing of secondary bars =                           260 mm
113  Recommended max. for main bars =                          300 mm       O.K.
114  Recommended for secondary bars =                          400 mm       O.K.
115 Conditions
116            Depth                 0
117            Rho                   0                 0 %Rho               1.60
118                                                     Case                   1
119 Clear spacing between main bars =                          147 mm
120    Main bar spacing also less than =                       147 mm       Still O.K.
121 ------------------------------------------------------------------------
122 Fire Resistance
123 ----------------
124            Min Thickness                 Cover to main Bars
125 Rating     SS          CO                SS          CO
126            1           95          95    20          20
127            2           125         125   35          25
128            3           150         150   45          35
129            4           170         170   55          45
130
131 Minimum Thickness Required =                                95 mm       O.K.
132 Minimum Cover to Main Bars =                                20 mm       O.K.
```

Condition (iii) $v > (v_c + 0.4)$ for which the minimum area of shear links required is given by:

$$A_{sv} \geq (v - v_c) b s_v / 0.87 f_{yv} \qquad [5.4]$$

These limiting equations are used in rows 79 to 84 to check if shear reinforcement is required and to allow the user to provide sufficient if required. The formulae in the key cells to achieve this are:

```
E79: @IF(D74<E78,"Not required",@IF(D74
     >=E78#AND#D74<=(E78+0.4),
     "Minimum links",
     "Provide reinforcement"))
H79: @IF(D74<E78,0,@IF(D74
     >=E78#AND#D74<=(0.4+E78),1,2))
E80: @IF(H79=0,"Not required",0.75*E77)
E82: @IF(H79=1,(0.4*E81*C9/(0.87*C15)),
```

@IF(H79=2,((D74−E78)
*C9*E81)/(0.87*C15),"Not required"))

E84: **@IF(H79=0,"Not required",
(@PI*(E83*E83)/4)*(C9/E81))**

In the above the formula in E79 determines which of three possible limits apply and the formula in H79 converts the result into a numerical value to simplify subsequent input. The maximum link spacing should not exceed $0.75d$ and this is calculated in E80.

The user is then required to select a value for the spacing in E81 and this value used in E82 to determine the minimum area required for the links using either Eqn 5.3 or 5.4.

Similarly the user is required to select a suitable link diameter in E83 and this value is used in E84 to calculate the area which would be provided if using a spacing defined by E81 and a diameter defined by E83. In practice the user would have to use a few trial values in E81 and E83 in order to find a satisfactory design. For cosmetic purposes the following formulae have been placed in cells G82 and G84:

G82: **@IF(H79=0," ","Area required")**
G84: **@IF(H79=0," ","Area provided")**

to act as a reminder. If links are not required then no output appears in the above cells.

Check for deflection

For this check additional information is required with regard to the span and type of slab support and these are entered in C97 and C98 using the abbreviations SS, CO and CA for simply supported, continuous and cantilever slabs respectively.

The limiting depth of slab required to ensure that the deflection is not excessive can now be determined by first assuming that the basic span/depth ratio is 20 for SS, 26 for CO and 7 for CA and then multiplying the value that applies by a modification factor to allow for the tension reinforcement.

The modification factor is defined by:

$$\text{Factor} = 0.55 + (477 - f_s)/(120[0.9 + K_m]) \leq 2 \qquad [5.5]$$

where $K_m = M/bd^2$ and f_s is defined by:

$$f_s = 0.625 f_y A_{s(reqd)}/(A_{s(prov)}\beta_b) \qquad [5.6]$$

where β_b is equal to (100 − %moment redistribution)/100. These equations have been inserted into the range F97 to F100 which contain:

F97: **(0.625*C14*G25*100)/(@IF(C16>F12,
G40,G32)*(100−C18))**
F98: **@MIN(0.55+((477−F97)/(120*(0.9+((C16
*1000000)/(C19*C12*C12))))),2)**
F99: **@IF(C98="SS",20,
@IF(C98="CO",26,7))*F98**
F100: **(C97*1000)/F99**
G100: **@IF(F100>=C12,"Depth O.K.",
"Check depth")**

The formula in G100 is used for checking that the depth provided is suitable and thus allowing the user to change the value if necessary.

Secondary reinforcement

Details about the requirements for the minimum area of reinforcement in each direction were given in the preliminary calculations above and a control constant placed in cell G18 to distinguish between (1) normal requirements and (2) special requirements. This information was used in cell F42 to determine the minimum area of reinforcement required and this value is now copied to F104.

The user now enters a trial diameter in F105 and a trial spacing in F106 so that the area of secondary reinforcement can be calculated and compared with that provided. Cells F107 and G107 contain:

F107: **@PI*(F105^2)*C9/(4*F106)**
G107: **@IF(F107>=F104,"O,K.",
"Check steel")**

Cracking

The rules governing the control for cracking vary from one code to another but whatever limits are specified they can usually be readily included into the spreadsheet. The limits used below are a mixture of those given in BS 8110 and those given in *Institution of Structural Engineers Manual*.

(1) Clear spacing between bars $\leq 3d$ or 300 mm for main bars.
(2) Clear spacing for distribution bars $\leq 3d$ or 400 mm.
(3) For $f_y = 250\ \text{N mm}^{-2}$, h must be ≤ 250 mm and for $f_y = 460\ \text{N mm}^{-2}$, h must be ≤ 200 mm.
(4) The steel ratio $A_s/bd < 0.3\%$.

Conditions 1 and 2 must be satisfied. Then, if either of 3 or 4 are satisfied, the design is satisfactory with regard to cracking control. If, however, neither of the conditions 3 or 4 are met then a further limitation is placed on the allowable clear spacing of the main bars as follows:

(5) If $A_s/bd \geq 1\%$ then the clear spacing must not exceed the value given by

$$\text{Clear spacing} = 75000(100 - \beta\%)/(100 f_y) \leq 300 \qquad [5.7]$$

where $\beta\%$ represents the moment redistribution.

(6) If A_s/bd *is* $< 1\%$ then the clear spacing is calculated by dividing the value obtained from equation 5.7 by the value $100A_s/bd$.

These conditions and the checks to establish if they are satisfied have been entered in rows 111 to 120 using the following entries:

F111: **@IF(F11<=F10,G30−G31,G38−G39)**
F112: **+F106**

in order to calculate the actual spacing of the main and secondary bars.

F113: **@MIN(3*C12,300)**
F114: **@MIN(3*C12,400)**

in order to calculate the recommended spacing in accordance with conditions 1 and 2 above. Cells H113 and H114 are then used to check if each spacing is satisfactory using:

H113: **@IF(F113>=F111,"O.K.", "Check Steel")**
H114: **@IF(F114>=F112, "O.K.","Check Steel")**

The conditions stated in 3 and 4 are now tested in cells D116 and D117 and a value 1 output if the condition is true and 0 if it is not. The appropriate cell formulae are:

D116: **@IF((C14=250#AND#C10<=250) #OR#(C14=460#AND#C10<=200),1,0)**
D117: **@IF(100*@IF(C16<F12,G32, (D40+G40))/(C9*C12)<0.3,1,0)**

Since only one of these conditions need be satisfied it is convenient to test if either of D116 or D117 contain 0 and this has been done in cell E117:

E117: **@IF(D116=0#OR#D117=0,0,1)**

so that if E117 outputs a zero, then neither of conditions 3 and 4 are satisfied whilst if it outputs a 1 then at least one is satisfied.

For convenience the percentage value of A_s/bd is calculated in G117 and the case which applies (5 or 6) in cell G118.

G117: **@IF(E117=0,100*(@IF(C16<F12,G32, (D40+G40))/(C9*C12)),"")**
G118: **@IF(E117=0,@IF(G117>=1,1,2),"")**

where the first @IF function, in each case, is used to suppress any output if the value in E117 is 1. This procedure is used for all the remaining cells in this section, both labels and formulae, so that output will only appear if the value in E117 is 0. If the value in E117 is 1 then the design is satisfactory as far as crack control is concerned and this information is output to C118 using:

C118: **@IF(E117=1,"Crack design O.K.","")**

and if the value of E117 is 0 then nothing would appear in cell C118, but information would appear in the lower cells in rows 119 and 120 using the following formulae:

F119: **@IF(E117=0,@MIN((75000 *(100−C18)/(C14*100)),300),"")**
F120: **@IF(E117=0,@IF(G118=1,F119, F119/G117),"")**
H120: **@IF(E117=0,@IF(F111<=F120, "Still O.K.","Check Steel"),"")**

The labels in this section are treated similarly so that they appear on the spreadsheet only if E117 = 0.

Fire resistance

To check if the effective depth of the slab and the cover to the reinforcement is sufficient to satisfy the required fire rating the British Standard code uses a table of values and a selection of these are shown in rows 126 to 129.

The appropriate values of effective depth and cover, for the fire rating specified in G17, are output to F131 and F132 using:

F131: **@VLOOKUP(G17,B126..F129,1)**
F132: **@VLOOKUP(G17,B126..F129, @IF(C98="CO",4,3))**

where a 1 is used at the end of the formula in F131 since columns 1 and 2 are identical. The @IF function in F132 is used to check if the slab is continuous and if so use column 4 (i.e. column F of the spreadsheet).

Finally the required values are checked with those provided in cells H131 and H132:

H131: **@IF(C12>=F131,"O.K.", "Check depth")**
H132: **@IF(C60>=F132,"O.K.", "Check cover")**

5.3 Use of spreadsheet 5.1

Once the input has been completed the spreadsheet can be globally PROTECTED after which the cells which require input from the user can be UNPROTECTED. This safeguards the cells which contain formulae and which can so easily be accidentally erased. The cells which have to be UNPROTECTED are: C9 to C18, G16 to G18, G30, G31, D38, D39, G38, G39, E61, E81, E83, C97, C98, F105, and F106.

Once values have been entered in the above cells the user will know if any changes are required in order to satisfy the limiting conditions. Different codes may require different conditions to be satisfied but the basic principles remain the same and there should be no difficulty in modifying the spreadsheet to cope with other conditions.

The calculation of the bending moment and shear force from the basic characteristic dead and imposed loads has not been included in this spreadsheet since this would tend to restrict its use. Different methods are available for calculating the moments and shears in slabs depending on the shape and method of support and examples of some of these are given below. Once these moments and shears have been calculated, by whatever method, then the design could be continued by entering the values in Sheet 5.1.

5.4 Determination of the bending moments

There are a number of different methods for determining the bending moments in slabs and of these the three most generally used are:

(a) Johansen's yield line methods;
(b) Hillerborg strip method;
(c) Use of moment coefficients (BS 8110).

Examples of how spreadsheets can be used for each of these methods is given below. Explanations of the method are confined to the particular example used and for a more general study readers should refer to texts which deal exclusively with the subject in Johansen (1972), Hognestad (1953), Jones and Wood (1967), Armer (1968), Wood and Armer (1968) and Hillerborg (1982).

5.5 Johansen's yield line method

Basically in this method the designer has to assume a pattern of yield lines along which the slab collapses by bending. If the yield line is caused by a sagging moment it is considered positive and vice versa. There is no guarantee that the yield line pattern selected by the designer will be identical with the pattern that occurs at failure and it is usual to calculate the moments for a number of different patterns so as to estimate the worst case.

Patterns of yield lines depend on the shape of the slab, the support conditions and the loading and although a large number of patterns are available designers can usually obtain a satisfactory solution by the study of only a few variations. In this method a collapse mechanism is assumed. Other collapse mechanisms may be possible but for each mechanism the solution obtained for the collapse load will be either greater than or equal to the true collapse load. Thus the solution will be an upper bound solution.

For example to study the collapse of the square panel supported on edge beams and columns shown in Fig. 5.1 the designer would have to analyse the slab for the various possible failure modes shown.

Having selected the pattern a unit virtual displacement is assumed for one of the yield lines and an equation representing the equality between the external work done by the loading and the energy dissipation of the yield lines resulting from the displacement.

The energy dissipation of a yield line is the product of yield moment of resistance m, times the projection of the yield line on to the axis, times the rotation of the rigid region about that axis. If the yield line rotates about more than one axis then the dissipation is obtained by taking the sum of all products.

The yield moment of resistance m is the moment per unit of width. If there are two bands of reinforcement perpendicular to each other then the slab is orthotropically reinforced with yield moments of say m_1 and m_2. If m_1 and m_2 are equal then the slab is reinforced isotropically. Examples are given below.

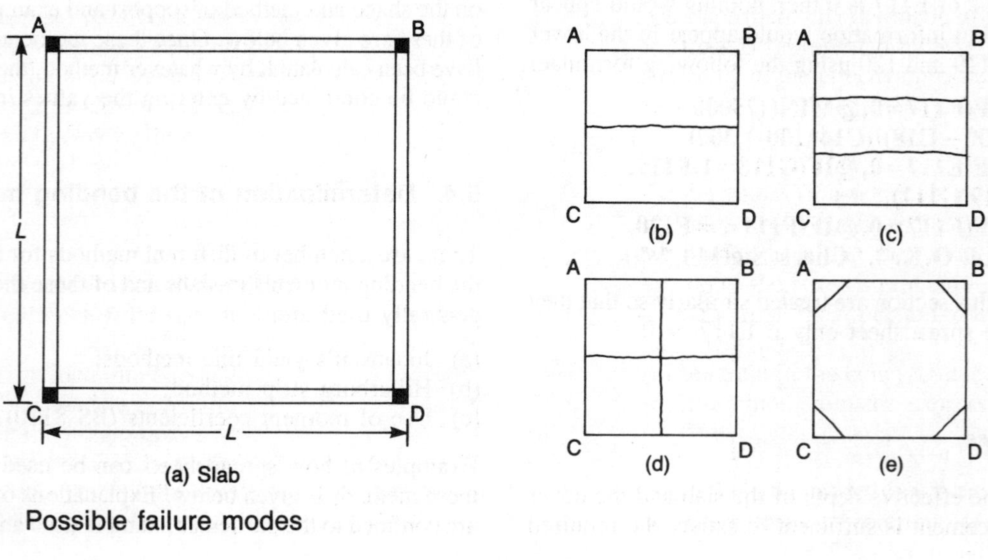

Figure 5.1 Possible failure modes

Example 1

Consider a rectangular slab of length L and width B which is simply supported on three sides and fixed on the remaining side AB. The loading is uniformly distributed and of intensity q kN m^{-2} and the slab is isotropically reinforced with steel top and bottom.

Assume that the yield line pattern is as shown in Fig. 5.2 and that the yield line EF is subjected to unit deflection. Note that the broken line along AB represents a negative yield line due to a hogging moment about the fixed support and requiring top steel.

The external work done can be calculated by dividing the area into convenient triangles and rectangles as shown in Fig. 5.2 and for each basic figure we can list as follows:

No.	*Area*	*Centroid movement*
1	$Bx/2$	$\frac{1}{3}$
2	$x(B-y)/2$	$\frac{1}{3}$
3	$xy/2$	$\frac{1}{3}$
4	$(B-y)(L-2x)$	$\frac{1}{2}$
5	$y(L-2x)$	$\frac{1}{2}$

The external work is the load times the volume swept by each area resulting from the unit deflection of EF. Therefore:

$$\begin{aligned}\text{External work} &= q[\tfrac{1}{3}(Bx + x(B-y) + xy] + \\ &\quad q\tfrac{1}{2}[(B-y)(L-2x) + y(L-2x)] \\ &= qB[L/2 - x/3] \qquad [5.7]\end{aligned}$$

Energy dissipation. AE rotates about AB and AC. The projection of AE on AB is x and on AC is y. The rotation of AE about AB is $1/y$ and about AC is $1/x$. Therefore the energy dissipation of AE (and BF) is given by:

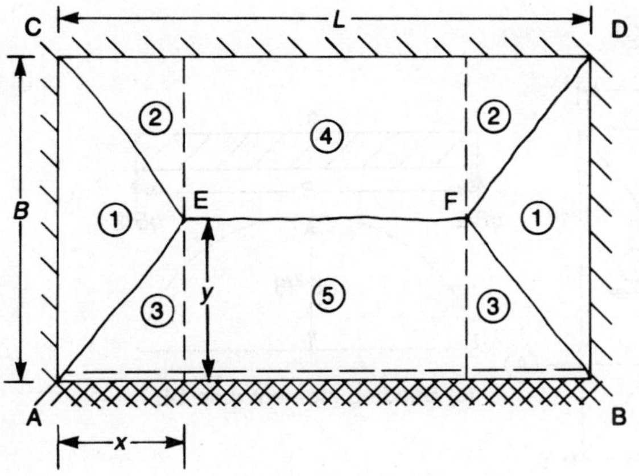

Figure 5.2 Example 1. Johansen's yield line method

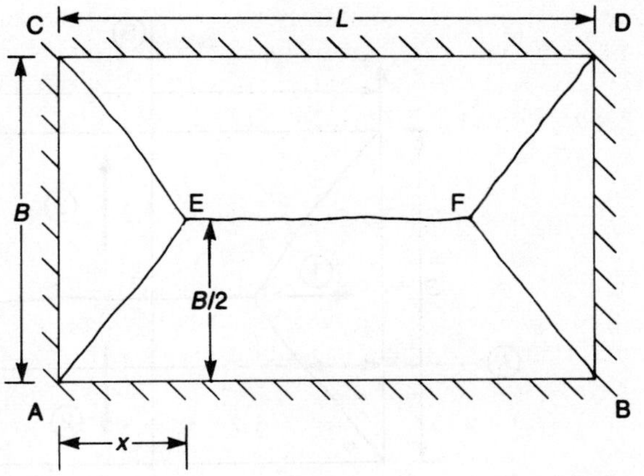

Figure 5.3 Example 2. Johansen's yield line method

AE and BF $= 2m[x/y + y/x]$

Similarly the energy dissipation along the other yield lines are:

CE and DF $= 2m[x/(B-y) + (B-y)/x]$
EF $= m[(L-2x)/y + (L-2x)/(B-y)]$
AB $= m[L/y]$

Equating the external work done to the total energy dissipation results in the equation:

$$q/m = [2B/x + 2L/y + L/(B-y)]/[B(L/2 - x/3)] \quad [5.8]$$

Values for x and y can now be obtained by equating the partial derivatives of q/m with regard to x and y to zero and then solving the resulting equations. This results in:

$$y = B(2-\sqrt{2}) \quad [5.9]$$

$$x = -1.333B^2 + \sqrt{(1.778B^4)} + 7.771L^2B^2/3.886L \quad [5.10]$$

Given any values of B and L there would be no difficulty in placing these equations into a spreadsheet so that the values of x and y are determined from Eqns 5.9 and 5.10 and the value m determined for any given distributed load using Eqn 5.8. See spreadsheet Sheet 5.2.

Example 2

If the slab of Example 1 had been simply supported on all four edges then it could be assumed that the yield line EF is on the centre line and the solution would only require the determination of x. There are no hogging moments so that no top steel is required.

Following a similar method used for Example 1 it can be shown that:

$$q/m = 6[2B/x + 4L/B]/(3LB-2Bx) \quad [5.11]$$

and

$$x = (-B^2/L + \sqrt{(B^4/L^2)} + 3B^2)/2 \quad [5.12]$$

5.6 Hillerborg strip method

In this method the analysis starts with an assumption regarding the way the loads are transmitted to the boundary. For example, if the rectangular slab shown in Fig. 5.4 is divided into the four areas shown then the areas marked 1 can be assumed to carry the load by bending in the X direction and those marked 2 by bending in the Y direction. Based on this assumption the bending moments and support reactions for sections in the X and Y direction, assuming a uniformly distributed load q, is as shown in Fig. 5.4 *b,c,d*.

Some skill is required initially in the selection of the way the load is divided, especially for non-rectangular slabs, but once an equilibrium distribution has been assumed, even though it is not the one that actually exists, then a solution can be obtained which is safe. In other words it satisfies the lower bound theorem and will be less than or equal to the true solution.

To analyse the rectangular slab shown in Fig. 5.4 it would be necesary to first assume a value for the angle ϕ so that the dimensions x and y for any section can be calculated and the maximum bending moment in that section determined. For the development given below ϕ is assumed to be 45°. Also since the maximum bending moment varies with the section it is more convenient to divide the slab into rectangular strips as shown in Fig. 5.5, see Kong and Evans (1987) and Armer (1968).

Example

To illustrate the method and the use of a spreadsheet solution consider a simply supported rectangular slab of length L and breadth B, Fig. 5.6.

Assume that the shorter side B is divided into N equal strips where N is an odd integer. Since ϕ is assumed to be 45° there will also be N strips in the longer direction and all strips, except the middle strip perpendicular to the longer side, will be of equal width B/N. The middle strip will be of width $L - (N-1)B/N$.

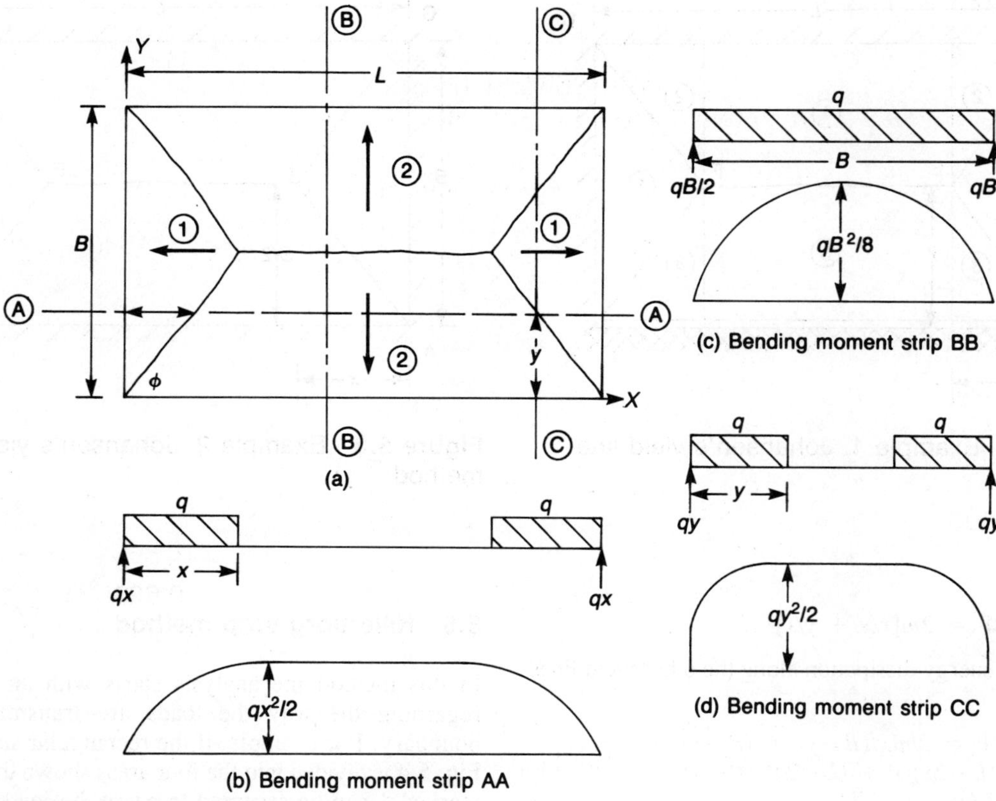

Figure 5.4 Hillerborg strip method

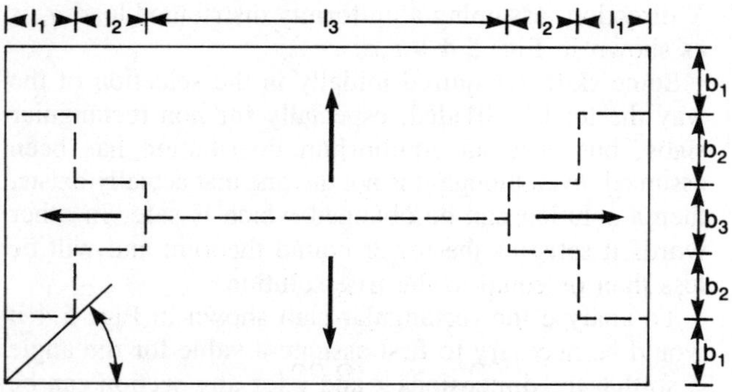

Figure 5.5 Hillerborg rectangular strip method

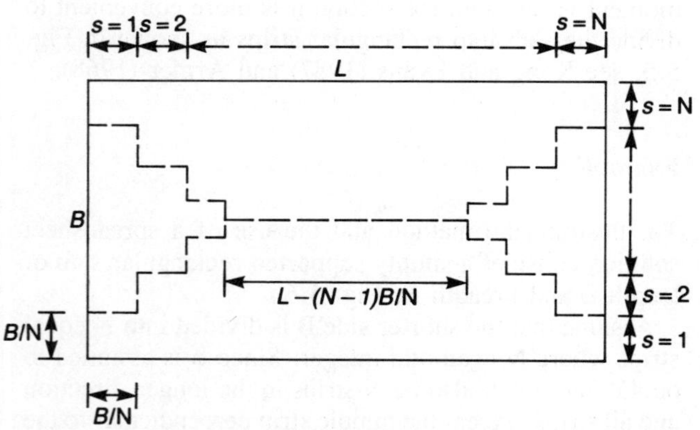

Figure 5.6 Hillerborg strip method for spreadsheet (ϕ = 45°)

For any number of strips S, expressions can be derived, by simple statics, for the bending moment at the centre and the reaction at the support. These expressions are:

X direction (parallel to the longer side)

Reaction $= (S-1)qB/N$ [5.13]

Bending Moment $= (S-1)^2q(B/N)^2/2$ [5.14]

Y direction (parallel to the shorter side) for (N − 1) strips

Reaction $= SqB/N$ [5.15]

Bending Moment $= S^2q(B/N)^2/2$ [5.16]

Y direction (for the middle strip)

Reaction $= qB/2$ [5.17]

Bending Moment $= qB^2/8$ [5.18]

In the above, S takes the values 1, 2, 3 ... depending on which strip is being considered and in the spreadsheet solution the value of N is limited to a maximum of 11. Because of symmetry the above equations need only be applied to $(N+1)/2$ strips in each direction and this will be illustrated on Sheet 5.2.

The method used for this simple example could be extended to slabs with other support conditions and using other values of ϕ but this would require some additional algebraic manipulation to derive the relevant equations.

Note that although this method of design provides a safe solution it does not necessarily follow that it is also a serviceable solution. Cracking could occur in some regions if the initial assumptions regarding the distribution are not valid.

Sheet 5.2 (Page 1)

1/A	B	C	D	E	F	G	H

2 Johansens's Method For
3 Determination of Moment in Rectangular Slab
4 --

	Initial Data			Preliminary calculations	
5	Initial Data			Preliminary calculations	
6	------------			------------------------	
7	Density	**26** kN/m3		Total(q)	16.88 kN/m2
8	Imposed	**6** kN/m2			
9	Thickness	**200** mm			
10	Breadth	**4.8** m			
11	Length	**9.2** m			
12					
13	Example 1 (Fig 5.2)			Example 2 (Fig 5.3)	
14	------------------------			------------------	
15	(3 sides SS 1 Fixed)			(4 sides SS)	
16	y	2.812 m		x	3.089 m
17	x	2.690 m		q/m	0.629
18	q/m	0.829		Moment	26.849 kNm/m
19	Moment	20.357 kNm/m			
20	--				

21 Hillerborg Strip Method For
22 Determination of Moment in Rectangular Slab
23 --
24 Example 2 (Fig 5.3)
25 -------------------
26 (4 sides SS)
27 Initial Data
28 ------------
29 Long side 9.2 m Shortside 4.8 m
30 No. of Strips(Odd) 5 Strip Width 0.960
31 --
32 Strips parallel to the X–axis
33 --

	No	Strip	Width	Reaction	Moment	Span
34	No	Strip	Width	Reaction	Moment	Span
35			m	kN	kNm	m
36	1	1	0.960	0.000	0.000	9.20
37	2	2	0.960	16.205	7.778	
38	3	3	0.960	32.410	31.113	
39	4	0				
40	5	0				
41	6	0				
42						

43 Strips parallel to the Y–axis
44 --

	No	Strip	Width	Reaction	Moment	Span
45	No	Strip	Width	Reaction	Moment	Span
46			m	kN	kNm	m
47	1	1	0.960	16.205	7.778	4.80
48	2	2	0.960	32.410	31.113	
49	3	3	5.360	40.512	48.614	
50	4	0				
51	5	0				
52	6	0				

Sheet 5.2 (Page 2)

```
59      B           C           D           E           F           G           H
60                              BS8110 Coefficient Method For
61                   Determination of Moment in Rectangular Slab
62                   -----------------------------------------------------------
63                    Suitable for All support conditions
64                              ----------------------
65 Example illustrated = (Fig 5.2)              (3 sides SS 1 Fixed)
66 Definition
67 -----------
68       SS =              1        CO =          0      Fixed =          0
69
70 Initial Conditions                   Preliminary Calculations
71 -------------------------            ------------------------------------
72      Side       Support              No of Discontinuous Edges                  3
73              1          1            Beta-y          0.0435
74              2          1            Beta-1           0.000 Beta-2          0.000
75              3          1            Gamma            0.461
76              4          0            B-3/Bx           0.000 B-4/Bx          1.333
77 AlphaSx           0.116              Beta-x           0.072
78 AlphaSy           0.032              Beta-3           0.000 Beta-4          0.096
79 Design Moments           (kNm/m)
80 ----------------
81                                                   Side 3     Side 4
82 X-Direction              Mid Span       28.094
83                          Support                   0.000       37.459
84                                                   Side 1     Side 2
85 Y-Direction              Mid Span       16.918
86                          Support                   0.000       0.000
```

5.7 BS 8110 moment coefficient method

If the bending moments at the supports and mid-section of a rectangular slab are defined in accordance with Fig. 5.7 then the bending moments can be expressed in the form:

M = Moment coefficient × the total load

and all that is required is the determination of the moment coefficients for each of the different locations and directions.

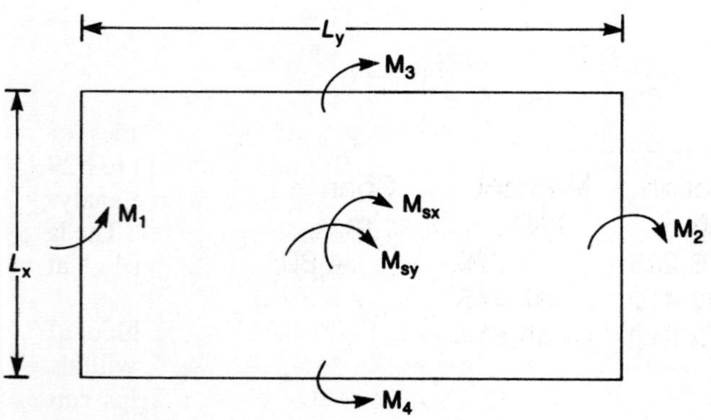

Figure 5.7 Notation for moment coefficient method ($L_y > L_x$)

All sides simply supported

Taking $L_y/L_x = L_a$ the coefficients for this case are given by:

$$\alpha_{sx} = (L_a)^4/[8(1+L_a^4)] \quad [5.19]$$

$$\alpha_{sy} = (L_a)^2/[8(1+L_a^4)] \quad [5.20]$$

The bending moments at the mid-section are then given by:

$$M_{sx} = \alpha_{sx} n L_x^2 \quad [5.21]$$

$$M_{sy} = \alpha_{sy} n L_x^2 \quad [5.22]$$

where n is the design ultimate load. Note that both of the above equations use L_x.

Restrained slabs

The moment coefficient β_y is found from:

$$\beta_y = (24+2S_d+1.5S_d^2)/1000 \quad [5.23]$$

where S_d is the number of discontinuous edges. The moment coefficients β_1 and β_2 are equal to $4\beta_y/3$ at a continuous support and equal to 0 at a simple support. A parameter γ is now determined using:

$$\gamma = (2/9)[3-(\sqrt{(18(1/L_a)}(\sqrt{(\beta_y+\beta_1)}+\sqrt{(\beta_y+\beta_2)}))] \quad [5.24]$$

Now using the equation:

$$\sqrt{(\beta_x+\beta_3)} + \sqrt{(\beta_x+\beta_4)} = \sqrt{\gamma} \quad [5.25]$$

and the fact that β_3 and β_4 are equal to either $4\beta_x/3$ or 0 depending on the type of support, the value of β_x can be determined. Using β_x found from above the values of β_3 and β_4 can be found.

The moments at the supports or mid-section can be calculated from expressions of the type:

$$M = \beta n L_x^2 \quad [5.26]$$

where β represents any of the coefficients β_1 to β_4, β_x and β_y.

Having calculated the moments the design is continued by first assuming that each span is divided into three strips taking the width of the middle strip as $\frac{3}{4}$ of the length of the side and the other two strips as being equal and of width $\frac{1}{8}$ the side length.

The calculated moments are used for determining the reinforcement required for the middle strip only and no allowance is made for any redistribution of moment. The edge strips are reinforced with the minimum percentage of reinforcement.

If the slab is simply supported then reinforcement is required both at the top and bottom of the slab near the corners in order to provide for torsional resistance. This reinforcement which runs parallel to the sides should extend a distance of at least $Lx/5$ from each edge and the area of each of the four layers should be $\frac{3}{4}$ of the area required for the maximum mid-span moment.

If the slab is continuous over one edge and simply supported over the adjacent edge then only half the above area of reinforcement is required. If the slab is continuous over both adjacent edges then no torsional reinforcement is required.

5.8 Spreadsheet layout for moments

In this spreadsheet, Sheet 5.2, examples are given of how the methods described above can be incorporated. The three approaches considered are:

(1) *Johansen's method*. This method is applied to two examples. A rectangular slab with three sides simply supported and the other side fixed and a rectangular slab with all sides simply supported. To apply the method to rectangular slabs with other support conditions would require the derivation of alternative equations.

(2) *Hillerborg strip method*. As presented this method is applied to a rectangular slab simply supported on all sides. The number of strips used can be varied up to a maximum of eleven. To apply the method to rectangular slabs with other support conditions would require the derivation of alternative equations.

(3) *BS 8110 moment coefficient method*. This method is applied to a rectangular slab with three sides simply supported and the other side fixed but the formulation is quite general so that any support conditions can be used.

Common initial data and preliminary calculation

Values of the density, imposed load, slab thickness, breadth and length are entered in the range C7 to C11.

The load intensity q is calculated in cell F7 using load factors of 1.4 and 1.6 for dead and imposed loads respectively. The formula in F7 is:

F7: **(1.4*C7*C9/1000)+(1.6*C8)**

Example 1 (Johansen's method)

Using Eqns 5.8 to 5.10 values of y, x and q/m are calculated in cells C16 to C18. The design moment is then determined in C19 using q/m and the value of q previously calculated in F7. The required cell formulae are:

```
C16: +C10*(2-@SQRT(2))
C17: (-1.333*(C10^2)+@SQRT((1.778
     *(C10^4)+7.771*(C10^2)
     *(C11^2))))/(3.886*C11)
C18: ((2*C10/C17)+(2*C11/C16)
     +C11/(C10-C16))/(C10
     *((C11/2)-(C17/3)))
C19: (F7/C18)
```

Example 2 (Johansen's method)

Using Eqns 5.11 and 5.12 values of x and q/m are calculated in cells G16 and G17. The design moment is then determined in G18 using q/m and the value of q previously calculated in F7. The required cell formulae are:

```
G16: ((-C10*C10/C11)
     +(@SQRT((C10^4)/C11^2)+(3*(C10^2))))/2
G17: (6*((2*C10/G16)+(4*C11/C10)))/((3
     *C10*C11)-(2*C10*G16))
G18: (F7/G17)
```

Example 3 (Hillerborg method)

For convenience the lengths of the two sides of the rectangle have been copied from C10 and C11 to F29 and C29. The number of strips to be used for the analysis is entered in cell D30 using an odd integer ≤ 11. The larger this integer the more accurate the solution but at the expense of requiring more detailing.

Cell G30 has been used to calculate the width of all the strips parallel to the longer side. This value will also be the width of all but the central strip of strips running parallel to the shorter side. The cell G30 contains:

G30: **+F29/D30**

The output in rows 34 to 52 has been arranged so as to allow for the symmetry of some of the strips. For example if five strips are selected in D30 then Strip 5 is the same as Strip 1 and Strip 4 the same as Strip 2 so that information is only required for Strips 1, 2 and 3. Similarly if eleven strips are selected then information will only appear for six strips since information regarding the other five will be known from symmetry. The ranges B36 to B41 and B47 to B52 are numbered 1 to 6 indicating that this is the maximum number of rows of strip information which will be revealed.

Strips are divided into those which are parallel to the X axis (the longer side) and those parallel to the Y axis. For each row information is given regarding the width of the strip, the support reaction and the mid-span bending moment using equations 5.13 to 5.18. The appropriate cell formulae are:

```
C36: @IF(B36<=($D$30+1)/2,B36,0)
D36: @IF(C36<>0,$G$30,"")
E36: @IF(C36<>0,(C36-1)*$F$7*$G$30,"")
F36: @IF(C36<>0,(((C36-1)^2)*$F$7
     *($G$30^2))/2,"")
C47: @IF(B47<=($D$30+1)/2,B47,0)
D47: @IF(C47<>0,(@IF(C47=($D$30+1)/2,
     ($C$11-(($D$30-1)
     *$G$30)),$G$30)),"")
E47: @IF(C47<>0,(@IF(C47=($D$30+1)/2,
     $F$7*$C$10/2,C47*$F$7*$G$30)),"")
F47: @IF(C47<>0,(@IF(C47=($D$30+1)/2,
     $F$7*($C$10^2)/8,((C47^2)*$F$7
     *($G$30^2))/2)),"")
```

Example 4 (BS 8110 moment coefficient method)

This is the same example as that solved in Example 1 above using Johansen's method. Initially the user has to select the support conditions using 1 for simply supported edges and 0 for continuous or fixed edges and this information is input into cells C73 to C76. Sides 1 and 2 refer to the short sides.

If the slab is simply supported on all sides then the coefficients α_{sx} and α_{sy} defined in Eqns 5.19 and 5.20 are required and these have been calculated in C77 and C78 which contain:

```
C77: ((C11/C10)^4)/(8*(1+((C11/C10)^4)))
C78: ((C11/C10)^2)/(8*(1+((C11/C10)^4)))
```

The remaining coefficients β_x, β_y, β_1, β_2, β_3 and β_4 are now calculated in rows 72 to 78 using equations 5.23 to 5.25. The formulae in the relevant cells are:

```
H72: @SUM(C73..C76)
F73: (24+(2*H72)+(1.5*H72*H72))/1000
F74: @IF(C73=1,0,4*$F$73/3)
H74: @IF(C74=1,0,4*$F$73/3)
F75: (2/9)*(3-((@SQRT(18))*(C10/C11)
     *((@SQRT(F73+F74))
     +(@SQRT(F73+H74)))))
F76: @IF(C75=1,0,4/3)
H76: @IF(C76=1,0,4/3)
F77: (F75)/((@SQRT(1+F76)
     +@SQRT(1+H76))^2)
F78: @IF(C75=1,0,4*F77/3)
H78: @IF(C76=1,0,4*F77/3)
```

Using these values of β the bending moments at the supports and mid-span, for both the X and Y directions can now be calculated from equations of the type represented by Eqn 5.26. Typically the mid-span moment in the X direction is given by:

```
E82: @IF(H72=4,C77,F77)*F7*C10*C10
```

where the @IF function allows for the use of either the α coefficients for the simply supported case or the β for mixed support conditions. The support moments are calculated from cell formula of the type:

```
F83: (F78*F7*C10*C10)
```

5.9 Use of Spreadsheet 5.2

The only unprotected cells required for Sheet 5.2 are:

- C7 to C11. The content of these cells are used by all three methods given on the spreadsheet to calculate the value of the uniformly distributed load and define the slab dimensions.
- D30. Used to define the number of strips for the Hillerborg method.
- C73 to C76. Used to define the type of support for the moment coefficient method of BS 8110.

6 Prestressed beams

The stresses arising in beams of reinforced concrete or masonry, due to the application of an imposed load, can be reduced by introducing initial stresses of opposite direction prior to the application of the load. These initial stresses can be introduced by means of a longitudinal force applied to reinforcement which is placed eccentrically to the centre of gravity of the section in the form of wires, cables or bars (tendons). Practically the longitudinal force can be applied before (pre-tensioned) or after (post-tensioned) the concrete has set and, for the latter case, ducts have to be provided to carry the reinforcement. These ducts may or may not be grouted after the beams are post-tensioned so that the beams are referred to as either bonded or unbonded.

6.1 Types of beams

For design purposes beams can therefore be classified as:

Pre-tensioned;
Post-tensioned (bonded)
Post-tensioned (unbonded).

When the initial stressing force is released a proportion of the load is transferred to the beam and the designer must ensure that the stresses induced at transfer are acceptable. Application of any imposed load results in a further loss in the transfer force and the stresses induced at these service conditions must also be checked. At the ultimate limit state a further set of conditions apply and sections must also be checked to ensure that these are satisfied.

In this chapter only the bending of simply supported beams of reinforced concrete is considered. Extensions of the approach to include topics such as reinforced masonry, shear, deflection and loss of pre-stress have not been included although examples of some of these topics have been given in other chapters.

Three spreadsheet examples are developed:

(a) *Ultimate load analysis or design* (Sheet 6.1). Given the dimensions of the section and the area of the reinforcement to determine the moment of resistance or given the ultimate moment to determine the area of reinforcement required.
(b) *Service and transfer analysis* (Sheet 6.2). Given the dimensions of the section, the area of the reinforcement, the magnitude of the jacking force and the service loading to check if the stresses at transfer and in service are acceptable.
(c) *Service and ultimate design* (Sheet 6.3). Given the service loading to determine the optimum values of the section moduli and the second moment of areas. Additionally to determine the limits of the magnitude of the jacking force and the corresponding eccentricity.

Generally, the method used for obtaining the solution to the above is quite tedious and requires the use of a trial-and-error approach. Tables of related parameters have been produced to simplify the process but the calculations are still cumbersome. It will be shown that the spreadsheet solution avoids these routine calculations and enables the designer to make any required adjustments to the initial assumptions with ease. Additional information on alternative methods of solution are given in Kong and Evans (1987) and Martin *et al.* (1989).

6.2 Principles of design and analysis at ultimate

Consider the cross section of the simply supported beam shown in Fig. 6.1.

If a prestressing force P_o is applied at a distance e_s from the centroid of the section, the strain distribution will be as shown by the broken line in Fig. 6.1(b). The prestress strain in the concrete at the level of the steel is denoted by ϵ_e. If the beam is now loaded to ultimate the strain diagram will change until the value at the top becomes 0.0035. The value at the bottom at ultimate is denoted by ϵ_u. During this loading process the loads first have to cancel the prestress strain ϵ_e before increasing to the ultimate value ϵ_u. The strain ϵ_{pa} in the concrete, at the level of the steel, resulting from the ultimate loads is therefore represented by:

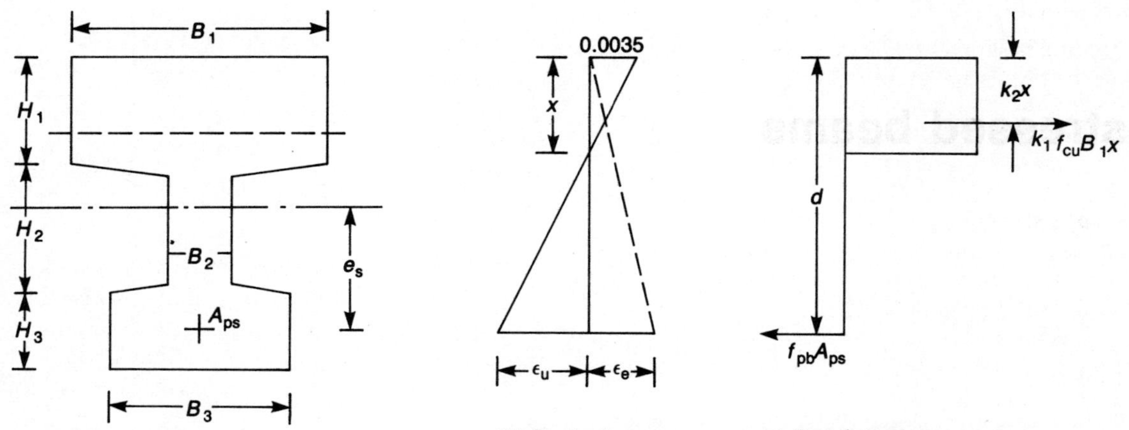

(a) Cross-section (b) Strain diagram (c) Stress diagram

Figure 6.1 Strains and stresses in prestressed beam

$$\epsilon_{pa} = \epsilon_e + \epsilon_u \qquad [6.1]$$

but this equation assumes that an effective bond exists between the steel and concrete. More generally the equation can be rewritten in the form:

$$\epsilon_{pa} = \beta_1\epsilon_e + \beta_2\epsilon_u \qquad [6.2]$$

where the bond factors β_1 and β_2 can be taken as unity for beams with an effective bond. For the case of unbonded post-tensioned beams typical values are 0.5 for β_1 and 0.1 to 0.25 for β_2.

Consider now the strain ϵ_{pb} in the tendon at the ultimate state. This strain is made up of the initial prestrain, ϵ_{pe}, plus the additional strain due to the applied loading. The additional strain in the tendon is the same as the strain in the surrounding concrete and is therefore as given by Eqn 6.2.

The required strain ϵ_{pb} is therefore defined by the equation:

$$\epsilon_{pb} = \epsilon_{pe} + \beta_1\epsilon_e + \beta_2\epsilon_u \qquad [6.3]$$

and it is necessary to re-express each of the strains on the right-hand side of Eqn 6.3 in terms of other known or assumed parameters before proceeding.

The prestress applied to the tendons must obviously be less than their characteristic strength f_{pu}. If the initial prestress is represented by αf_{pu} and the loss of stress in service by θ, the effective prestress f_{pe} can be represented by:

$$f_{pe} = \alpha(1-\theta)f_{pu} \qquad [6.4]$$

Typical values of α and θ are 0.7 and 0.25. Knowing f_{pe} the value of ϵ_{pe} can be expressed as f_{pe}/E_s. There is an assumption here that the stress f_{pe} is within the elastic portion of the stress/strain curve so that it is assumed to be on the first linear portion of the tri-linear stress/strain curve (Fig. 6.2).

The strain in the concrete due to the prestress, ϵ_e, can be determined from the expression:

$$\epsilon_e = [(P_s/A_c) + (P_s e_s/Z_1)]/E_c \qquad [6.5]$$

where P_s is the effective prestressing force applied at a distance e_s from the centroid, A_c the area of the concrete, Z_1 the section modulus referred to the bottom of the section and E_c Young's modulus of elasticity for concrete.

Finally the ultimate strain in the concrete, e_u can be expressed in the form:

$$\epsilon_u = (d/x - 1)0.0035 \qquad [6.6]$$

The usual procedure followed at this stage is to use the above equations, for the various strains, to obtain an expression for ϵ_{pb} as a function of x the depth to the neutral axis. Substituting this expression into the equation representing the balance of the compressive force in the concrete to the tensile force in the steel results in an equation for f_{pb} in terms of ϵ_{pb}. This equation must then be solved in conjunction with the stress/strain curve using a trial-and-error process.

For the case of unbonded tendons the values of f_{pb} and x can be calculated directly from equations given in BS 8110:

$$f_{pb} = f_{pe} + (7000d/L)[1 - 1.7(f_{pu}A_{ps}/f_{cu}B_1d)] \le 0.7f_{pu} \qquad [6.7]$$

$$x = 2.47[f_{pu}A_{ps}/f_{cu}B_1d][f_{pb}/f_{pu}]d \qquad [6.8]$$

The above process can be quite tedious but the spreadsheet solution avoids this by first calculating ϵ_e and ϵ_{pe} and then adding these values to a range of values of ϵ_u calculated for the full range of x/d values. This approach which has been used in other chapters amounts to solving the equations for the complete range of possible values of x and then determining the correct solution from inspection of the range of solutions. The decision as to which of the solutions is the correct one can usually be made automatically by the spreadsheet with very little additional input from the user. This results in a range of values of ϵ_{pb} and for each of these values the stress f_{pb} is calculated from the stress/strain curve. Then by checking the out-of-balance resulting from the equation:

$$\text{Out of balance} = f_{pb}A_{ps} - k_1 f_{cu}B_1x \qquad [6.9]$$

it is possible to find the values of x and f_{pu} which results in a zero out-of-balance and using these values determine the value of the ultimate moment of resistance M_u using:

$$M_u = f_{pb}A_{ps}[d - k_2x] \qquad [6.10]$$

This procedure is similar to that described previously for the design of beams in Chapter 3.

6.3 Spreadsheet layout for flexure at ultimate

Initial data. For this example the sections considered are of a rectangular form as defined by the six dimensions B_1 to B_3 and H_1 to H_3 and as illustrated in Fig. 6.1. Zero values can be used for some of these dimensions so that a range of different rectangular shapes is possible. It is assumed however that the neutral axis lies within the upper rectangle, i.e. $x < H_1$.

The area of the steel is entered in C19 of Sheet 6.1 and the values β_1 and β_2 in C20 and C21 respectively. The initial stress factor and the loss of prestress in service is entered in C22 and C23 since these quantities are used later in F13 to calculate P_s, the effective prestressing force, using equation 6.4 and the area of the steel.

Preliminary calculations. In the range F7 to F12 of Sheet 6.1, values of the area of the concrete, the position of the centroid, the second moment of area and the section moduli are determined using:

F7: **(C7*C10)+(C8*C11)+(C9*C12)**
F8: **((C10*C7*C10/2)+(C11*C8*(C10+(C11/2)))+(C12*C9*(C10+C11+(C12/2))))/F7**
F9: **(C10+C11+C12−F8)**
F10: **(C7*(C10^3)/12)+(C10*C7*((F8−(C10/2))^2))+(C9*(C12^3)/12)+(C12*C9*((C10+C11−F8+(C12/2))^2))+(C8*(C11^3)/12)+(C11*C8*((F8−C10−(C11/2))^2))**
F11: **(F10/F8)**
F12: **(F10/F9)**

The effective prestressing force can be obtained from the product of f_{pe} and A_{ps} and this value is calculated in F13. The value of the eccentricity e_s is determined in F14:

F13: **(C22*(1−C23)*C15*C19)/1000**
F14: **(C13−F8)**

so that the values of ϵ_e and ϵ_{pe} can be determined in F15 and F16:

F15: **(((F13/F7)+(F13*F14/F12))/C18)*C20**
F16: **(F13/(C17*C19))**

The trilinear stress/strain curve for the tendons is as represented in Fig. 6.2 and it is necessary to define the strains at A and B so that these values can be used later to define the limits of the linear ranges.

The strains at A and B are calculated in F17 and F18 using:

F17: **0.8*C15/(C16*C17*1000)**
F18: **0.005+(C15/(C16*C17*1000))**

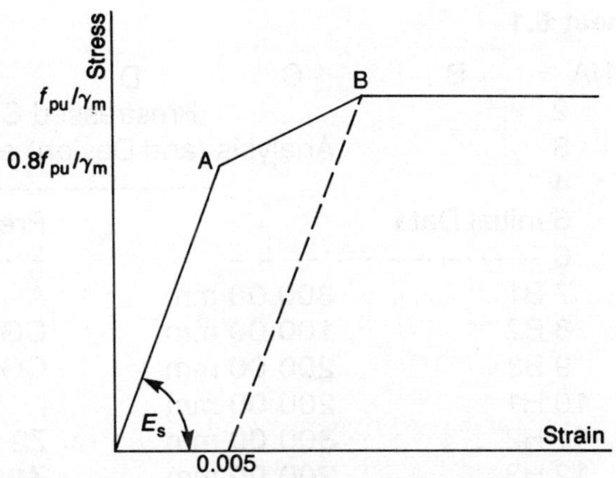

Figure 6.2 Short-term design stress/strain curve (BS 8110)

Trial values

A range of trial values of x/d from 0.05 to 0.95 in intervals of 0.05 is inserted in the range B34 to B52 and the values of x corresponding to each of the trial values determined in the range C34 to C52 using formulae of the type **(B34*$C$13)**.

The total strain, as given by Eqn 6.3, can now be calculated in the range D34 to D52 and the stresses corresponding to these total strains in E34 to E52 using equations representing the stress/strain relationship between the limits defined previously in cells F17 and F18. Typically D34 and E34 contain:

D34: **((0.0035*((1/B34)−1))*$C$21)+$F$15+$F$16**
E34: **@IF(D34<=$F$17,0.8*$C$15*D34/($C$16*$F$17),@IF(D34>=$F$18,$C$15/$C$16,(0.8*$C$15/$C$16)+(0.2*$C$15*(D34−$F$17))/($C$16*($F$18−$F$17))))**

The stresses in E34 to E52 can now be used to check for the out-of-balance defined by Eqn 6.9 and this calculation has been performed in F34 to F52 using formulae of the type:

F34: **(E34*$C$19)−(0.405*$C$7*$C$14*C34)**

The out-of-balance values output to cells F34 to F52 will range from positive to negative values and it is now necessary to determine the value of x/d which would result in a zero out-of-balance. To simplify this process the range G34 to G51 has been used to find the lowest positive value of the out-of-balance and the range H35 to H52 to find the lowest negative value. The value of x/d which produced the lowest positive out-of-balance has been found in I34 to I52. The required formulae are:

G34: **@IF(F34>0#AND#F35<0,F34,0)**
H35: **@IF(F35<0#AND#F34>0,F35,0)**
I34: **@IF(G34<>0,B34,0)**

This ensures that only the salient values appear in the block G34 to I52 and these can then be copied automatically to another part of the spreadsheet so that a trial

Sheet 6.1

1/A	B	C	D	E	F	G	H	I
2			Prestressed Concrete Beams					
3			Analysis (and Design) at Ultimate Load					
4			------------------------------					
5 Initial Data				Preliminary Calculations.				
6 ------------				------------------------------				
7 B1		**300.00** mm		A−conc	130000 mm2			
8 B2		**100.00** mm		CG top	312 mm			
9 B3		**200.00** mm		CG Bot	388 mm			
10 H1		**200.00** mm		I	6.6E+09 mm4			
11 H2		**300.00** mm		Z2	2.12E+07 mm3			
12 H3		**200.00** mm		Z1	1.70E+07 mm3			
13 d		**600.00** mm		Ps	536 kN			
14 fcu		**60** N/mm2		es	288 mm			
15 fpu		**1700** N/mm2		Stra ee	0.0004			
16 Gms		**1.15**		Stra epe	0.0045			
17 Es		**200** kN/mm2		StainA	0.0059			
18 Ec		**36** kN/mm2		StrainB	0.0124			
19 Aps		**600** mm2						
20 Beta1		**1.00**						
21 Beta2		**1.00**						
22 IntStress		**0.70** of fpu						
23 ServLoss		**0.25**						
24 ------------	------------	------------	------------	------------	------------	------------	------------	------------
25 Final Calculation								
26 ------------	------------	------------			Unbonded			
27 Bonded					Span	**10.00** m		
28 x		122 mm	O.K.		fpb	1190 N/mm2		
29 Mu		484 kNm			x	98 mm		O.K.
30					Mu	397 kNm		
31 ------------	------------	------------	------------	------------	------------	------------	------------	------------
32								
33	x/d	x	TotStrain	Stress	Check			
34	0.05	30	0.0713	1478	668257	0		0
35	0.10	60	0.0363	1478	449557	0	0	0
36	0.15	90	0.0247	1478	230857	0	0	0
37	0.20	120	0.0188	1478	12157	12157	0	0.2
38	0.25	150	0.0153	1478	−206543	0	−206543	0
39	0.30	180	0.0130	1478	−425243	0	0	0
40	0.35	210	0.0113	1430	−673036	0	0	0
41	0.40	240	0.0101	1373	−925964	0	0	0
42	0.45	270	0.0091	1328	−1171286	0	0	0
43	0.50	300	0.0083	1293	−1411284	0	0	0
44	0.55	330	0.0077	1264	−1647409	0	0	0
45	0.60	360	0.0072	1240	−1880630	0	0	0
46	0.65	390	0.0067	1219	−2111617	0	0	0
47	0.70	420	0.0063	1202	−2340849	0	0	0
48	0.75	450	0.0060	1186	−2568676	0	0	0
49	0.80	480	0.0057	1141	−2814737	0	0	0
50	0.85	510	0.0054	1089	−3064319	0	0	0
51	0.90	540	0.0052	1044	−3310470	0	0	0
52	0.95	570	0.0050	1003	−3553731		0	0
53	0.202779	122	0.0186	1478	0			
54	**0.202779**	122	0.0186	1478	1			
55			0.20	12157				
56			0.25	−206543				

interpolation can be performed in order to determine the approximate value of x/d required to produce a zero out-of-balance. The lowest positive value is transferred to E55 and the lowest negative value to E56. The interpolation block D55 to E56 contains:

D55: **@SUM(I34..I52)**
E55: **@SUM(G34..G51)**
D56: **0.05+D55**
E56: **@SUM(H35..H52)**

Using the above a value of x/d which will result in a near zero value for the out-of-balance can be found. This interpolation is performed in B53 using:

B53: **(D55)+(E55*0.05)/(E55−E56).**

Note that even if the 'check' value in F53 is zero the user **must** copy the interpolated value from B53 to B54 since subsequent calculations use values from row 54 and not row 53. The value in B53 is calculated automatically by the spreadsheet as a guide to a suitable value but the user must signify acceptance of this value by re-entering the same value or signifying non-acceptance by entering some other value. Although the entry in B54 is shown to six decimal places it will be found that such accuracy is not required. By observing the changes that occur to the values of x and f_{pb}, in C54 and E54, resulting from changes in the value input for x/d in cell B54, the user can decide on the degree of accuracy required.

The formulae in the remaining cells of rows 53 and 54 are copies of those in rows 34 to 52.

A reminder could be included so that the user is made aware if any changes are required to the value of x/d in B54 (see Sheet 4.8).

Final calculation

The ultimate moment of resistance of the given section is calculated in C29 for the case of bonded tendons and in G30 for unbonded tendons.

For the bonded case the value of x calculated previously in C54 has been copied to C28 and the value of M_u determined from equation 6.10. The formula entered in C29 is:

C29: **(C19*E54*(C13−(0.45*C54)))/1000000**

For the unbonded case values of f_{bp} and x are calculated from Eqns 6.7 and 6.8. Equation 6.7 requires the value of the span of the beam and this is entered by the user in G27. The remaining formulae for the unbonded case are:

G28: **@MIN((C22*C15*(1−C23))+(7000 *C13/(G27*1000))*(1−(1.7*C15 *C19)/(C14*C7*C13))),0.7*C15)**
G29: **(2.47*(C15*C19/(C14*C7*C13)) *(G28/C15))*C13**
G30: **(C19*G28*(C13−(0.45*G29)))/1000000**

Note that no direct comparison can be made between the values of M_u for the bonded case in C29 and the unbonded case in G30 since the values of β_1 and β_2 in C20 and C21 would normally be different for the two cases.

In cells E28 and I29 a check is made to determine if the value of x is within the upper flange and an appropriate message printed:

E28: **@IF(C28<=C10,"O.K.","Check C10")**

6.4 Use of spreadsheet 6.1

If the spreadsheet is now protected then the only cells that have to be unprotected would be: C7 to C23, G27 and B54. The entry to G27 is only required if the tendons are unbonded.

Use for analysis. Once the initial data has been entered all that remains is for the user to select a value of x/d in B54 which results in a near-zero value in F54. An indication of the value of x/d required to achieve this is given in B53.

Use for design. Once the initial data has been entered different values of A_{sp} can be tried in cell C19 until the value of M_u found in C29 (or G30 for unbonded) is equal to the ultimate moment determined from the ultimate loading.

Note that spreadsheet 6.1 has been developed for the purposes of 'analysis' but can also be used for 'design'. It is possible to do the opposite and develop a spreadsheet for 'design' which then can be used for 'analysis'. To achieve this three changes would have been required:

(1) Using the known value of M_u determine the values of the product $f_{pb}A_{ps}$ (from Eqn 6.10) for the full range of x/d values.
(2) Using the total strain value for the value of x/d which produces the zero 'check' value, determine the value of f_{pb} from the equation representing the stress/strain relationship.
(3) Calculate A_{ps} using the value of f_{pb} found in (2) and the value of $f_{pb}A_{ps}$ found in (1) at the zero 'check' value.

6.5 Principles of design and analysis in service

The maximum tensile and compressive stresses that arise when the prestressing force is transferred to the concrete and also due to the application of an imposed load must be checked to determine if they exceed specified limits. Limiting values for the tensile stresses, as given in the British Standard code, are divided into three classes defined by:

Class 1. In which no flexural tensile stresses are permitted.
Class 2. In which a limited amount of tensile stress is allowed but no visible cracking is permitted.
Class 3. In which tensile stresses are allowed but crack widths are not to exceed specified widths depending on the environment.

If only Class 1 and 2 members are considered the limiting design stresses given in BS 8110 can be summarised as:

(a) *Compressive stresses in bending for simply supported beams*
At transfer, limit = $0.5f_{ci}$
In-service, limit = $0.33f_{cu}$

(b) *Tensile stresses in bending*
Class 1. At transfer, limit = 1 Nmm^{-2}
In-service, limit = 0
Class 2, pre-tensioned. At transfer limit = $0.45\sqrt{f_{ci}}$
In-service, limit = $0.45\sqrt{f_{cu}}$
Class 2, post-tensioned. At transfer limit = $0.36\sqrt{f_{ci}}$
In-service, limit = $0.36\sqrt{f_{cu}}$

where f_{ci} is the cube strength and f_{cu} is the characteristic strength.

Therefore to define the limiting stresses it is necessary to include information on the spreadsheet regarding the class of the member and also whether it is pre- or post-tensioned.

Bending stresses at transfer and in service

If a jacking force P_o is applied to the tendons then during transfer the resulting transfer force P_t will be less than P_o due to losses. Further losses occur on application of the service loads so that the resulting service load on the tendons P_s, the effective load, will be reduced even further.

At transfer the critical stresses at the top and bottom of the section with the maximum bending moment are defined by:

$$f_{t1} = -P_t/A_c - P_t e_s/Z_1 + M_t/Z_1 \quad [6.11]$$

$$f_{t2} = -P_t/A_c + P_t e_s/Z_2 - M_t/Z_2 \quad [6.12]$$

where the suffix t refers to 'transfer', the suffix 1 the bottom of the section and the suffix 2 the top of the section. The bending moment M_t at transfer represents the bending moment at the critical section. If the prestressing force is not parallel to the centroidal axis the critical section is at mid-span and M_t is the bending moment due to the self weight, if the prestressing force is parallel to the centroidal axis then the critical section is at the support and M_t is ignored as in Martin *et al.* (1989).

In service the critical stresses at the top and bottom of the section with the maximum bending moment are defined by:

$$f_{s1} = -P_s/A_c - P_s e_s/Z_1 + M_s/Z_1 \quad [6.13]$$

$$f_{s2} = -P_s/A_c + P_s e_s/Z_2 - M_s/Z_2 \quad [6.14]$$

where the suffix s refers to 'service'. The bending moment M_s in service represents the bending moment due to the self weight and the imposed loads.

The limiting stresses are represented by f_{tc}, f_{sc}, f_{tt} and f_{st} where the first suffix represents 'transfer' or 'service' and the second suffix represents 'compression' or 'tension'. Substituting this notation into Eqns 6.11 to 6.14 and rearranging yields expressions for the optimum values for Z_1, Z_2, the minimum and maximum values of P_o and e_s:

$$Z_1 = \frac{[M_s - (\eta M_t)]}{[f_{st} - (\eta f_{tc})]} \quad [6.15]$$

$$Z_2 = \frac{[M_s - (\eta M_t)]}{[(\eta f_{tt}) - f_{sc}]} \quad [6.16]$$

$$P_{o(min)} = \frac{A_c(f_{st}Z_1 + \eta f_{tt}Z_2 - M_s + \eta M_t)}{[-\eta_s(Z_1 + Z_2)]} \quad [6.17]$$

$$e_{s(max)} = \frac{[Z_1Z_2(f_{st} - \eta f_{tt}) - \eta M_t Z_1 - M_s Z_2]}{[A_c(f_{st}Z_1 + \eta f_{tt}Z_2 - M_s + \eta M_t)]} \quad [6.18]$$

$$P_{o(max)} = \frac{A_c(f_{sc}Z_2 + \eta f_{tc}Z_1 + M_s - \eta M_t)}{[-\eta_s(Z_1 + Z_2)]} \quad [6.19]$$

$$e_{s(min)} = \frac{[Z_1Z_2(-f_{sc} + \eta f_{tc}) - \eta M_t Z_2 - M_s Z_1]}{[A_c(f_{sc}Z_2 + \eta f_{tc}Z_1 + M_s - \eta M_t)]} \quad [6.20]$$

where $\eta = P_s/P_t$ and $\eta_s = P_s/P_o$.

Although the four Eqns 6.11 to 6.14 have been rearranged into the six Eqns 6.15 to 6.20 these rearrangements are not mutually exclusive and if the values of Z_1 and Z_2 found from Eqns 6.15 and 6.16 are used in the remaining equations then it will be found that $P_{o(max)}$ would be the same as $P_{o(min)}$ and $e_{s(max)}$ the same as $e_{s(min)}$. However Eqns 6.15 to 6.20 are convenient since it is unlikely that the exact values found for Z_1 and Z_2 from Eqns 6.15 and 6.16 are used in practice so Eqns 6.17 and 6.19 will result in different values for P_o and similarly with e_s. This also implies that if different values of Z_1 and Z_2, from those found using the above equations, are used then the four stress equations 6.11 to 6.14 are not exactly satisfied.

6.6 Spreadsheet layout for in-service loads

Initial data. For this example the sections considered are of a rectangular form as defined by the six dimensions B_1 to B_3 and H_1 to H_3 and as illustrated in Fig. 6.1. Zero values can be used for some of these dimensions so that a range of different rectangular shapes is possible. It is assumed however that the neutral axis lies within the upper rectangle, i.e. $x < H_1$.

The value of the preload and the percentage losses both at transfer and also in-service are entered in the range C15 to C17 in Sheet 6.2. In C18 and C19 values of the dead and imposed load have been inserted since these will be required to calculate the bending moments M_t and M_s. The characteristic and cube strengths have been placed in C20 and C21.

Finally information is entered in C22 to C24 about the class of the beam (1 or 2), whether it is pre-tensioned or post-tensioned (1 or 2), and whether the prestressing force is parallel to the centroidal axis (Y or N).

Preliminary calculations. For the given section the values of the area, position of the centroid, second moment of area and the section moduli can be calculated using similar formulae to those given previously for Sheet 6.1.

Sheet 6.2

1/A		B	C	D	E	F	G	H
2			Prestressed Concrete Simply Supported Beams					
3			Analysis and Design at Service Load					
4					------------------------------			
5	Initial Data				Preliminary Calculations			
6	------------				------------------------			
7	B1		350 mm		A–Conc	80000 mm2		
8	B2		150 mm		CG Top	144 mm		
9	B3		300 mm		CG Bot	156 mm		
10	H1		100 mm		I	7.1E+08 mm4		
11	H2		100 mm		Z2	4.96E+06 mm3		
12	H3		100 mm		Z1	4.57E+06 mm3		
13	d		222 mm		es	78 mm		
14	Span		10.00 m		Pt	576 kN		
15	Preload		640 kN		Ps	480 kN		
16	TrLoss		10 %		Mt	38 kNm		
17	SeLoss		25 %		Ms	75 kNm		
18	DeadLd		3 kN/m		eta	0.8333		
19	ImpLoad		6 kN/m		Mmin	0 kNm		
20	fcu		50 N/mm2					
21	fci		35 N/mm2					
22	Class1/2		2 (1 or 2)					
23	Pre/Post		1 (1 or 2)					
24	Parallel	Y	(Y or N)					
25	--							
26	Check Stresses							
27	--------------							
28		At				At		
29		Transfer				Service		
30		--------				-------		
31	Top		1.88 O.K.			–13.54 O.K.		
32	Bottom		–17.07 O.K.			2.20 O.K.		
33	Allowable							
34	Tension		2.66 ftt			3.18 fst		
35	Compn.		–17.50 ftc			–16.50 fsc		
36	--							
37	Check Section							
38	-------------							
39	Z1Reqd		4.22E+06 mm3		Z1Prov	4.57E+06 O.K.		
40	Z2Reqd		4.01E+06 mm3		Z2Prov	4.96E+06 O.K.		
41								
42	Check Jacking Force and eccentricity							
43	--							
44	Using required values of Z							
45	MinPo		683 kN	at		es–max		67 mm
46	MaxPo		683 Kn	at		es–min		67 mm
47								
48	Using provided values of Z							
49	MinPo		554 kN	at		es–max		89 mm
50	MaxPo		823 kN	at		es–min		51 mm
51								
52	Po–prov		640 kN			es–prov		78 mm
53			O.K.				O.K.	

Values of the eccentricity e_s, the force in the cable at transfer P_t and in-service P_s, are calculated in cells F13 to F15 using:

F13: **(C13 − F8)**
F14: **(C15*(100 − C16)/100)**
F15: **(C15*(100 − C17)/100)**

Values of M_t and M_s are then calculated in F16 and F17 using:

F16: **(C18*(C14^2))/8**
F17: **(C19*(C14^2))/8**

The moments have been calculated assuming the beam is simply supported. Other coefficients would have to be introduced into the formulae of F16 and F17 to allow for continuous beams.

In cell F18 the ratio η defined as P_s/P_t or F15/F14 is entered and in cell F19 the value of M_{min} to be used at transfer. As stated above this value could be equal to M_t or zero.

F19: **@IF(C24 = "Y",0,F16)**

Calculation of the stresses

Four stresses have to be calculated and compared with the allowable stresses. Two at transfer, top and bottom fibres, and two in-service. These have been calculated as follows:

At transfer, top fibre.

C31: **(−(F14/F7) + (F14*F13/F11) − (F19*1000/F11))*1000**

At transfer, bottom fibre.

C32: **(−(F14/F7) − (F14*F13/F12) + (F19*1000/F12))*1000**

In-service, top fibre.

F31: **(−(F15/F7) + (F15*F13/F11) − (F17*1000/F11))*1000**

In-service, bottom fibre.

F32: **(−(F15/F7) − (F15*F13/F12) + (F17*1000/F12))*1000**

The allowable stresses at transfer are given by:

Tension (+)

C34: **@IF($C$22 = 1,1,@IF($C$22 = 2#AND#$C$23 = 1,0.45 *@SQRT($C$21),0.36*@SQRT($C$21)))**

Compression (−)

C35: **−0.5*C21**

The allowable stresses in-service are given by:

Tension (+)

F34: **@IF($C$22 = 1,0,@IF($C$22 = 2#AND#$C$23 = 1,0.45*@SQRT($C$20), 0.36*@SQRT($C$20)))**

Compression (−)

F35: **−0.33*C20.**

Finally checks are made in cells D31, D32, G31 and G32 to ascertain if the calculated stresses are acceptable using:

D31: **@IF(C31 > 0#AND#C31 < = C34) #OR#(C31 < 0#AND#C31 > = C35, "O.K.", "Check")**
D32: **@IF(C32 > 0#AND#C32 < = C34) #OR#(C32 < 0#AND#C32 > = C35, "O.K.", "Check")**
G31: **@IF(F31 > 0#AND#F31 < = F34) #OR#(F31 < 0#AND#F31 > = F35, "O.K.", "Check")**
G32: **@IF(F32 > 0#AND#F32 < = F34) #OR#(F32 < 0#AND#F32 > = F35, "O.K.", "Check")**

If the checks are not satisfied then alterations to the design are required.

Check on section moduli

The values of the required section moduli Z_1, referred to the bottom, and Z_2, referred to the top, are calculated in C39 and C40 using:

C39: **((F17 − (F18*F19))*1000000)/(F34 − (F18 *C35)**
C40: **((F17 − (F18*F19))*1000000)/((F18 *C34) − F35)**

The values of Z_1 and Z_2 provided have been copied from F12 and F11 to F39 and F40 so that they can be conveniently compared. Cells G39 and G40 contain formulae which indicate the result of the comparison:

G39: **@IF(F39 > = C39, "O.K.", "Check")**
G40: **@IF(F40 > = C40, "O.K.", "Check")**

Check on jacking force and eccentricity

Equations 6.17 to 6.20 are now used to calculate the values of the jacking force, P_o, and the eccentricity.

In rows 45 and 46 these calculations use the required values of Z_1 and Z_2 found previously in C39 and C40. Equations 6.17 and 6.18 are used in row 45 and Eqns 6.19 and 6.20 in row 46. The values in both these rows should be identical and the two sets of calculations have only been included since they serve as a check on the formulae entered. The above calculations are repeated in rows 49 and 50 but this time using the values of Z_1 and Z_2 provided. Formulae entered for these operations are:

C45: **((((F34*C39) + (F18*C34*C40) − (F17*1000000) + (F18*F19*1000000))/ (−(F15/C15)*(C39 + C40)))/1000)*F7**

```
C46: (((F35*C40)+(F18*C35*C39)
     +(F17*1000000)-(F18*F19*1000000))/
     (-(F15/C15)*(C39+C40)))/1000)*F7
G45: ((C39*C40*((F34)-(F18*C34)))
     -(F18*F19*C39*1000000)
     -(F17*C40*1000000))/(((F34*C39)
     +(F18*C34*C40)-(F17*1000000)
     +(F18*F19*1000000))*F7)
G46: ((C39*C40*((-F35)+(F18*C35)))
     -(F18*F19*C40*1000000)
     -(F17*C39*1000000))/(((F35*C40)
     +(F18*C35*C39)+(F17*1000000)
     -(F18*F19*1000000))*F7)
C49: (((F34*F39)+(F18*C34*F40)
     -(F17*1000000)+(F18*F19*1000000))/
     (-(F15/C15)*(F39+F40)))/1000)*F7
C50: (((F35*F40)+(F18*C35*F39)
     +(F17*1000000)-(F18*F19*1000000))/
     (-(F15/C15)*(F39+F40)))/1000)*F7
G49: ((F39*F40*((F34)-(F18*C34)))
     -(F18*F19*F39*1000000)
     -(F17*F40*1000000))/(((F34*F39)
     +(F18*C34*F40)-(F17*1000000)
     +(F18*F19*1000000))*F7)
G50: ((F39*F40*((-F35)+(F18*C35)))
     -(F18*F19*F40*1000000)
     -(F17*F39*1000000))/(((F35*F40)
     +(F18*C35*F39)+(F17*1000000)
     -(F18*F19*1000000))*F7)
```

With a suitable choice of initial data the value of P_o in row 49 should be less than the value of P_o in row 50 but it is possible to obtain the reverse effect or even negative values. Similarly with the values of e_s in cells G49 and G50. Usually when this occurs it will be found that the stress limits, in rows 31 to 35, have not been satisfied or that the magnitude of the imposed load is insufficient to cancel the transfer stresses. A more judicious choice of initial data is required.

For purposes of comparison the value of P_o entered in C15 and the value of e_s from F13 are copied to C52 and G52. Checks are then made in D53 and G53 to indicate if the values of P_o and e_s are within the limits.

```
D53: @IF(C52>=C49#AND#C52<=C50,
     "O.K.","Check")
G53: @IF(G52<=G49#AND#G52>=G50,
     "O.K.","Check")
```

Note that the values of f_{pu} and A_{ps} are not required for the analysis and design at the serviceability limit state. The stresses in the concrete at transfer and in service have been checked but not the stress in the cable. It is assumed that a sufficient area of cable, A_{ps}, is provided to ensure that the stress in the cable is below the value of f_{pu} by at least 70 per cent.

For example if the value of P_o is 640 kN and f_{pu} equals 1700 N mm^{-2} then there should be sufficient wires in the cable to provide an area, A_{ps}, of at least 640000/(0.7 × 1700), i.e. 538 mm^2. The required value of A_{ps} is also determined at ultimate, Sheet 6.1, to ensure that there is sufficient steel to carry the ultimate loads.

6.7 Use of spreadsheet 6.2

If the spreadsheet is now protected then the only cells that have to be unprotected would be: C7 to C24. Initial data has to be entered into these cells and the output to the cells listed below will then indicate if the design is acceptable with regard to that particular check:

D31, D32, G31, G32 (check for stresses at transfer and in service).
G39, G40 (check for section moduli).
D53, G53 (check for P_o and e_s).

The spreadsheet can now be used either for analysis in which all the parameters are known but have to be checked to ensure that they are within prescribed limits or for design involving a selection of suitable initial data.

6.8 Combined spreadsheet for serviceability and ultimate design

The spreadsheets developed for design at the serviceability state, Sheet 6.1, and the ultimate state, Sheet 6.2, can now be combined to cover both states. This has been done in Sheet 6.3 but in order to avoid repetition the two sheets have been combined with the minimum of change so that explanations do not have to be repeated. The result of this is that there are two sets of initial data with some degree of duplication. Where such duplication occurs the relevant cell in the second set of data is connected to the corresponding cell of the first set by entering the address.

For example cell C63 contains the value of B_1 and so does cell C7. These two cells are connected by entering +C7 in cell C63. All cells which are connected in this way are then protected to avoid illegal entry, i.e. to enter a value for B_1 entry must be via cell C7 and not C63.

The data entered on the first page of Sheet 6.3 was initially identical with that given on Sheet 6.2 so that all the constraints were known to be satisfied. However on checking the second page of Sheet 6.3 it was found from E84 that the depth of 100 mm in cell C10 was insufficient since x was equal to 142. The depth in C10 was therefore increased to 150 mm so that all the conditions would be satisfied.

6.9 Use of spreadsheet 6.3

If the spreadsheet is now protected then the only cells that have to be unprotected would be: C7 to C24, C71 to C77 and B110. Note that a value **must** be entered in B110 and that this value will be one that results in the output to F110 being small.

Initial data has to be entered into the above cells and the output to the cells listed below will then indicate if the design is acceptable with regard to that particular check:

D31, D32, G31, G32 Check for stresses at transfer and in service.
G39, G40 (check section moduli).
D53, G53 (check for P_o and e_s).
E78 (check that stress in tendon is below $0.7f_{pu}$).
E84 (check if x is within upper flange).

Sheet 6.3 (Page 1)

1/A	B	C	D	E	F	G	H
2		Prestressed Concrete Simply Supported Beams					
3		Design in Service and at Ultimate					
4			———————	———————	———————	———————	
5 Initial Data				Preliminary Calculations			
6 ———————	———————			———————	———————	———————	
7 B1		350	mm	A−Conc	97500 mm2		
8 B2		150	mm	CG Top	163 mm		
9 B3		300	mm	CG Bot	187 mm		
10 H1		150	mm	I	1.1E+09 mm4		
11 H2		100	mm	Z2	6.89E+06 mm3		
12 H3		100	mm	Z1	6.04E+06 mm3		
13 d		222	mm	es	59 mm		
14 Span		10.00	m	Pt	576 kN		
15 Preload		640	kN	Ps	480 kN		
16 TrLoss		10	%	Mt	38 kNm		
17 SeLoss		25	%	Ms	75 kNm		
18 DeadLd		3	kN/m	eta	0.8333		
19 ImpLoad		6	kN/m	Mmin	0 kNm		
20 fcu		50	N/mm2				
21 fci		35	N/mm2				
22 Class1/2		2	(1 or 2)				
23 Pre/Post		1	(1 or 2)				
24 Parallel	Y		(Y or N)				
25 ———————	———————	———————	———————	———————	———————	———————	———————
26 Check Stresses							
27 ———————	———————						
28	At				At		
29	Transfer				Service		
30	———————				———————		
31 Top		−1.01 O.K.			−11.73 O.K.		
32 Bottom		−11.49 O.K.			2.85 O.K.		
33 Allowable							
34 Tension		2.66 ftt			3.18 fst		
35 Compn.		−17.50 ftc			−16.50 fsc		
36 ———————	———————	———————	———————	———————	———————	———————	———————
37 Check Section							
38 ———————	———————						
39 Z1Reqd		4.22E+06 mm3		Z1Prov	6.04E+06 O.K.		
40 Z2Reqd		4.01E+06 mm3		Z2Prov	6.89E+06 O.K.		
41							
42 Check Jacking Force and eccentricity							
43 ———————	———————	———————	———————	———————	———————		
44 Using required values of Z							
45 MinPo		832 kN		at	es−max		55 mm
46 MaxPo		832 Kn		at	es−min		55 mm
47							
48 Using provided values of Z							
49 MinPo		407 kN		at	es−max		121 mm
50 MaxPo		1274 kN		at	es−min		30 mm
51							
52 Po−prov		640 kN			es−prov		59 mm
53		O.K.				O.K.	

Sheet 6.3 (Page 2)

58	B	C	D	E	F	G	H	I
59			Ultimate Limit State					
60			------------------------					
61	Initial Data			Preliminary Calculations.				
62	------------			------------------------				
63	B1	350.00 mm		A−conc	97500 mm2			
64	B2	150.00 mm		CG top	163 mm			
65	B3	300.00 mm		CG Bot	187 mm			
66	H1	150.00 mm		I	1.1E+09 mm4			
67	H2	100.00 mm		Z2	6.89E+06 mm3			
68	H3	100.00 mm		Z1	6.04E+06 mm3			
69	d	222.00 mm		Ps	640 kN			
70	fcu	50 N/mm2		es	59 mm			
71	fpu	**1700** N/mm2		Stra ee	0.0004			
72	Gms	**1.15**		Stra epe	0.0037			
73	Es	**205** kN/mm2		StainA	0.0058			
74	Ec	**36** kN/mm2		StrainB	0.0122			
75	Aps	**846** mm2						
76	Beta1	**1.00**						
77	Beta2	**1.00**						
78	IntStress	0.59 of fpu		O.K.				
79	ServLoss	0.25						
80	------------	------------	------------	------------	------------	------------	------------	------------
81	Final Calculation							
82	------------	------------			Unbonded			
83	Bonded				Span	10.00 m		
84	x	142 mm		O.K.	fpb	814 N/mm2		
85	Mu	159 kNm			x	97 mm		O.K.
86					Mu	123 kNm		
87	------------	------------	------------	------------	------------	------------	------------	------------
88								
89	x/d	x	TotStrain	Stress	Check			
90	0.05	11	0.0705	1478	1171937	0		0
91	0.10	22	0.0355	1478	1093266	0	0	0
92	0.15	33	0.0239	1478	1014595	0	0	0
93	0.20	44	0.0180	1478	935924	0	0	0
94	0.25	56	0.0145	1478	857252	0	0	0
95	0.30	67	0.0122	1478	778581	0	0	0
96	0.35	78	0.0105	1402	635225	0	0	0
97	0.40	89	0.0093	1344	508022	0	0	0
98	0.45	100	0.0083	1300	391603	0	0	0
99	0.50	111	0.0075	1264	282735	0	0	0
100	0.55	122	0.0069	1235	179356	0	0	0
101	0.60	133	0.0064	1211	80096	80096	0	0.6
102	0.65	144	0.0059	1190	−15997	0	−15997	0
103	0.70	155	0.0055	1137	−139732	0	0	0
104	0.75	167	0.0052	1068	−276213	0	0	0
105	0.80	178	0.0049	1009	−405468	0	0	0
106	0.85	189	0.0047	956	−528772	0	0	0
107	0.90	200	0.0044	909	−647117	0	0	0
108	0.95	211	0.0042	867	−761286		0	0
109	0.641676	142	0.0060	1193	−188			
110	0.641577	142	0.0060	1193	0			
111			0.60	80096				
112			0.65	−15997				
113	------------	------------	------------	------------	------------	------------	------------	------------

7
Analysis of frame structures

In this chapter the analysis of a selection of frames of either steel or reinforced concrete is described. Such frames have to be analysed in order to determine the moments and shears resulting from the application of the selected loading. They can be considered to be either braced or unbraced depending on the amount of sidesway that occurs and different codes specify different limits to distinguish between the two cases.

7.1 Allowance for stiffness

The sidesway for braced frames is usually small because of the stiffening effects resulting from components such as shear walls, lifts or staircases and such frames are analysed by considering only a subframe of the whole structure. Different codes may use different subframes or different definitions of the factored loading on each span but the principles of analysis are similar for all and this chapter contains examples which illustrate how a spreadsheet can be used to obtain solutions. An important parameter in the analysis is the value taken for the stiffness of each member (EI/L). A reduced stiffness is taken for some members and codes specify the method of calculating this reduction. In the examples that follow it is assumed that the value of the stiffness reduction factor is known.

For unbraced frames design usually consists of both subframe analysis using only factored vertical loading on each span and an analysis of the whole frame using only specified horizontal forces. In BS 8110 the analysis of the whole frame is achieved by first assuming that points of contraflexure exist at defined locations in the beams and columns so that the frame becomes statically determinate.

7.2 Braced frames of reinforced concrete

As an example of the method of analysis consider the frame shown in Fig. 7.1 consisting of three bays and four storeys.

Letters A to P have been used for the joints of this frame starting at the top and letters Q to T for the supports. The beams and columns have been numbered sequentially from the bottom as shown. Typical subframes that can be used for braced frame analysis are shown in Fig. 7.2. The 0.5 values shown in (a) and (b) represent the reduction factor to be used for determining the stiffness. In fact the subframe shown in Fig. 7.2(c) is representative of all the other subframes as long as the appropriate stiffness and lengths are used. It is therefore convenient to use this subframe in preference to the others since no additional principles are introduced when using the spreadsheet approach for obtaining a solution. The same spreadsheet solution can be used for any subframe of the frame shown in Fig. 7.2(c) by inserting zero lengths for any member which is not present.

7.3 Basic equations

Consider the subframe shown in Fig. 7.3 in which the columns and beams have been numbered 1 to 11, the joints lettered E to H and the supports lettered A to D and I to L.

Assuming that each of the beams carries a uniformly distributed load then the fixed end moments at the end of each beam are given by:

Member 5	$-w_5L_5^2/12$	$+w_5L_5^2/12$
Member 6	$-w_6L_6^2/12$	$+w_6L_6^2/12$
Member 7	$-w_7L_7^2/12$	$+w_7L_7^2/12$

where w_5, w_6 and w_7 are the loads on the beams, in kN m^{-1}, and clockwise moments are considered as positive. Other types of loading could be allowed for by using the appropriate expressions for the fixed end moments but the above equations represent the normal design requirements.

Using the method of slope deflection expressions for the four unknowns θ_E, θ_F, θ_G and θ_H can be written in matrix form as:

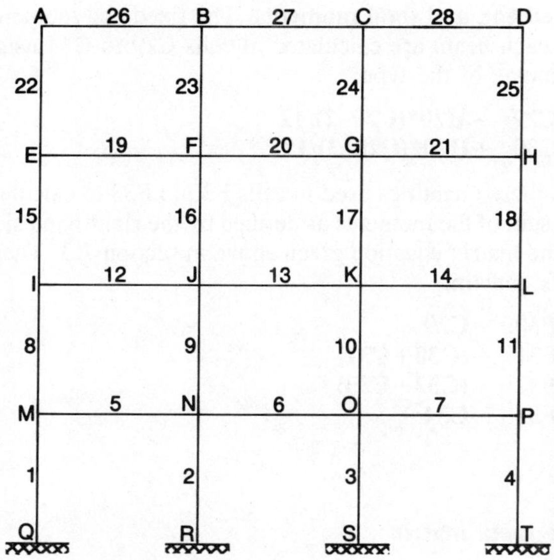

Figure 7.1 Multi-storey frame

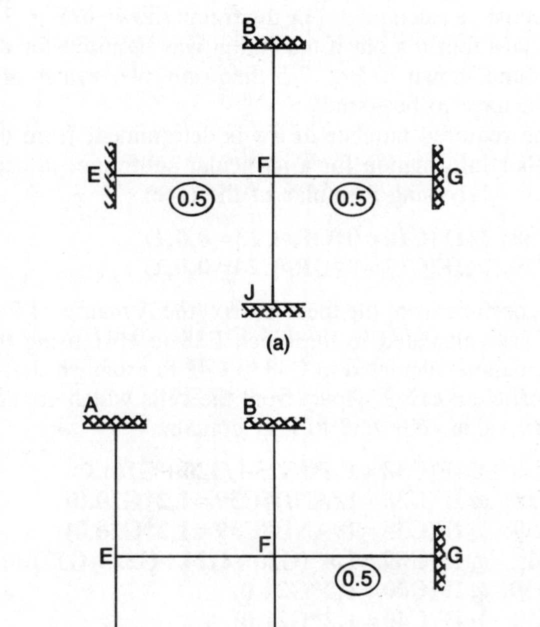

Figure 7.2 Typical subframes

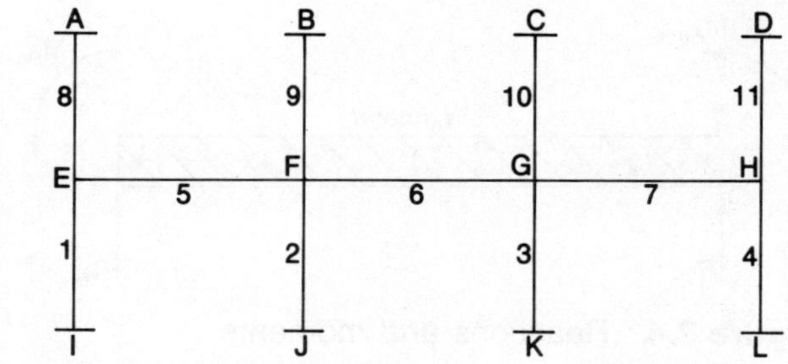

Figure 7.3 Subframe used for analysis

$$\begin{bmatrix} 4k(851) & 2k(5) & 0 & 0 \\ 2k(5) & 4k(9625) & 2k(6) & 0 \\ 0 & 2k(6) & 4k(10{,}736) & 2k(7) \\ 0 & 0 & 2k(7) & 4k(11{,}47) \end{bmatrix} \begin{Bmatrix} \theta_E \\ \theta_F \\ \theta_G \\ \theta_H \end{Bmatrix} = \begin{Bmatrix} +F_5 \\ -F_5+F_6 \\ -F_6+F_7 \\ -F_7 \end{Bmatrix}$$

where the numbers in brackets represent the sum of the stiffnesses of the members framing into a particular joint. For example 4k(10,736) represents $4(k_{10}+k_7+k_3+k_6)$. The symbol F has been used to replace $wL^2/12$ so that F_6 represents $w_6L_6^2/12$.

The terms on the right-hand side of the matrix equation and the value of the stiffness of each member are known so that the values of θ's can be determined by first inverting the stiffness matrix and then multiplying by the load vector. Algebraically this process is represented by the equations:

$$[A]\{\theta\} = \{B\} \qquad [7.1]$$

$$[A^{-1}][A]\{\theta\} = [A^{-1}]\{B\} \qquad [7.2]$$

$$\{\theta\} = [A^{-1}]\{B\} \qquad [7.3]$$

where Eqn 7.1 represents the above matrix equation and Eqns 7.2 and 7.3 illustrate the effect of pre-multiplying by the inverse matrix A^{-1}. The product of A^{-1} and A is equal to the unit matrix, thus giving a solution for the unknown θ values.

Once the values of the θ's have been determined the values of the moments and reactions at the ends of each member can be readily determined from the standard slope deflection equations. For example the moment M_{E5} representing the moment at E in beam 5 is given by:

$$M_{E5} = 4\theta_E k_5 + 2\theta_F k_5 - w_5L_5^2/12 \qquad [7.4]$$

and the moment M_{G3} by:

$$M_{G3} = 4\theta_G k_3 \qquad [7.5]$$

Equation 7.5 consists of only one term because the far end of the member is fixed and there is no load.

The equations for the reactions are illustrated in Fig. 7.4 and are represented by the equations:

$$R_{An} = -(M_{An} + M_{Bn})/L + w_n/2 \qquad [7.6]$$

$$R_{Bn} = +(M_{An} + M_{Bn})/L + w_n/2 \qquad [7.7]$$

Reactions are defined as positive upwards or from left to right.

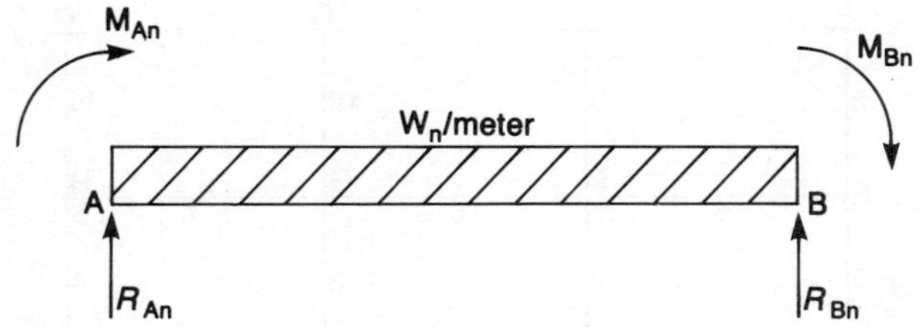

Figure 7.4 Reactions and moments

7.4 Spreadsheet layout for solution of subframes

Many spreadsheet packages contain a facility for both matrix inversion and matrix multiplication and in this example it is assumed that such facilities are available to the user. However it will be found that a more efficient spreadsheet can be produced by working from first principles so that solutions can be obtained automatically without any additional input from the user. This alternative approach will be illustrated in later examples.

Initial data. This consists of the values of the dead and live load, in kN m^{-1}, entered in cells C9 and C10 in Sheet 7.1 and the value of the modulus of elasticity in cell C11. If the value of the modulus is unknown then any value (such as 1) could be entered since this will not affect the values calculated for the moments and reactions.

Preliminary calculations. The factored dead load and the sum of the factored dead plus live loads are calculated in cells F9 and F10 using:

F9: **1*C9**
F10: **1.6*C10+1.4*C9**

using partial factors of 1.0, 1.6 and 1.4.

Member information. This subframe consists of eleven members and details about each of these members are listed in the block B16 to G26. Lengths and second moment of areas are entered in columns C and D and any required reduction factor for the element stiffness in column F. Using these values the quantities *EI/L* and *k* are calculated in columns E and G using, for example:

E16: **@IF(C16=0,0,+$C$11*D16/(C16 *1000000))**
G16: **E16*F16.**

If the factor in column F, for any member, is made equal to 1 then the value in columns E and G for that member would be the same. If the length of any member is taken as zero then the value of *k* for that member would be zero and this enables the spreadsheet to be used for a range of other subframes consisting of less than eleven members. This will be illustrated later.

Finally in cells H20 to H22 the value of the loading to be used for each member is entered. Analysis might be required using the maximum loading on each span or alternatively using the maximum loading only on alternate spans and the dead load on the remaining spans. The user has to enter the appropriate value in cells H20 to H22.

Fixed end and total moments. The fixed end moments for each beam are calculated in cells C29 to C34 using formulae of the type:

C29: **−H20*(C20^2)/12**
C30: **+H20*(C20^2)/12**

and these quantities used in cells F30 to F33 to calculate the sum of the moments as defined by the right-hand side of the matrix equation given above in section 7.3. These cells contain:

F30: **−C29**
F31: **−(C30+C31)**
F32: **−(C32+C33)**
F31: **−C34**

The theta matrix

Since the spreadsheet is designed to solve not only the frame shown in Fig. 7.3 but also subframes of this frame, it is necessary to determine how many unknown values of θ must be calculated. For the frame shown in Fig. 7.3 there are four θ's but if a solution was required for the subframe shown in Fig. 7.5 then only two values of θ would have to be found.

The required number of θ's is determined, from the member information for a particular subframe, in cells C38 to C41 using formulae of the type:

C38: **@IF(C16=0#OR#C23=0,0,1)**
C39: **@IF(C17=0#OR#C24=0,0,1)**

The coefficients of the theta matrix (the A matrix of Eqn 7.1) are calculated in the block E38 to H41 using the information calculated in C38 to C41 to establish first if a coefficient exists. Apart from the cells which contain a zero value cells E38 to H41 contain:

E38: **@IF(C38=1,4*(G23+G20+G16),0)**
F38: **@IF(C38=1#AND#C39=1,2*G20,0)**
E39: **@IF(C38=1#AND#C39=1,2*G20,0)**
F39: **@IF(C39=1,4*(G20+G24+G21+G17),0)**
G39: **@IF(C40=1,2*G21,0)**
F40: **@IF(C40=1,2*G21,0)**
G40: **@IF(C40=1,4*(G21+G25+G22+G18),0)**
H40: **@IF(C41=1,2*G22,0)**
G41: **@IF(C40=1,2*G22,0)**
H41: **@IF(C41=1,4*(G26+G19+G22,0)**

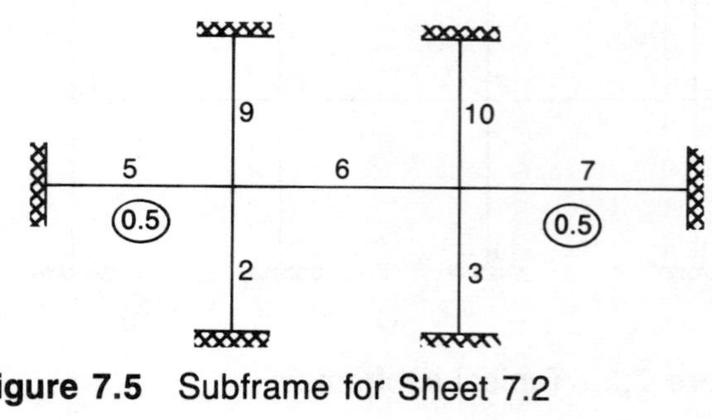

Figure 7.5 Subframe for Sheet 7.2

Sheet 7.1 (Page 1)

1/A		B	C	D	E	F	G	H
2				Analysis of Braced Frames				
3				------------------------------				
4	Note that ROWS 46 to 49 should be cleared of any numerical							
5	values, left over from a previous example, before starting.							
6								
7	Initial Data				Preliminary Calculations			
8	-------------				--------------------------			
9	DeadLd		**45** kN/m		DeadLd	45 kN/m		
10	LiveLd		**30** kN/m		Dd+Le	111 kN/m		
11	Ec		**26** kN/mm2					
12								
13	Member Information							
14	------------------							
15			L	I	EI/L	Factor	k	
16		1	**2.85**	**2.00E+00**	1.82E−05	**1**	1.82E−05	
17		2	**2.85**	**2.00E+00**	1.82E−05	**1**	1.82E−05	
18		3	**2.85**	**2.00E+00**	1.82E−05	**1**	1.82E−05	Loading
19		4	**2.85**	**2.00E+00**	1.82E−05	**1**	1.82E−05	kN/m
20		5	**6.5**	**2.00E+00**	8.00E−06	**1**	8.00E−06	**45**
21		6	**8.2**	**2.00E+00**	6.34E−06	**1**	6.34E−06	**111**
22		7	**6.2**	**2.00E+00**	8.39E−06	**1**	8.39E−06	**45**
23		8	**2.85**	**2.00E+00**	1.82E−05	**1**	1.82E−05	
24		9	**2.85**	**2.00E+00**	1.82E−05	**1**	1.82E−05	
25		10	**2.85**	**2.00E+00**	1.82E−05	**1**	1.82E−05	
26		11	**2.85**	**2.00E+00**	1.82E−05	**1**	1.82E−05	
27	Fixed Moments							
28	--------------					Sum of Moments		
29	m5a		−158 kNm			--------------		
30	m5b		158 kNm		E	158 kNm		
31	m6a		−622 kNm		F	464 kNm		
32	m6b		622 kNm		G	−478 kNm		
33	m7a		−144 kNm		H	−144 kNm		
34	m7b		144 kNm					
35								
36					Theta Matrix			
37					E	F	G	H
38	ThetaE		1	E	1.78E−04	1.60E−05	0.00E+00	0.00E+00
39	ThetaF		1	F	1.60E−05	2.03E−04	1.27E−05	0.00E+00
40	ThetaG		1	G	0.00E+00	1.27E−05	2.05E−04	1.68E−05
41	ThetaH		1	H	0.00E+00	0.00E+00	1.68E−05	1.80E−04
42								
43								
44					Inverse Matrix			
45					E	F	G	H
46	ThetaE		6.8E+05	E	5.7E+03	−4.5E+02	2.8E+01	−2.6E+00
47	ThetaF		2.4E+06	F	−4.5E+02	5.0E+03	−3.1E+02	2.9E+01
48	ThetaG		−2.4E+06	G	2.8E+01	−3.1E+02	4.9E+03	−4.6E+02
49	ThetaH		−5.8E+05	H	−2.6E+00	2.9E+01	−4.6E+02	5.6E+03
50					--------------------------			
51								

Sheet 7.1 (Page 2)

55	B	C	D	E	F	G	H
56		Bending Moments kNm (Clockwise Positive)					
57		----------	----------	----------	----------	----------	----------
58	Upper Columns						
59	8		9		10		11
60	----------	----------	----------	----------	----------	----------	----------
61	25	Mb	87	Mb	−89	Mb	−21
62	49	Ma	174	Ma	−178	Ma	−42
63	Beams						
64	5	5	6	6	7	7	
65	----------	----------	----------	----------	----------	----------	----------
66	Ma	Mb	Ma	Mb	Ma	Mb	
67	−99	245	−592	590	−235	84	
68	238	<−FreeM−	933	<−FreeM−	216 kNm		
69	Lower Columns						
70	1		2		3		4
71	----------	----------	----------	----------	----------	----------	----------
72	49	Mb	174	Mb	−178	Mb	−42
73	25	Ma	87	Ma	−89	Ma	−21
74	Check						
75	0		8.5E−14		2.8E−14		0
76	O.K.		O.K.		O.K.		O.K.
77		----------	----------	----------	----------	----------	
78							
79		Reactions kN (Positive Left to Right)					
80		----------	----------	----------	----------	----------	
81	Upper Columns						
82	8		9		10		11
83	----------	----------	----------	----------	----------	----------	----------
84	−26	Rb	−91	Rb	93	Rb	22
85	26	Ra	91	Ra	−93	Ra	−22
86	Beams						
87	5	5	6	6	7	7	
88	----------	----------	----------	----------	----------	----------	----------
89	Ra	Rb	Ra	Rb	Ra	Rb	
90	124	169	455	455	164	115	
91	Lower Columns						
92	1		2		3		4
93	----------	----------	----------	----------	----------	----------	----------
94	−26	Rb	−91	Rb	93	Rb	22
95	26	Ra	91	Ra	−93	Ra	−22
96							
97			----------	----------	----------	----------	
98							
99							

Inversion and multiplication

The matrix in block E38 to H41 is now inverted, using the special facility provided by the spreadsheet package, and the result placed in the block E46 to H49. It is advisable to clear any numerical values from block E46 to H49, which result from a previous calculation, before starting the process of inversion. Note that if a subframe is being analysed then the matrix A will not be a 4 × 4 matrix.

The appropriate moments from the range F30 to F33 are then pre-multiplied by the inverse matrix and the result placed in the range C46 to C49. This calculation is under the control of the user and care must be taken that the correct values of the moment vector are used and that the results are placed in the correct theta location. An indication of which values to use is given by the output to the range C38 to C41. For example if the subframe is such that these cells contain 0110 then the theta matrix will only contain non-zero values in the reduced block F39 to G40. The inverted matrix should then be placed in F47 to G48 and the moments F31 and F32 used for the multiplication process. The results would be placed in C47 and C48. (It is advisable to clear any residual numerical values from the range C46 to C49 before starting the multiplication process.)

Bending moments and reactions

The bending moments and reactions can now be determined from the values of θ using equations similar to Eqns 7.4 to 7.7. Typical examples of some of the bending moment formulae are:

B67: **4*C46*G20+2*C47*G20+C29**
C67: **4*C47*G20+2*C46*G20+C30**
D67: **4*C47*G21+2*C48*G21+C31**
D68: **(H21*C21*C21/8)**

where the values in row 68 are the free bending moments in the beams.

Rows 75 and 76 are used to check that the bending moments at the joints are in balance using:

B75: **@IF(C38=1,(B62+B67+B72),0)**
B76: **@IF(@ABS(B75)<0.01,"O.K.","Check")**

where the first formula calculates the out-of-balance at joint E and the second checks if this out-of-balance is acceptable. A tolerance of ±0.01 has been defined as acceptable. Similar checks are made at the other joints.

7.5 Use of spreadsheet 7.1

If the spreadsheet is now protected then the only cells which have to be unprotected are: C9 to C11, C16 to C26, D16 to D26, F16 to F26, H20 to H22, the block E46 to H49 and the range C46 to C49. There are no particular difficulties with the input of the information required in these cells although it is advisable to delete any residual numerical values from rows 46 to 49 before starting the process of inversion and multiplication.

Note that the values used in the current example have no real significance and arbitrary values have been used for the second moment of area for each member.

7.6 Additional example (spreadsheet 7.2)

In order to illustrate how the method applies when working with a subframe of the frame shown in Fig. 7.3 consider the subframe shown in Fig. 7.5.

The results are given on Sheet 7.2.

Compared with Sheet 7.1 changes have been made to the lengths of columns 1, 4, 8 and 11 and reduced stiffness factors of 0.5 used for beams 5 and 7. The maximum load has been placed on the outer spans and only the dead load placed on the inner span. No change has been made to the values of the second moment of areas since the zero value entered for the length automatically makes the value of k equal to zero for that member.

For this example there are only two thetas so that the theta matrix is reduced to a 2 × 2. Similarly only two values of theta are obtained after multiplication and these must be placed in cells C47 and C48.

The method is quite versatile and can be applied to any subframe of the original frame. There are 10 different possibilities:

(a) the original frame, a 4 × 4 matrix;
(b) two subframes each using a 3 × 3 matrix;
(c) three subframes each using a 2 × 2 matrix;
(d) four subframes each using a 1 × 1 matrix.

There is of course an assumption that the user has available a spreadsheet package which can invert and multiply matrices but it will be shown in a later section that this difficulty can be overcome. In fact it will be shown that there are advantages in working from first principles with regard to the manipulation of matrices.

7.7 Analysis of unbraced frames

For unbraced frames it is usual to assume points of contraflexure at the mid-sections of all beams and columns of the complete two-dimensional structure as shown in Fig. 7.6. Factored wind loads are then applied at the joints A, E, I, and M and the resulting forces and moments calculated by simple statics.

In the British Standard code the partial factor is given as 1.2 times the characteristic wind load as determined using the code for wind loads as in BS Code of Practice *CP3: Chapter V: Part 2: Wind Loads* (1972).

Prior to applying simple statics for the determination of the forces and moments at each storey level it is first necessary to apportion the applied joint load between the columns in such a way that there is horizontal equilibrium. Various methods exist for this initial sharing of the applied load but one in common use is to apportion in proportion to the stiffnesses of the columns. Using this approach then, if the columns are of equal stiffness, the horizontal load at each column is equal to the applied load divided by the number of columns. This assumption will be made in the example that follows although there would be no great difficulty in working with another set of assumptions.

In Fig. 7.6 it is further assumed that the supports are all fixed. If a particular support was pinned then the point of contraflexure would be at the support rather than the mid section of the lower column.

Forces and moments in the upper storey

Using the exploded view of the upper storey shown in Fig. 7.7 it can be shown that:

1. The horizontal forces h_1 to h_4 all take an equal share of the force at the joint A resulting from the factored wind load.

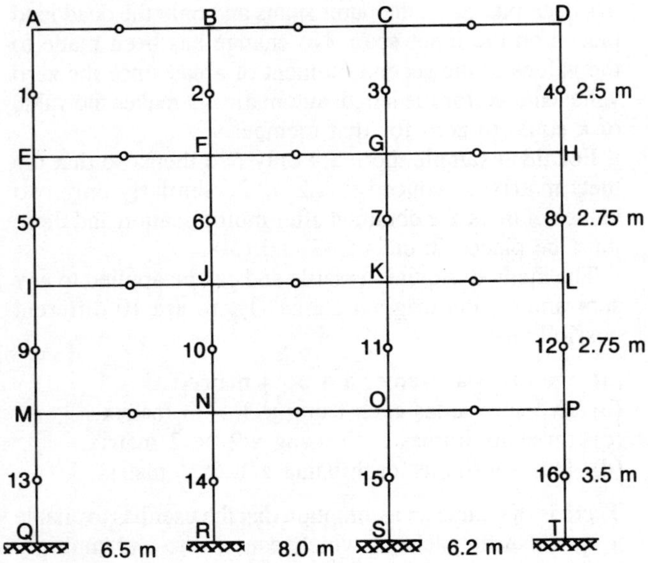

Figure 7.6 Unbraced frame

$$h_1 \text{ to } h_4 = (1.2W_kH_1/2)/4 = (1.2W_kH_1/8) \quad [7.8]$$

2. Taking moments about a, b, c and d the vertical forces at the midsections of the upper columns are given by:

$$v_1 = h_1H_1/L_1 \quad [7.9]$$

$$v_2 = H_1(h_2/L_2 - h_1/L_2 - h_1/L_1) \quad [7.10]$$

$$v_3 = -H_1(h_3/L_2 - h_4/L_2 - h_4/L_3) \quad [7.11]$$

$$v_4 = -h_4H_1/L_3 \quad [7.12]$$

3. The column bending moment at a particular joint is given by the product of the horiztonal force in the column times half the column height. For example the bending moment at A is given by the equation:

$$M_A = h_1H_1/2 \quad [7.13]$$

and similarly for the other upper storey joints.

4. The beam bending moments for the roof beams will be equal and opposite to the column bending moments for the outer beams and zero for the central beam.

Forces and moments in the 3rd storey

Using the exploded view of the 3rd storey shown in Fig. 7.8 it can be shown that:

1. The horiztonal forces h_5 to h_8 all take an equal share of the sum of the forces at the joint A and E resulting from the factored wind load.

$$h_5 \text{ to } h_8 = (1.2w_k(2H_1+H_2)/2)/4$$
$$= [1.2w_k(2H_1+H_2)/8] \quad [7.14]$$

2. Taking moments about e, f, g and h the vertical forces at the mid sections of the upper columns are given by:

$$v_5 = h_1H_1/L_1 + h_5H_2/L_1 + v_1 \quad [7.15]$$

$$v_6 = h_6H_2/L_2 + h_2H_1/L_2 + v_2 - (v_5 - v_1)(L_1+L_2)/L_2 \quad [7.16]$$

$$v_7 = h_7H_2/L_3 + h_3H_1/L_3 + v_3 - (v_6+v_5-v_1-v_2)(L_2+L_3)/L_3 \quad [7.17]$$

$$v_8 = -h_8H_2/L_3 - h_4H_1/L_3 + v_4 \quad [7.18]$$

3. The column bending moment at a particular joint is given by the product of the horizontal force in the column times half the column height. For example the bending moment at E in column EI is given by the equation:

$$M_E = h_5H_2/2 \quad [7.19]$$

and similarly for the other 3rd storey joints.

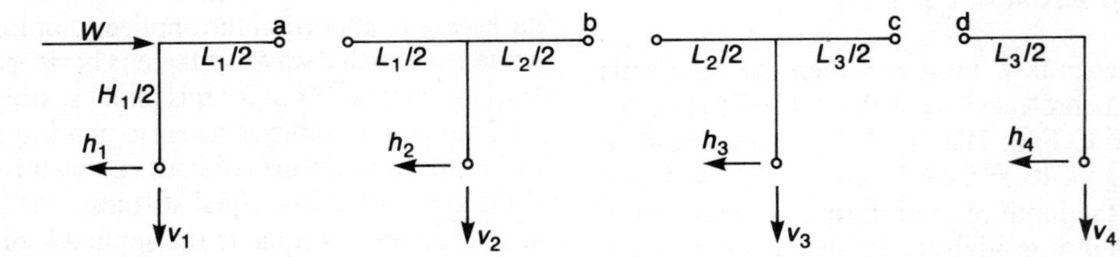

Figure 7.7 Upper storey forces

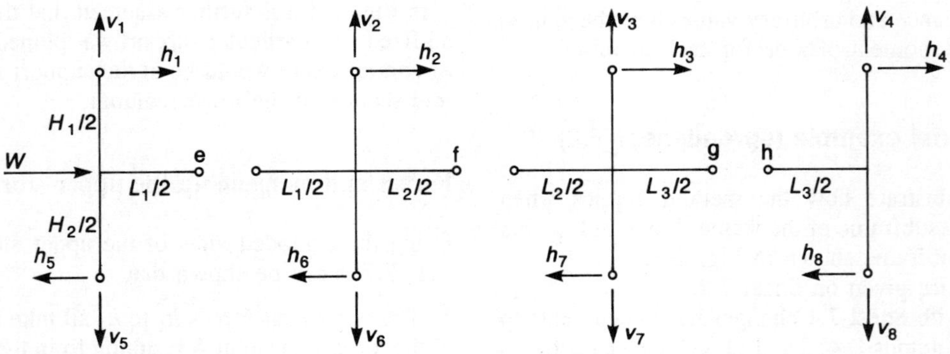

Figure 7.8 3rd storey forces

Sheet 7.2 (Page 1)

1	A	B	C	D	E	F	G	H
2				Analysis of Braced Frames				
3				------------------------------				
4	Note that ROWS 46 to 49 should be cleared of any numerical							
5	values, left over from a previous example, before starting.							
6								
7	Initial Data				Preliminary Calculations			
8	-------------				--------------------------			
9	DeadLd		45	kN/m	DeadLd	45	kN/m	
10	LiveLd		30	kN/m	Dd+Le	111	kN/m	
11	Ec		26	kN/mm2				
12								
13	Member Information							
14	-------------------							
15			L	I	EI/L	Factor	k	
16		1	0	2.00E+00	0.00E+00	1	0.00E+00	
17		2	2.85	2.00E+00	1.82E−05	1	1.82E−05	
18		3	2.85	2.00E+00	1.82E−05	1	1.82E−05	Loading
19		4	0	2.00E+00	0.00E+00	1	0.00E+00	kN/m
20		5	6.5	2.00E+00	8.00E−06	0.5	4.00E−06	111
21		6	8.2	2.00E+00	6.34E−06	1	6.34E−06	45
22		7	6.2	2.00E+00	8.39E−06	0.5	4.19E−06	111
23		8	0	2.00E+00	0.00E+00	1	0.00E+00	
24		9	2.85	2.00E+00	1.82E−05	1	1.82E−05	
25		10	2.85	2.00E+00	1.82E−05	1	1.82E−05	
26		11	0	2.00E+00	0.00E+00	1	0.00E+00	
27	Fixed Moments							
28	--------------					Sum of Moments		
29	m5a		−391	kNm		--------------		
30	m5b		391	kNm	E	391	kNm	
31	m6a		−252	kNm	F	−139	kNm	
32	m6b		252	kNm	G	103	kNm	
33	m7a		−356	kNm	H	−356	kNm	
34	m7b		356	kNm				
35								
36					Theta Matrix			
37					E	F	G	H
38	ThetaE		0	E	0.00E+00	0.00E+00	0.00E+00	0.00E+00
39	ThetaF		1	F	0.00E+00	1.87E−04	1.27E−05	0.00E+00
40	ThetaG		1	G	0.00E+00	1.27E−05	1.88E−04	0.00E+00
41	ThetaH		0	H	0.00E+00	0.00E+00	8.39E−06	0.00E+00
42								
43								
44					Inverse Matrix			
45					E	F	G	H
46	ThetaE			E				
47	ThetaF		−7.8E+05	F		5.4E+03	−3.6E+02	
48	ThetaG		6.0E+05	G		−3.6E+02	5.3E+03	
49	ThetaH			H				
50					----------------------------			
51								

Sheet 7.2 (Page 2)

	A	B	C	D	E	F	G	H
55		B	C	D	E	F	G	H
56			Bending Moments kNm (Clockwise Positive)					
57			- - - - - - - -	- - - - - - - -	- - - - - - - -	- - - - - - - -	- - - - - - - -	- - - - - - - -
58	Upper Columns							
59		8		9		10		11
60		- - - - - - - -	- - - - - - - -	- - - - - - - -	- - - - - - - -	- - - - - - - -	- - - - - - - -	- - - - - - - -
61		0	Mb	−28	Mb	22	Mb	0
62		0	Ma	−57	Ma	44	Ma	0
63	Beams							
64		5	5	6	6	7	7	
65		- - - - - - - -	- - - - - - - -	- - - - - - - -	- - - - - - - -	- - - - - - - -	- - - - - - - -	- - - - - - - -
66		Ma	Mb	Ma	Mb	Ma	Mb	
67		−397	378	−264	258	−345	361	
68		586	<−FreeM−	378	<−FreeM−	533 kNm		
69	Lower Columns							
70		1		2		3		4
71		- - - - - - - -	- - - - - - - -	- - - - - - - -	- - - - - - - -	- - - - - - - -	- - - - - - - -	- - - - - - - -
72		0	Mb	−57	Mb	44	Mb	0
73		0	Ma	−28	Ma	22	Ma	0
74	Check							
75		0		1.4E−14		−4.3E−14		0
76	O.K.		O.K.		O.K.		O.K.	
77			- - - - - - - -	- - - - - - - -	- - - - - - - -	- - - - - - - -	- - - - - - - -	
78								
79			Reactions kN (Positive Left to Right)					
80			- - - - - - - -	- - - - - - - -	- - - - - - - -	- - - - - - - -	- - - - - - - -	- - - -
81	Upper Columns							
82		8		9		10		11
83		- - - - - - - -	- - - - - - - -	- - - - - - - -	- - - - - - - -	- - - - - - - -	- - - - - - - -	- - - - - - - -
84		0	Rb	30	Rb	−23	Rb	0
85		0	Ra	−30	Ra	23	Ra	0
86	Beams							
87		5	5	6	6	7	7	
88		- - - - - - - -	- - - - - - - -	- - - - - - - -	- - - - - - - -	- - - - - - - -	- - - - - - - -	- - - - - - - -
89		Ra	Rb	Ra	Rb	Ra	Rb	
90		364	358	185	184	342	347	
91	Lower Columns							
92		1		2		3		4
93		- - - - - - - -	- - - - - - - -	- - - - - - - -	- - - - - - - -	- - - - - - - -	- - - - - - - -	- - - - - - - -
94		0	Rb	30	Rb	−23	Rb	0
95		0	Ra	−30	Ra	23	Ra	0
96								
97				- - - - - - - -	- - - - - - - -	- - - - - - - -	- - - -	
98								
99								

4. The beam bending moments in the beams will be equal and opposite to the sum of the bending moments in the columns framing into the joint for the outer beams and zero for the central beam.

Forces and moments in lower storeys

The horizontal forces in the lower storeys are calculated using similar equations to 7.8 and 7.14 but using the appropriate height above the pin joint being considered. At foundation level the horiztonal force at each support is equal to $\frac{1}{4}$ of the total wind force.

The vertical forces, column bending moments and beam bending moments are calculated using formulae similar to those used for the third storey using the relevant forces and dimensions for that level. At foundation level the values of these quantities are the same as those calculated for the first storey.

7.8 Spreadsheet layout for unbraced frames

Initial data and preliminary calculations. The value of the wind load is entered in cell C6 in Sheet 7.3 and the factored wind load calculated in F6 using 1.2*C6.

Member information. This consists of the spans of the beams entered in C11 to E11 and the height of the storeys entered in G11 to G14.

Horizontal forces

C19: (top storey) **(F6*G11/8)**
C27: (3rd storey) **((F6*((2*G11)+G12))/8)**
C35: (2nd storey) **((F6*((2*G11)+ (2*G12)+G13))/8)**
C43: (1st storey) **((F6*((2*G11)+(2*G12) +(2*G13)+G14))/8)**
C52: (foundation) **+C43+(F6*G14/8)**

The horizontal forces at each column for a particular level are equal and can be obtained directly by entering +C19 into cells D19 to F19 and similarly for the other rows containing horizontal forces.

Vertical forces

The formulae required to calculate the vertical forces are based on equations 7.9 to 7.12 and equations similar to 7.15 to 7.18.

For the top storey:

C20: **(C19*G11/C11)**
D20: **(D19*G11/D11)−(C19*G11/D11) −(C19*G11/C11)**
E20: **(G11*((F19/E11)+(F19/D11)−(E19/D11)))**
F20: **−(F19*G11/E11).**

For the 3rd storey:

C28: **(C27*G12/C11)+(C19*G11/C11)+C20**
D28: **(D27*G12/D11)−((C28−C20) *(C11+D11)/D11)+(D19*G11/D11)+D20**
E28: **(E27*G12/E11)−((D28+C28−C20−D20) *(D11+E11)/E11)+(E19*G11/E11)+E20**
F28: **−(F27*G12/E11)−(F19*G11/E11)+F20**

For the 2nd storey:

C36: **(C35*G13/C11)+(C27*G12/C11)+C28**
D36: **(D35*G13/D11)−((C36−C28) *(C11+D11)/D11)+(D27*G12/D11)+D28**
E36: **(E35*G13*E11)−((D36+C36−C28−D28) *(D11+E11)/E11)+(E27*G12/E11)+E28**
F36: **−(F35*G13/E11)−(F27*G12/E11)+F28**

For the 1st storey:

C44: **(C43*G14/C11)+(C35*G13/C11)+C36**
D44: **(D43*G14/D11)−((C44−C36) *(C11+D11)/D11)+(D35*G13/D11)+D36**
E44: **(E43*G14/E11)−((D44+C44−C36−D36) *(D11+E11)/E11)+(E35*G13/E11)+E36**
F44: **−(F43*G14/E11)−(F35*G13/E11)+F36**

At foundation level the vertical forces are the same as those calculated at the first storey level so that, for example, the entry for cell C53 is +C44.

Column bending moments

These moments are the product of the horizontal force times half the column height. Typical examples of the required formulae are:

Top storey:

C22: **(C19*G11/2)**
D22: **(D19*G11/2)**

3rd storey:

C30: **(C27*G12/2)**
D30: **(D27*G12/2)**

and similarly for the other storeys. At foundation level the value of the bending moment is equal to the value found for the first storey level.

Beam bending moments

These moments are the product of the vertical force times half the beam length. At any storey level the required vertical force is the algebraic sum of all vertical forces to one side of the section. For example at the third storey level, working from left to right the vertical force in beam EF is (C28−C20) and in FG (C28+D28−C20−D20). For beam GH it is more convenient to work from right to left since less forces are involved. Typical examples of the required formulae are:

Top storey

C24 & D24: **−(C20*C11/2)**
E24 & F24: **−(C20+D20)*D11/2**
G24 & H24: **+F20*E11/2**

3rd storey

C32 & D32: **−(C28−C20)*C11/2**

Sheet 7.3

1/A	B	C	D	E	F	G	H	I
2			Analysis of Unbraced Frames					
3			------------------------					
4 Initial Data				Preliminary Calculations				
5 ------------				------------------------				
6 WindLd		60	kN/m	WindLd	72 (1.2Wk)			
7								
8 Member Information								
9 ------------------								
10		1stBay	2ndBay	3rdBay		StoreyHt		
11 Length		6.5	8	6.2 m		2.5 m		
12						2.75 m		
13						2.75 m		
14						3.5 m		
15 Results								
16 -------								
17			Top Storey					
18			----------					
19 Hor.Frce		23	23	23	23 kN			
20 Ver.Frce		9	−9	9	−9 kN			
21 Joints		A	B	C	D			
22 BM.Cols		28	28	28	28 kNm			
23			1stBay		2ndBay		3rdBay	
24 BM.Beams		−28	−28	0	0	−28	−28 kNm	
25			3rd Storey					
26			----------					
27 Hor.Frce		70	70	70	70 kN			
28 Ver.Frce		47	−47	49	−49 kN			
29 Joints		E	F	G	H			
30 BM.Cols		96	96	96	96 kNm			
31			1stBay		2ndBay		3rdBay	
32 BM.Beams		−124	−124	0	0	−124	−124 kNm	
33			2nd Storey					
34			----------					
35 Hor.Frce		119	119	119	119 kN			
36 Ver.Frce		127	−127	133	−133 kN			
37 Joints		I	J	K	L			
38 BM.Cols		164	164	164	164 kNm			
39			1stBay		2ndBay		3rdBay	
40 BM.Beams		−260	−260	0	0	−260	−260 kNm	
41			1st Storey					
42			----------					
43 Hor.Frce		176	176	176	176 kN			
44 Ver.Frce		272	−272	285	−285 kN			
45 Joints		M	N	O	P			
46 BM.Cols		241	241	241	241 kNm			
47			1stBay		2ndBay		3rdBay	
48 BM.Beams		−471	−471	0	0	−471	−471 kNm	
49			Foundation Level					
50			----------------					
51 Support		Q	R	S	T			
52 Hor.Frce		207	207	207	207 kN			
53 Ver.Frce		272	−272	285	−285 kN			
54 BM.Cols		241	241	241	241 kNm			

E32 & F32: $-(C28+D28-C20-D20)*D11/2$
G32 & H32: $(F28-F20)*E11/2$

2nd storey

C40 & D40: $-(C36-C28)*C11/2$
E40 & F40: $-(C36+D36-C28-D28)*D11/2$
G40 & H40: $(F36-F28)*E11/2$

and similarly for the other storeys. At foundation level the moment is equal to the value calculated at the first storey level.

Note that both the beam and column bending moments calculated above follow the clockwise positive convention.

7.9 Use of spreadsheet 7.3

If the spreadsheet is now protected then the only cells which have to be unprotected are: C6, C11, D11, E11 and G11 to G14.

This simple spreadsheet could easily be extended to frames with more storeys and bays.

7.10 Automatic matrix inversion and multiplication

Inversion

On spreadsheets 7.1 and 7.2 matrix inversion and multiplication were performed using the special facility available in most spreadsheet packages. To use Sheets 7.1 and 7.2 it was necesary for the user to call up the inversion and multiplication processes and follow the instructions output by the program. In this section it will be shown how the process can be automated so that inversion and multiplication are achieved without any additional input from the user.

The procedure is illustrated using a 4 × 4 matrix but there would be no difficulty in extending this to cater for larger matrices. For convenience the matrix to be inverted will be defined by the cell addresses rather than the numerical values within those cells.

One of the simplest methods for inversion is by means of Gauss Jordan elimination method in which the unit matrix is placed alongside the matrix to be inverted and operations on each row performed until the original matrix is converted to a unit matrix. At this stage the original unit matrix will have been converted to the inverse of the original matrix, see Eqns 7.1 to 7.3. As an example consider the matrix occupying block B8 to I11 consisting of the original matrix and the unit matrix, Sheet 7.4:

	B	C	D	E	F	G	H	I
8	**B8**	**C8**	**D8**	**E8**	**1**	**0**	**0**	**0**
9	**B9**	**C9**	**D9**	**E9**	**0**	**1**	**0**	**0**
10	**B10**	**C10**	**D10**	**E10**	**0**	**0**	**1**	**0**
11	**B11**	**C11**	**D11**	**E11**	**0**	**0**	**0**	**1**

A new matrix is produced in block B13 to I16 using the operations described in Table 7.1.

Table 7.1

Row	*Operation*	*1st elimination* B	C	D	E	F	G	H	I
13	Divide by $B$8	1	C13	D13	E13	F13	0	0	0
14	[B9-(B13*$B$9)]	0	C14	D14	E14	F14	1	0	0
15	[B10-(B13*$B$10)]	0	C15	D15	E15	F15	0	1	0
16	[B11-(B13*$B$11)]	0	C16	D16	E16	F16	0	0	1

Note that some terms of the operator are anchored, using the dollar sign, since these terms remain the same for the operation on all values in the row. As written the operators apply to elements in column B only since all B terms, except the appropriate anchored term, will change as the operator is applied to different elements along the row. The second elimination matrix is then produced in block B18 to I21 using the operations described in Table 7.2.

Table 7.2

Row	*Operation*	*2nd elimination* B	C	D	E	F	G	H	I
18	[B13-(B19*$C$13)]	1	0	D18	E18	F18	G18	0	0
19	Divide by $C$14	0	1	D19	E19	F19	G19	0	0
20	[B15-(B19*$C$15)]	0	0	D20	E20	F20	G20	1	0
21	[B16-(B19*$C$16)]	0	0	D21	E21	F21	G21	0	1

The third elimination matrix is produced in block B23 to I26 using the operations described in Table 7.3.

Table 7.3

Row	Operation	3rd elimination B	C	D	E	F	G	H	I
23	[B18 – (B25*$D$18)]	1	0	0	E23	F23	G23	H23	0
24	[B19 – (B25*$D$19)]	0	1	0	E24	F24	G24	H24	0
25	Divide by $D$20	0	0	1	E25	F25	G25	H25	0
26	[B21 – (B25*$D$21)]	0	0	0	E26	F26	G26	H26	1

The fourth elimination matrix is produced in block B28 to I31 using the operations described in Table 7.4.

Table 7.4

Row	Operation	4th elimination B	C	D	E	F	G	H	I
28	[B23 – (B31*$E$23)]	1	0	0	0	F28	G28	H28	I28
29	[B24 – (B31*$E$24)]	0	1	0	0	F29	G29	H29	I29
30	[B25 – (B31*$E$25)]	0	0	1	0	F30	G30	H30	I30
31	Divide by $E$26	0	0	0	1	F31	G31	H31	I31

The unit matrix has now changed position and the matrix in block F28 to I31 is the inverse of the original matrix. This inverse matrix has been copied to block F35 to I38 in Sheet 7.3 for convenience.

Multiplication

Pre-multiplication of the vector in range B35 to B38 by the matrix in block F35 to I38 can be achieved using the following formulae:

D35: **(F35*$B$35)+(G35*$B$36)+(H35 *$B$37)+(I35*$B$38)**
D36: **(F36*$B$35)+(G36*$B$36)+(H36 *$B$37)+(I36*$B$38)**
D37: **(F37*$B$35)+(G37*$B$36)+(H37 *$B$37)+(I37*$B$38)**
D38: **(F38*$B$35)+(G38*$B$36)+(H38 *$B$37)+(I38*$B$38)**

7.11 Spreadsheet layout for inversion and multiplication

Inversion

The initial matrix has been placed in block B8 to E11 in Sheet 7.4 and the unit matrix in block F8 to I11.

The four eliminations are then set out in rows 13 to 31 using the operating formulae given above in Section 7.10. The formulae used are identical with the operating formulae except for those which require division, namely rows 13, 19, 25 and 31. The formulae in these rows take the form:

B13: **@IF($B$8< >0,B8/$B$8,0)**
B19: **@IF($C$14< >0,B14/$C$14,0)**
B25: **@IF($D$20< >0,B20/$D$20,0)**
B31: **@IF($E$26< >0,B26/$E$26,0**

to allow for the possibility that the pivot has a zero value when the matrix being inverted is smaller than 4 × 4.

Multiplication

If the B matrix is placed in B35 to B38 and the inverted A matrix in F35 to I38 then the formulae given above in Section 7.10 define the multiplication process. The result is shown in D35 to D38.

7.12 Use of spreadsheet 7.4

This spreadsheet was designed to be used in conjunction with other spreadsheets requiring inversion and multiplication and an example of this will be given in the next section. However, it could if necessary be used alone.

If the spreadsheet is protected then the only cells which need to be unprotected are block B8 to E11, for the original matrix, and B35 to B38 for the B matrix.

The original matrix can be 4 × 4, 3 × 3, 2 × 2 or 1 × 1 but the first pivot must be on the leading diagonal. For example a 2 × 2 matrix in which the terms had been placed in block D8 to E9 would not be acceptable.

7.13 Spreadsheet layout using automatic matrix operations

As an example of the use of automatic inversion and multiplication the process will be applied to the problem previously solved on Sheet 7.1 and illustrated in Fig. 7.3

A comparison of page 1 of Sheet 7.1 with page 1 of Sheet 7.5 will show that the two pages contain identical numerical values. The only difference between the two is that for Sheet 7.1 inversion and multiplication required input from the user whilst for Sheet 7.5 the process was automatic.

The calculations shown on page 2 of Sheet 7.1 have been repeated on Sheet 7.5, and are shown on the third page. The second page of Sheet 7.5 is a copy of the inversion and multiplication process to block K8 to R31.

Sheet 7.4

1/A	B	C	D	E	F	G	H	I
2								
3			Matrix Inversion and Multiplication					
4			--------------	--------------	--------------	--------------	--------------	--------------
5								
6		Initial Matrix					Unit Matrix	
7		--------------	--------------	--------------			--------------	--------------
8	**15.00**	**3.00**	**0.00**	**0.00**	1	0	0	0
9	**6.00**	**12.00**	**5.00**	**0.00**	0	1	0	0
10	**0.00**	**4.00**	**10.00**	**3.00**	0	0	1	0
11	**0.00**	**0.00**	**5.00**	**15.00**	0	0	0	1
12								
13	1.00	0.20	0.00	0.00	0.07	0.00	0.00	0.00
14	0.00	10.80	5.00	0.00	−0.40	1.00	0.00	0.00
15	0.00	4.00	10.00	3.00	0.00	0.00	1.00	0.00
16	0.00	0.00	5.00	15.00	0.00	0.00	0.00	1.00
17								
18	1.00	0.00	−0.09	0.00	0.07	−0.02	0.00	0.00
19	0.00	1.00	0.46	0.00	−0.04	0.09	0.00	0.00
20	0.00	0.00	8.15	3.00	0.15	−0.37	1.00	0.00
21	0.00	0.00	5.00	15.00	0.00	0.00	0.00	1.00
22								
23	1.00	0.00	0.00	0.03	0.08	−0.02	0.01	0.00
24	0.00	1.00	0.00	−0.17	−0.05	0.11	−0.06	0.00
25	0.00	0.00	1.00	0.37	0.02	−0.05	0.12	0.00
26	0.00	0.00	0.00	13.16	−0.09	0.23	−0.61	1.00
27								
28	1.00	0.00	0.00	0.00	0.08	−0.02	0.01	0.00
29	0.00	1.00	0.00	0.00	−0.05	0.12	−0.06	0.01
30	0.00	0.00	1.00	0.00	0.02	−0.05	0.14	−0.03
31	0.00	0.00	0.00	1.00	−0.01	0.02	−0.05	0.08
32								
33	B Matrix		Multiplication			Inversion		
34	--------------		--------------	--------------		--------------		
35	**10.00**		0.77		0.08	−0.02	0.01	0.00
36	**5.00**		−0.53		−0.05	0.12	−0.06	0.01
37	**6.00**		1.35		0.02	−0.05	0.14	−0.03
38	**−20.00**		−1.78		−0.01	0.02	−0.05	0.08

This process is identical with that described for Sheet 7.4 except for the change in cell location and the resulting changes in the formulae due to this translation. Column B becomes Column K, etc. Note that the matrix placed in the block K8 to N11 is not entered by the user but is a copy of the matrix calculated previously in E37 to H40.

The results of the inversion are shown in block M35 to P38 and these values are copied to block E45 to H48 using for example:

E45: +**M35**
F45: +**N35**

Similarly the results of the multiplication of the inverse matrix and the B matrix which are calculated in K35 to K38, using the method described with reference to Sheet 7.4, are copied to the range C45 to C48.

The net result of this is that the user need never refer to the existing page 2 of Sheet 7.5 but would work entirely within page 1. It will of course be necessary to add formulae similar to those from the second page of Sheet 7.1 in order to calculate the moments and reactions.

7.14 Use of spreadsheet 7.5

If the spreadsheet is now protected then the only cells which have to be unprotected are: C7 to C9, C14 to C24, D14 to D24, F14 to F24, H18 to H20. Note that this list

Sheet 7.5 (Page 1)

1/A		B	C	D	E	F	G	H
2				Analysis of Braced Frames				
3				------------------------------				
4								
5	Initial Data				Preliminary Calculations			
6	-------------				-----------------------------			
7	DeadLd		**45**	kN/m	DeadLd	45	kN/m	
8	LiveLd		**30**	kN/m	Dd+Le	111	kN/m	
9	Ec		**26**	kN/mm2				
10								
11	Member Information							
12	-------------------							
13			L	I	EI/L	Factor	k	
14		1	**2.85**	**2.00E+00**	1.82E−05	**1**	1.82E−05	
15		2	**2.85**	**2.00E+00**	1.82E−05	**1**	1.82E−05	
16		3	**2.85**	**2.00E+00**	1.82E−05	**1**	1.82E−05	Loading
17		4	**2.85**	**2.00E+00**	1.82E−05	**1**	1.82E−05	kN/m
18		5	**6.5**	**2.00E+00**	8.00E−06	**1**	8.00E−06	**45**
19		6	**8.2**	**2.00E+00**	6.34E−06	**1**	6.34E−06	**111**
20		7	**6.2**	**2.00E+00**	8.39E−06	**1**	8.39E−06	**45**
21		8	**2.85**	**2.00E+00**	1.82E−05	**1**	1.82E−05	
22		9	**2.85**	**2.00E+00**	1.82E−05	**1**	1.82E−05	
23		10	**2.85**	**2.00E+00**	1.82E−05	**1**	1.82E−05	
24		11	**2.85**	**2.00E+00**	1.82E−05	**1**	1.82E−05	
25	Fixed Moments							
26	--------------					Sum of Moments		
27	m5a		−158	kNm		---------------		
28	m5b		158	kNm	E	158	kNm	
29	m6a		−622	kNm	F	464	kNm	
30	m6b		622	kNm	G	−478	kNm	
31	m7a		−144	kNm	H	−144	kNm	
32	m7b		144	kNm				
33								
34								
35					Theta Matrix			
36					E	F	G	H
37	ThetaE		1	E	1.78E−04	1.60E−05	0.00E+00	0.00E+00
38	ThetaF		1	F	1.60E−05	2.03E−04	1.27E−05	0.00E+00
39	ThetaG		1	G	0.00E+00	1.27E−05	2.05E−04	1.68E−05
40	ThetaH		1	H	0.00E+00	0.00E+00	1.68E−05	1.80E−04
41								
42								
43					Inverse Matrix			
44					E	F	G	H
45	ThetaE		6.8E+05	E	5.7E+03	−4.5E+02	2.8E+01	−2.6E+00
46	ThetaF		2.4E+06	F	−4.5E+02	5.0E+03	−3.1E+02	2.9E+01
47	ThetaG		−2.4E+06	G	2.8E+01	−3.1E+02	4.9E+03	−4.6E+02
48	ThetaH		−5.8E+05	H	−2.6E+00	2.9E+01	−4.6E+02	5.6E+03
49					-----------------------------			

Sheet 7.5 (Page 2)

J	K	L	M	N	O	P	Q	R
2								
3	Matrix Inversion and Multiplication							
4	--							
5								
6	A–Matrix					Unit–Matrix		
7	--------					------------		
8	1.8E–04	1.6E–05	0.0E+00	0.0E+00	1	0	0	0
9	1.6E–05	2.0E–04	1.3E–05	0.0E+00	0	1	0	0
10	0.0E+00	1.3E–05	2.0E–04	1.7E–05	0	0	1	0
11	0.0E+00	0.0E+00	1.7E–05	1.8E–04	0	0	0	1
12								
13	1.0E+00	9.0E–02	0.0E+00	0.0E+00	5.6E+03	0.0E+00	0.0E+00	0.0E+00
14	0.0E+00	2.0E–04	1.3E–05	0.0E+00	–9.0E–02	1.0E+00	0.0E+00	0.0E+00
15	0.0E+00	1.3E–05	2.0E–04	1.7E–05	0.0E+00	0.0E+00	1.0E+00	0.0E+00
16	0.0E+00	0.0E+00	1.7E–05	1.8E–04	0.0E+00	0.0E+00	0.0E+00	1.0E+00
17								
18	1.0E+00	0.0E+00	–5.6E–03	0.0E+00	5.7E+03	–4.5E+02	0.0E+00	0.0E+00
19	0.0E+00	1.0E+00	6.3E–02	0.0E+00	–4.5E+02	5.0E+03	0.0E+00	0.0E+00
20	0.0E+00	0.0E+00	2.0E–04	1.7E–05	5.6E–03	–6.3E–02	1.0E+00	0.0E+00
21	0.0E+00	0.0E+00	1.7E–05	1.8E–04	0.0E+00	0.0E+00	0.0E+00	1.0E+00
22								
23	1.0E+00	0.0E+00	0.0E+00	4.6E–04	5.7E+03	–4.5E+02	2.8E+01	0.0E+00
24	0.0E+00	1.0E+00	0.0E+00	–5.2E–03	–4.5E+02	5.0E+03	–3.1E+02	0.0E+00
25	0.0E+00	0.0E+00	1.0E+00	8.2E–02	2.8E+01	–3.1E+02	4.9E+03	0.0E+00
26	0.0E+00	0.0E+00	0.0E+00	1.8E–04	–4.6E–04	5.2E–03	–8.2E–02	1.0E+00
27								
28	1.0E+00	0.0E+00	0.0E+00	0.0E+00	5.7E+03	–4.5E+02	2.8E+01	–2.6E+00
29	0.0E+00	1.0E+00	0.0E+00	0.0E+00	–4.5E+02	5.0E+03	–3.1E+02	2.9E+01
30	0.0E+00	0.0E+00	1.0E+00	0.0E+00	2.8E+01	–3.1E+02	4.9E+03	–4.6E+02
31	0.0E+00	0.0E+00	0.0E+00	1.0E+00	–2.6E+00	2.9E+01	–4.6E+02	5.6E+03
32								
33 Multiplication			Inversion					
34	-------------------------							
35	6.8E+05		5.7E+03	–4.5E+02	2.8E+01	–2.6E+00		
36	2.4E+06		–4.5E+02	5.0E+03	–3.1E+02	2.9E+01		
37	–2.4E+06		2.8E+01	–3.1E+02	4.9E+03	–4.6E+02		
38	–5.8E+05		–2.6E+00	2.9E+01	–4.6E+02	5.6E+03		

is shorter than the equivalent list for the use of Sheet 7.1 so that less user interaction is required for Sheet 7.5.

As stated with reference to Sheet 7.1 the actual value used for the Modulus of Elasticity in cell C11 is not important as far as calculations for the moments and reactions are concerned. However the values of the rotation in C46 to C49 will not be the true values when using an assumed value for E.

7.15 Spreadsheet for sub-frames (column design)

Multi-storey frames are subjected to vertical and horizontal loads and it is usual to design for these loads separately. When designing for horizontal loading it is necessary to consider the whole frame and an example of a spreadsheet which can be used for this case was illustrated in Sheet 7.3. When dealing with vertical loading then the analysis can be based either on the whole structure or on the use of subframes. Illustrations of the use of subframe analysis using spreadsheets were given in Sheets 7.1, 7.2 and 7.3.

The subframes recommended in the BS code for steelwork design, BS 5950 *Part 1: Code of Practice for Design in Simple and Continuous Construction* (1990), are illustrated in Fig. 7.9. The upper two diagrams of Fig. 7.9 are used for the design of the beam indicated and as shown in Sheet 7.2 this subframe could be analysed by treating it as a subset of the subframe solved on Sheet 7.1.

Sheet 7.5 (Page 3)

56		B	C	D	E	F	G	H
57			Bending Moments kNm (Clockwise Positive)					
58			--------	--------	--------	--------	--------	--------
59	Upper Columns							
60		8		9		10		11
61		--------	--------	--------	--------	--------	--------	--------
62		25	Mb	87	Mb	−89	Mb	−21
63		49	Ma	174	Ma	−178	Ma	−42
64	Beams							
65		5	5	6	6	7	7	
66		--------	--------	--------	--------	--------	--------	--------
67		Ma	Mb	Ma	Mb	Ma	Mb	
68		−99	245	−592	590	−235	84	
69		237.6562	<−FreeM−	932.955	<−FreeM−	216.225	kNm	
70	Lower Columns							
71		1		2		3		4
72		--------	--------	--------	--------	--------	--------	--------
73		49	Mb	174	Mb	−178	Mb	−42
74		25	Ma	87	Ma	−89	Ma	−21
75	Check							
76		0		0		0		0
77	O.K.		O.K.		O.K.		O.K.	
78			--------	--------	--------	--------	--------	--------
79								
80			Reactions kN (Positive Left to Right)					
81			--------	--------	--------	--------	--------	--------
82	Upper Columns							
83		8		9		10		11
84		--------	--------	--------	--------	--------	--------	--------
85		−26	Rb	−91	Rb	93	Rb	22
86		26	Ra	91	Ra	−93	Ra	−22
87	Beams							
88		5	5	6	6	7	7	
89		--------	--------	--------	--------	--------	--------	--------
90		Ra	Rb	Ra	Rb	Ra	Rb	
91		124	169	455	455	164	115	
92	Lower Columns							
93		1		2		3		4
94		--------	--------	--------	--------	--------	--------	--------
95		−26	Rb	−91	Rb	93	Rb	22
96		26	Ra	91	Ra	−93	Ra	−22
97								
98				--------	--------	--------		

Compare Fig. 7.9(a) with Fig. 7.5. The subframe shown in Fig. 7.9(b) could also be analysed using a similar approach; in fact, all that is required is to ensure that the length of member 5 is entered as zero and that there is no load in cell H20.

For the design of the columns, BS 5950 recommends the use of subframes of the type shown in Fig. 7.9(c) and (d) and an example of a suitable spreadsheet layout for these subframes is given on Sheet 7.6.

Basic equations

The columns of the subframe are numbered 1, 4 and 7 as shown in Fig. 7.10. The beams are numbered 2, 3, 5 and 6. Joints are lettered C and F and the supports A, B, D, E, G and H. Factored loading would be placed on the beams and since there are only two joints the theta matrix will be a 2×2. Using the slope deflection method the following matrix equation can be derived:

Sheet 7.6 (Page 1)

1/A		B	C	D	E	F	G	H
2				Analysis of Braced Frames				
3				----------------------------				
4								
5	Initial Data				Preliminary Calculations			
6	--------------				----------------------------			
7	DeadLd		**45**	kN/m	DeadLd	45	kN/m	
8	LiveLd		**30**	kN/m	Dd+Le	111	kN/m	
9	E		**1**	kN/mm2				
10								
11	Member Information							
12	-------------------							
13			L	I	EI/L	Factor	k	
14		1	**3**	**1.00E+02**	3.33E-05	1	3.33E-05	
15		4	**3**	**1.00E+02**	3.33E-05	1	3.33E-05	Loading
16		7	**3**	**1.00E+02**	3.33E-05	1	3.33E-05	kN/m
17		2	**4**	**1.00E+02**	2.50E-05	1	2.50E-05	**45**
18		3	**4**	**1.00E+02**	2.50E-05	1	2.50E-05	**111**
19		5	**4**	**1.00E+02**	2.50E-05	1	2.50E-05	**111**
20		6	**4**	**1.00E+02**	2.50E-05	1	2.50E-05	**45**
21								
22	Fixed Moments							
23	--------------					Sum of Moments		
24	m2a		-60	kNm		--------------		
25	m2b		60	kNm	C	88	kNm	
26	m3a		-148	kNm	F	-88	kNm	
27	m3b		148	kNm		Theta Matrix		
28	m5a		-148	kNm		------------		
29	m5b		148	kNm		C	F	
30	m6a		-60	kNm	C	4.67E-04	6.67E-05	
31	m6b		60	kNm	F	6.67E-05	4.67E-04	
32								
33	A-Matrix			Unit-Matrix				
34	--------			-----------				
35		4.67E-04	6.67E-05	1	0			
36		6.67E-05	4.67E-04	0	1			
37								
38		1.0E+00	1.4E-01	2.1E+03	0.0E+00			
39		0.0E+00	4.6E-04	-1.4E-01	1.0E+00			
40								
41		1.0E+00	0.0E+00	2.2E+03	-3.1E+02			
42		0.0E+00	1.0E+00	-3.1E+02	2.2E+03			
43								
44			Multiplication		Inversion			
45			------------------------					
46	ThetaC		2.2E+05		2.2E+03	-3.1E+02		
47	ThetaF		-2.2E+05		-3.1E+02	2.2E+03		

Sheet 7.6 (Page 2)

	A	B	C	D	E	F	G	H
54		B	C	D	E	F	G	H
55			Bending Moments kNm (Clockwise Positive)					
56			----------	----------	----------	----------	----------	----------
57	Columns							
58	--------		7		4		1	
59			----------	----------	----------	----------	----------	
60	Top		−15		−15		29 kNm	
61	Bottom		−29		15		15 kNm	
62	Beams 5 and 6							
63	--------------							
64		E	5	F	F	6	G	
65		----------	----------	----------	----------	----------	----------	
66		Ma		Mb	Ma		Mb	
67		−159		126	−82		49 kNm	
68			222	<−Free Moment−>		90		
69	Beams 2 and 3							
70	--------------							
71		B	2	C	C	3	D	
72		----------	----------	----------	----------	----------	----------	
73		Ma		Mb	Ma		Mb	
74		−49		82	−126		159 kNm	
75			90	<−Free Moment−>		222		
76								
77	Moment Check			F		C		
78				−3.6E−15		3.6E−15		
79				O.K.		O.K.		
80			----------	----------	----------	----------	----------	
81								
82			Reactions kN (Positive Left to Right)					
83			----------	----------	----------	----------	----------	
84	Columns							
85	-------		7		4		1	
86			----------	----------	----------	----------	----------	
87	Top		15	Rb	0	Rb	−15 kN	
88	Bottom		−15	Ra	0	Ra	15 kN	
89	Beams 5 and 6							
90	--------------							
91			5	5	6	6		
92			----------	----------	----------	----------		
93			Ra	Rb	Ra	Rb		
94			230	214	98	82 kN		
95	Beams 2 and 3							
96	--------------							
97			2	2	3	3		
98			----------	----------	----------	----------		
99			Ra	Rb	Ra	Rb		
100			82	98	214	230 kN		

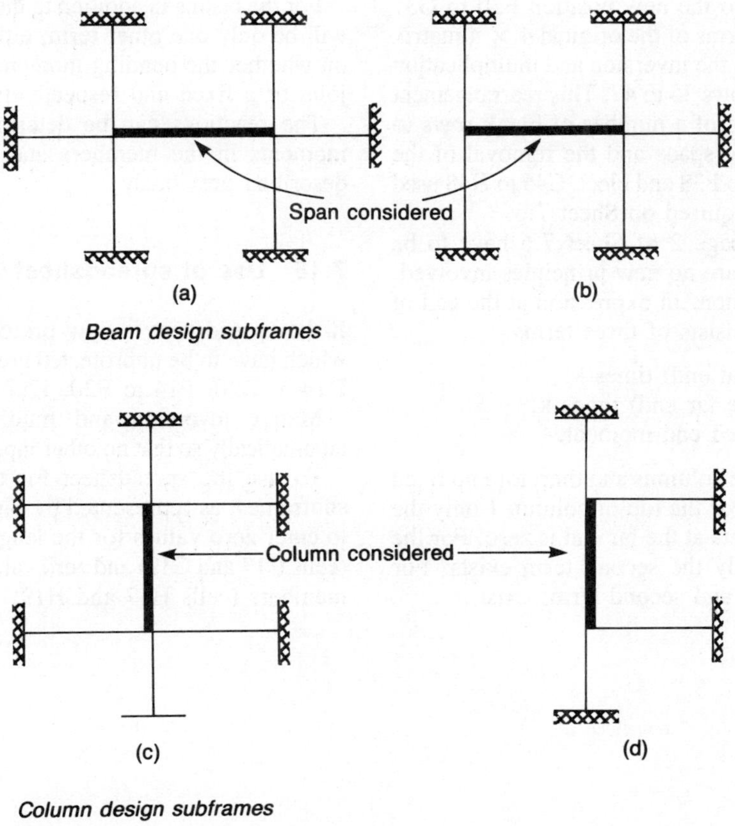

Figure 7.9 Subframes for design to BS 5950

$$\begin{bmatrix} 4k(1234) & 2k(4) \\ 2k(4) & 4k(4567) \end{bmatrix} \begin{Bmatrix} \theta_C \\ \theta_F \end{Bmatrix} = \begin{Bmatrix} +F3-F2 \\ -F5+F6 \end{Bmatrix}$$

where the terms in brackets represent the sum of the stiffnesses of the members framing into a particular joint. For example 4k(1234) represents $4(k_1+k_2+k_3+k_4)$. The symbol F has been used to replace $wL^2/12$ so that F6 represents $w_6L_6^2/12$.

The terms on the right-hand side of the matrix equation and the value of the stiffness of each member are known so that the values of θ's can be determined by first inverting the stiffness matrix as described by Eqns 7.1 to 7.3.

Once the values of the θ's have been determined the values of the moments and reactions at the ends of each member can be readily determined from the standard slope deflection equations. For example the moment M_{E5} representing the moment at E in beam 5 is given by:

$$M_{E5} = 2\theta_F k_5 - w_5L_5^2/12 \qquad [7.20]$$

and the moment M_{F4} by:

$$M_{F4} = 4\theta_F k_4 + 2\theta_C k_4 \qquad [7.21]$$

The reactions in each of the members can be determined from the end moments and the loading in the member using Eqn 7.6 and 7.7.

Spreadsheet layout (Sheet 7.6)

Although the layout of spreadsheet 7.6 is slightly different from that developed for Sheet 7.5 it will be found convenient to start with a copy of 7.5 and to work through the necessary changes.

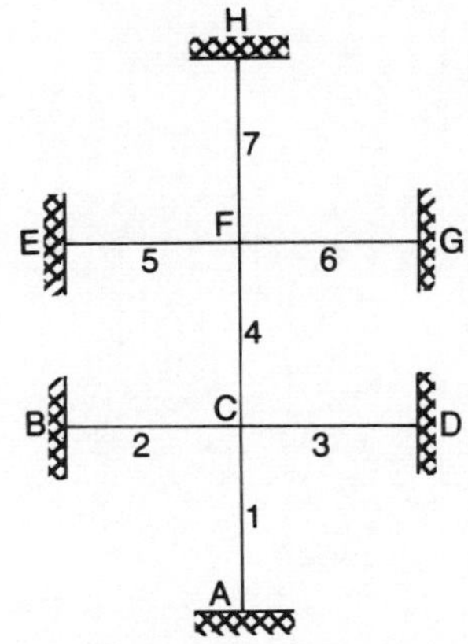

Figure 7.10 Subframe for column design

Member information has been reduced from eleven members to seven members grouping the columns and beams separately. There are now four beams so that the fixed end moments in column C have to be changed from six values to eight. On Sheet 7.5 there were four joints but now there are only two so that 'Sum of Moments' entry has to be reduced.

The theta matrix in block E38 to F39 of Sheet 7.5 should be changed in accordance with the matrix equation given above and then moved to the new position F30 to G31 of Sheet 7.6. The zero terms of the original 4 × 4 matrix can now be removed and the inversion and multiplication process repositioned to rows 35 to 47. This rearrangement will require the insertion of a number of blank rows to give some extra working space and the removal of the link between block K35 to P38 and block C45 to H48 used on Sheet 7.5 and not required on Sheet 7.6.

All the formulae of page 2 of Sheet 7.6 have to be changed although there are no new principles involved. In general the bending moment expression at the end of a particular member consists of three terms:

(1) 4 times (theta at that end) times k;
(2) 2 times (theta at the far end) times k;
(3) the value of the fixed end moment.

There are no loads on the columns and therefore no fixed end column moments. For the top of column 1 only the first term exists since theta at the far end is zero. For the bottom of column 1 only the second term exists. For column 4 both the first and second terms exist.

For the beams in addition to the fixed end moment there will be only one other term, either (1) or (2) depending on whether the bending moment is being calculated at a joint or a fixed end respectively.

The reactions can be determined from the bending moments in the members and any applied loading as described previously.

7.16 Use of spreadsheet 7.6

If the spreadsheet is now protected then the only cells which have to be unprotected are: C7 to C9, C14 to C20, D14 to D20, F14 to F20, H17 to H20.

Matrix inversion and multiplication is performed automatically so that no other input is required by the user.

To use the spreadsheet for the solution of external subframes, as represented by Fig. 7.9(d), it is necessary to enter zero values for the lengths of members 2 and 5 (cells C17 and C19) and zero values for the loads on these members (cells H17 and H19).

8 Single and continuous span beams

To simplify design procedures many handbooks of engineering contain formulae for the determination of the maximum bending moment and maximum deflection for single span beams loaded with simple or standard loading. Solutions for other standard forms of loading are also included in some texts but often the equations for the maximum deflection are not quoted because of their complexity. Solutions are also available for the redundant support reactions and bending moments for continuous beams but, for simplicity, only standard loading such as a point load acting at the centre section, or two point loads at the third or quarter points, or a uniformly distributed load over the full length are considered. Design involving non-standard loading must be solved from first principles.

8.1 Types of beams

The spreadsheets developed in this chapter are designed for the analysis of beams, either single span or continuous. Initially the procedure will be applied to simply supported beams carrying five point loads acting at any location and two uniformly distributed loads covering any portion of the span. The approach can easily be extended to include other forms of loading such as triangular loading. The value of the shear, bending moment and deflection are found at a number of sections along the span with very little input from the user.

Having illustrated the procedure using the simply supported beams it is then shown how the approach can be extended to determine the support reactions and bending moments for continuous beams of up to five spans.

The method could easily be extended to obtain solutions for rectangular or ridge portal frames subjected to various forms of loading and for different support conditions (Kleinlogel). However, to cover all the possible variations a large number of formulae would be required and the spreadsheet solution might have to be limited to symmetrical portals and to assume also that certain parameters are constant throughout.

8.2 Simply supported beams

Point loads. For a point load W acting at a distance a from the left support, Fig. 8.1, the left support reaction and the bending moment at any section x are given by:

$$R_1 = Wb/L \qquad [8.1]$$

$$M_{xx} = Wbx/lb/ - W[x-a] \qquad [8.2]$$

Using Macauley's method the deflected form of the beam due to bending can be expressed by the equation:

$$6EIy = Wx(L-a)(2aL-a^2-x^2)/L + W[x-a]^3 \qquad [8.3]$$

where the term in the square bracket is only evaluated when positive, i.e. when $x > a$.

Uniformly distributed load. The uniformly distributed load, of intensity w, shown in Fig. 8.2 covers only a portion of the span and for this case the reaction at the left support and the bending moment at any section are defined by:

$$R_1 = w(b-a)(1-(a+b)/2L) \qquad [8.4]$$

$$M_{xx} = R_1x - w[x-a]^2/2 + w[x-b]^2/2 \qquad [8.5]$$

where the square brackets are only evaluated when positive.

The equation representing the deflected form can be expressed, using Macauley brackets, as:

$$EIy = -R_1x^3/6 + w[x-a]^4/24 - w[x-b]^4/24 + Ax \qquad [8.6]$$

where A is a constant of integration given by:

$$A = R_1L^2/6 - w(L-a)^4/24L + w(L-b)^4/24L \qquad [8.7]$$

Discrete solution for deflections

In order that the deflections resulting from the application of different load systems can be combined it is convenient to apply Eqns 8.3 and 8.6 to a set of selected nodes along

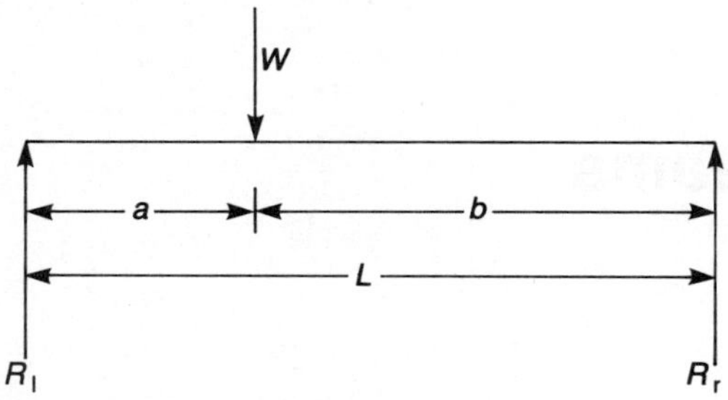

Figure 8.1 Point load

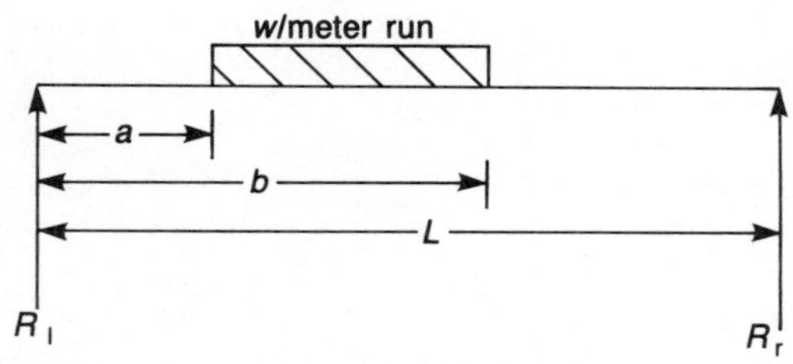

Figure 8.2 Uniformly distributed load

the beam and to facilitate this the beam is assumed to be divided into ten equal intervals, eleven nodes.

Defining the interval as u the deflection resulting at the ith node from the application of a point load at a distance a from the left support can be obtained from Eqn 8.3 by substituting ui for x:

$$y = (W/6EI)\{ui(L-a)(2aL-a^2-u^2i^2)/L + [ui-a]^3\} \quad [8.8]$$

where i is any nodal value between 0 and 11 and the square bracket is omitted if ui is $> a$.

Similarly, Eqn 8.6 for the uniformly distributed load can be re-written as:

$$EIy = -R_l(ui)^3/6 + w[ui-a]^4/24 - w[ui-b]^4/24 + Aui \quad [8.9]$$

Using this approach the deflection calculated at each node will be accurate but, since the position of the maximum deflection resulting from a particular loading may not be at a nodal point, the method will not necessarily predict the true maximum value. However with ten intervals the position and magnitude of the maximum deflection can be estimated with sufficient accuracy for most cases.

8.3 Spreadsheet layout for simply supported beams

Consider a simply supported beam which can be loaded with up to 5 point loads and two uniformly distributed loads acting on any part of the span.

Initial data

In addition to values of the span, modulus of elasticity and second moment of area entered in cells B7, D7 and F7, information is required about the location and magnitude of the applied loading. For the case being considered the magnitudes of the five loads are placed in cells D11 to H11 and their position relative to the left-hand support placed in D12 to H12. If less than five loads are active then zero values are inserted as appropriate in D11 to H11 but the spreadsheet requires that the loads entered must start in D11. Thus for three loads entries would be made to D11 to F11 and zero values entered in G11 and H11. Also, the lengths from the left support at which the loads act are entered in row 12 but these must be placed so that the smallest value is placed first and the remainder increasing progressively along the beam to the right. Residual values of position, left over from a previous run, do not affect the result as long as a zero value has been entered for the load. See block F11 to H12 of Sheet 8.1.

Allowance for up to two cases of uniformly distributed load has been included. Data for the first of these is entered in C18 to C20 which contains the magnitude and the start and end position of the load measured from the left support. Corresponding data for the second uniformly distributed load is entered in a similar manner into cells G18 to G20. Note that these two loadings can start and end at any position, not necessarily a node point, and can overlap each other.

Shear forces

Note that in what follows the calculation of the shear forces resulting from the action of any point loads is kept quite separate from the shear forces resulting from any uniformly distributed loads. The total shear can be obtained by simple addition of the two results.

The values of the shear force resulting from the action of the point loads are calculated in the block B26 to D32. The distances from the left support appear in the B column of this block and although these are copied from the information in row 12 it is necessary to allow for the fact that the number of active loads may vary from 0 to 5. With 5 loads 7 values should appear, two for the supports plus five for the loads. For no loads then no values should appear. This can be achieved by using:

B26: **0.0**
B27: **@IF(D11<>0,D12,0)**
B28: **@IF(E11<>0,E12,@IF(D11<>0,B7,0))**
B32: **@IF(B31=0,0,B7)**

the formula in B28 being repeated down the column to B31 but using appropriate relative CELL references. These formulae determine which value is to be output; either the value from row 12 or the span or zero. Values of the shear force are determined using:

C26: **0.0**
D26: **((D11*(B7−D12))+(E11*(B7−E12))+(F11*(B7−F12))+(G11*(B7−G12))+(H11*(B7−H12))/B7**
C27: **@IF(B27>0,D26," ---")**
D27: **@IF(B27=B7,0,@IF(B27>0#AND#B27<B7,(C27−D$11),"---"))**

and similarly for C28 to C32 and D28 to D31. Note that

Sheet 8.1 (Page 1)

```
1/A    B        C         D         E         F         G         H        I
 2                   Simply Supported Beams
 3                   ------------------------
 4 Initial Data
 5 --------------
 6 Span (m)             Mod.(E) N/mm2       I mm4
 7          20          2.0E+08             3.0E+06
 8
 9 PointLoads                   1         2         3         4         5
10 -------------      -----------------------------------------------------
11 Load (kn)                   10        10         0         0         0
12 Position (m)                 5        15         4         6         8
13
14 Uniformly Distributed Loads
15 --------------------------------
16 CASE 1                     CASE 2
17 ------                     ------
18 load           5 kn/m      load                3 kN/m
19 StartPos       0 m         StartPos           15 m
20 LastPos        5 m         LastPos            20 m
21
22 Results for Shear Force
23 ---------------------------
24            Point Loads            Distributed Loads              Dist to
25      Dist   Shear (kN)              Dist  Shear (kN)           Zero shear
26      0.00       0.00    10.00       0.00     23.75              for U.D.L
27      5.00      10.00     0.00       0.00     23.75                  0.00
28     15.00       0.00   -10.00       5.00     -1.25                  4.75
29     20.00     -10.00     0.00      15.00     -1.25                  0.00
30      0.00  ---        ---          20.00    -16.25                  0.00
31      0.00  ---        ---          20.00    -16.25                  0.00
32      0.00  ---        ---
33    -------------------------  -------------------------
34    Distributed Loads           Useful Extras
35      Dist  Shear (kN)          --------------
36      0.00     23.75              U D Loads
37      0.00     23.75                Dist  Shear
38      4.75      0.00                5.00    -1.25
39      5.00     -1.25               15.00    -1.25
40     15.00     -1.25               20.00   -16.25
41     20.00    -16.25                0.00
42     20.00    -16.25                0.00
43
44 Results for Bending Moment
45 ------------------------------
46            Point Loads            Distributed Loads
47      Dist  Bending (kNm)           Dist Bending (kNm)
48      0.00       0.00               0.00      0.00
49      5.00      50.00               0.00      0.00
50     15.00      50.00               4.75     56.41
51     20.00       0.00               5.00     56.25
52      0.00   ---                   15.00     43.75
53      0.00   ---                   20.00      0.00
54      0.00   ---                   20.00      0.00
```

Sheet 8.1 (Page 2)

59	B	C	D	E	F	G	H	I
60	Additional Calculations			Useful Extras				
61	------------------------------------							
62	Load1	25		Distributed Load Moments				
63	Load2	15		At Point Load Locations				
64	Dist1	2.5	No.	Dist	Moment (kNm)			
65	Dist2	17.5	1	5.00	56.25			
66			2	15.00	43.75			
67			0	---	---			
68			0	---	---			
69			0	---	---			
70								

71 Results for Point Load Deflection at 11 nodes.(mm)

72 --

73	Nodal Interval =		2 m				Total
74		1	2	3	4	5	Deflection
75	Node						
76	0	0.00	0.00	0.00	0.00	0.00	0.00
77	1	0.71	0.52	0.00	0.00	0.00	1.23
78	2	1.33	1.00	0.00	0.00	0.00	2.32
79	3	1.74	1.41	0.00	0.00	0.00	3.15
80	4	1.93	1.73	0.00	0.00	0.00	3.65
81	5	1.91	1.91	0.00	0.00	0.00	3.82
82	6	1.73	1.93	0.00	0.00	0.00	3.65
83	7	1.41	1.74	0.00	0.00	0.00	3.15
84	8	1.00	1.33	0.00	0.00	0.00	2.32
85	9	0.52	0.71	0.00	0.00	0.00	1.23
86	10	0.00	0.00	0.00	0.00	0.00	0.00
87							

88 Results for U.D.Load Deflection (mm)

89 --

90				Case1	Case2	
91		Reaction		21.875	1.875	
92	Constant of Integration =			319.010	121.093	
93				Total U D L		Overall
94		1	2	Deflection		Deflection
95	Node					
96	0	0.00	0.00	0.00		0.00
97	1	1.02	0.40	1.42		2.65
98	2	1.83	0.77	2.60		4.92
99	3	2.33	1.10	3.43		6.58
100	4	2.54	1.35	3.88		7.54
101	5	2.50	1.50	3.99		7.81
102	6	2.25	1.52	3.77		7.42
103	7	1.83	1.40	3.23		6.38
104	8	1.29	1.10	2.39		4.71
105	9	0.67	0.61	1.28		2.51
106	10	0.00	0.00	0.00		0.00

the term carrying the dollar sign in D27 will change to E$11 for cell D28 and similarly for the other cells D29 to D31. Cell D32 contains:

D32: **@IF(B32>0,0,"---")**

For the distributed loads the shear forces at selected sections are calculated in E26 to F31. The number of values output to the block E26 to E31 will vary with the number of UDLs (uniformly distributed loads) applied. For two distributed loads there are six values, two for the supports and the start and end position of each load. With one distributed load only four values should appear. Additionally it is convenient if the distances entered in E26 to E31 are in their correct ascending order and the formulae should allow for this. Required formulae are:

```
E26: 0.0
E27: @IF(C18=0#AND#F18<>0,F19,
     @IF(C18<>0#AND#F18=0,C19,
     @IF(C18=0#AND#F18=0,B7,
     @MIN(C19,F19))))
E28: @IF(C18=0#AND#F18<>0,F20,
     @IF(C18<>0#AND#F18=0,C20,
     @IF(C18=0#AND#F18=0,0,@IF(E27=C19,
     @MIN(F19,C20),@MIN(C19,F20)))))
E29: @IF(C18=0#AND#F18=0,0,
     @IF(C18=0#OR#F18=0,B7,
     @IF(E30=F20,@MAX(C20,F19),
     @MIN(C19,F20))))
E30: @IF(C18=0#OR#F18=0,0,
     @MAX(C20,F20))
E31: @IF(C18=0#OR#F18=0,0,B7)
```

Actual values of the shear force at the above locations are then calculated in F26 to F31 using:

```
F26: ((((C20-C19)*C18)
     *(((C20-C19)/2)+(B7-C20)))
     +((F20-F19)*F18)*(((F20-F19)/2)
     +(B7-F20)))/B7
F27: +F26
F28: @IF(C18=0#AND#F18=0," ---",
     (F27-((E28-E27)*@IF(E27=C19
     #AND#C18>0,C18,F18))))
F29: @IF(C18=0#AND#F18=0,
     "---",(F28)-((E29-E28)*@IF(E28=C20
     #OR#E28=F20),0,(C18+F18))))
F30: @IF(C18=0#OR#F18=0,"---",
     (F29)-((E30-E29)
     *@IF(E29=C20,C18,F18)))
F31: +F30
```

Since the load acting on the portion of the beam under consideration can be zero, or load 1 (C18), or load 2 (F18), or the sum of the loads (C18+F18), the calculations for shear in the above formulae must allow for these possible variations. The formulae first check if the load is zero and if not calculate the shear by subtracting the product of the load times the distance over which it acts from the shear value calculated at the previous section. These calculations yield the values of the shear forces at the supports and also at the start and end of the distributed loads. The location of zero shear has not yet been identified and since this is required for determining the maximum moment it is convenient to calculate this value, at this stage, in cells H27 to H31 using:

```
H27: @IF(F27<0#AND#F26>0,
     (((F26*(E27-E26))/(F26-F27))+E26),0)
```

and similarly for the other formulae but using the appropriate cells. In the above formula a test is first made to find the sections between which the shear force becomes negative and then uses a simple proportion equation to find the location within this section where the shear value becomes zero.

Once the position of zero shear force has been determined it can be included in a recalculated set of ordered distances along the beam. This has been done in B36 to B42 using:

```
B36: 0.0
B37: @IF(F27>=0#AND#F26>=0,E27,
     @IF(F27<=0#AND#F26<=0,E26,
     @SUM($H$27..$H$31)))
```

and similarly for the other cells using the appropriate cell changes. In these formulae a simple check is first made to determine if the relevant values are either both positive or both negative and if not the non zero value from the range H27 to H31 is inserted. Values of the shear force are entered alongside in C36 to C42 using:

```
C36: @IF(B36=@SUM($H$27..$H$31),0,F26)
C37: @IF(B37=@SUM($H$27..$H$31),0,
     @IF(B37>@SUM($H$27..$H$31),
     F26,F27))
```

and similarly for the other cells C38 to C41 using the appropriate changes. The formula in cell C42 is similar to that in cell C36 after making the necessary relative changes. Note that the @SUM term in the above represents the position of zero shear due to the distributed loading. The formulae in cells C37 to C41 allow for the three possibilities for the distance being considered. This distance can be less than the distance to zero shear, equal to it or greater than it. Having determined which case applies the value of the shear is either copied from the appropriate cell from the range F26 to F31 or equated to zero.

The graphical facility of the spreadsheet could be used at this stage to plot the shear force diagram for the uniformly distributed loading using B36 to B42 for the X axis and C36 to C42 for the Y axis. To obtain a graphical output of the shear force diagram for the point loads would require some additional spreadsheet formulae but, since the diagram is a series of rectangles and salient values have already been calculated, these formulae are not included.

Similarly the combined shear force diagram for both forms of loading has not been determined since this can be sketched from the values calculated above. To assist with this some useful extra information has been calculated in block E38 to F42. Since the sections loaded with point loads may not be the same as the start and end positions of the uniformly distributed loads it is convenient, when sketching the combined shear force, to know the value of the uniformly distributed shear at the point load sections. The values entered in E38 to E42 are the same as those in B27 to B31 and the remaining cells of the block contain formulae of the type:

```
F38: @IF(E38< =E$217,F$26,@IF((E38>E$27
     #AND#E38< =E$28),((F$27-F$28)
     *(E$28-E38)/(E$28-E$27))+F$28,
     @IF((E38>E$28#AND#E38< =E$29),
     ((F$28-F$29)*(E$29-E38)/(E$29-E$28))
     +F$29,@IF((E38>E$29
     #AND#E38< =E$30),((F$29-F$30)
     *(E$30-E38)/(E$30-E$29))
     +F$30,F$30))))
F39: @IF(E39=0," ",@IF(E39< =E$27,F$26,
     @IF((E39>E$27#AND#E39< =E$28),
     ((F$27-F$28)*(E$28-E39)/(E$28-E$27))
     +F$28,@IF((E39>E$28
     #AND#E39< =E$29),((F$28-F$29)
     *(E$29-E39)/(E$29-E$28))+F$29,
     @IF((E39>E$29#AND#E39< =E$30),
     ((F$29-F$30)*(E$30-E39)/(E$30-E$29))
     +F$30,F$30)))))
```

and similarly for the other cells but allowing for the appropriate changes. As usual those cell references containing the $ sign remain unchanged. Although the expression looks fairly complicated it is in fact a series of checks to find which two values of the previously calculated UDL shear straddle the particular value of the distance being considered and, having found them, applying the rules of simple proportion.

Bending moments

The bending moments at the sections carrying the point loads are calculated in block B48 to C54 with cells B48 to B54 being copied from B26 to B32. Moments are calculated using:

```
C48: 0.0
C49: @IF(B49=0," ---",C48+(B49-B48)
     *C27+0.00001)
C54: @IF(B54=0," ---",0)
```

The formulae in C50 to C53 are a repeat of the formula in C49 after allowing for appropriate cell changes. The value 0.00001 has been added to the formula in C49 to ensure that the computer calculation of the zero value is positive.

Bending moments for the distributed loads are calculated in the block E48 to F54 in which the range E48 to E54 is copied from B36 to B42 and cells F48 to F54 contain formulae of the type:

```
F48: 0.0
F49: @IF(E49> =0,((C37+C36)/2)
     *(E49-E48)+F48," ---")
F50: @IF(E50>0,((C38+C37)/2)
     *(E50-E49)+F49," ---")
```

and similarly for the remaining cells except that a small rounding value of 0.00001 is added to the calculation to ensure that zero values appear as positive rather than negative. For similar cosmetic reasons the additional = sign has been included in F49 to improve the output when there are no uniformly distributed loads.

The graphical facility could now be used to plot the bending moment diagrams resulting from the action of the point loads or from the action of the uniformly distributed loads. Since the salient values in these diagrams may be at different locations additional spreadsheet formulae would be required to draw the combined diagram. Since the combined bending moment at any location can readily be obtained by inspection the additional formulae are not included but for convenience some additional calculations are given in block D65 to F69.

These additional calculations represent the values of the bending moments resulting from uniformly distributed loads at the section where the point loads act. The formulae required to calculate these moments are quite lengthy and it is convenient to make some preliminary additional calculations. This has been done in C62 to C65 which contain:

```
C62: +C18*(C20-C19)
C63: +F18*(F20-F19)
C64: (C19+C20)/2
C65: (F19+F20)/2
```

A count of the number of active loads has been made in D65 to D69 using formulae of the type:

```
D65: @IF(D11>0,1,0)
D66: @IF(E11>0,2,0) etc.,
```

and the distance to each active load entered in E65 to E69 using:

```
E65: @IF(D65>0,B49," ---")
```

Note that, except for LOTUS 1-2-3 (Release 2), the formulae in D65 to D69 can be entered as:

```
D65: @IF(D11>0,1," ")
```

in order to output the null string if there is no point load.

The formulae in E66 to E69 are relatively similar. The actual bending moment at each location can now be determined using:

```
F65: @IF(D65=1,($C$36*E65)
       -@IF($C$20< =E65,$C$62
       *(E65-$C$64),@IF($C$20>E65
       #AND#$C$19<E65,$C$18*(E65-$C$19)
       *(E65-($C$19+E65)/2),0))
       -@IF($F$20< =E65,$C$63
       *(E65-$C$65),@IF($F$20>E65
       #AND#$F$19<E65,$F$18*(E65-$F$19)
       *(E65-($F$19+E65)/2),0)),"---")
```

and relatively similar formulae in cells F66 to F69. The formulae are lengthy but can easily be explained by noting that if the first cosmetic @IF is ignored the remainder consists of three parts:

```
(C36*E65) - @IF(@IF) - @IF(@IF)
```

The first term is the product of reaction at the support and the distance to the load position and, from this product, the influence of the uniformly distributed loads must be subtracted. Each @IF(@IF) term represents the effect of one of the uniformly distributed loads. A check is made if the distance E65 is greater than the 'last' position of the load being considered or between the 'start' and 'last' position and using this information the appropriate moment equation used.

Point load deflections

These have been calculated in rows 76 to 86 for the eleven nodes 0 to 10 using the nodal interval, represented by (B7/10), calculated in D73. The formulae for determining the deflections are based on Eqn 8.8 and appear on the spreadsheet in the form:

C76: **((B76*$D$73/$B$7)*(($D$11*($B$7−$D$12)*((2*D$12*$B$7)−(D$12^2)−((B76*$D$73)^2)))−@IF(B76*$D$73)>$D$12,$D$11*(($D$12−(B76*$D$73))^3),0))*(1000^4)/(6*$D$7*$F$7)**

and similarly for the other formulae in cells C77 to C86 except that B76 changes progressively to B77, B78 ... etc.

The formula for D76 is similar:

D76: **((B76*$D$73/$B$7)*(($E$11*($B$7−$E$12)*((2*E$12*$B$7)−(E$12^2)−((B76*$D$73)^2)))−@IF(B76*$D$73)>$E$12,$E$11*(($E$12−(B76*$D$73))^3),0))*(1000^4)/(6*$D$7*$F$7)**

except that references to cells D11 and D12 must be changed to E11 and E12. Similar changes must be made for the other columns, i.e. the substitution of (F11, F12), (G11, G12) and (H11, H12) in cells E76 to G76 respectively. Once the required changes have been made to the formulae in row 76 the formulae in the other rows can be obtained by copying. Finally the total deflection at each node is determined in H76 to H86.

Uniformly distributed load deflections

Prior to calculating the deflections it is convenient to determine the values of the reactions resulting from each of the two loads and the constant of integration as defined by Eqn 8.7. The reaction and constant of integration for the first uniformly distributed load (Case 1) are calculated in E91 and E92 and for Case 2 loading in F91 and F92.

E91: **+C18*(C20−C19)*((2*B7)−C19−C20)/(2*B7)**

E92: **((E91*4*(B7^3))−(C18*((B7−C19)^4))+(C18*((B7−C20)^4)))/(24*B7)**

and similarly for F91 and F92 but using F18 to F20 to replace C18 to C20 and F91 to replace E91.

The deflections at each node are calculated in rows 96 to 106 using formulae, based on equation 8.9, of the type:

C96: **(((−$E$91*((B96*$D$73)^3))/6)+($E$92*(B96*$D$73))+(@IF(B96*$D$73>=$C$19,$C$18*(((B96+$D$73)−$C$19)^4),0)−@IF(B96*$D$73>$C$20,$C$18*(((B96*$D$73)−$C$20)^4),0))/24)*((1000^4)/($D$7*$F$7))**

and similarly for the other cells of column C but using B97, B98 ... etc, progressively down the column instead of B96. A similar formula is also used for Case 2 deflections in column D except that: E91 and E92 would be replaced by F91 and F92 and C18 to C20 would be replaced by F18 to F20.

The total uniformly distributed load deflection has been entered in column E and the overall deflection due to all active loading in column G. Using the graphical facility the user can now plot graphs of the variation of the deflection along the span for either Case 1, Case 2 or the combined loading and thus estimate the magnitude and position of the maximum deflection. Such an estimate can also be carried out visually.

8.4 Use of spreadsheet 8.1

If the spreadsheet is now protected then the only cells which have to be unprotected so as to allow entry by the user are: B7, D7, F7, D11 to H11, D12 to H12, C18 to C20, and F18 to F20.

It has been assumed that any zero entries in the range D11 to H12 appear at the end of the range and not at the beginning so that, for example, with only two point loads the load values would be entered in D11 and D12. Similarly for C18 and F18, if there is only one distributed load then the value of this load would be placed in C18 and not F18. Also the positions entered in row 12 must start with the one nearest to the left support and increase progressively along the beam to the right.

8.5 Single span fixed end beams

The method given above for simply supported beams can be readily adapted to solve for beams with fixed ends and it is possible to use one spreadsheet for both cases.

Equations for the fixed end moments for a single point load positioned as shown in Fig. 8.1 are:

$$\text{Left-hand moment} = -Wab^2/L^2 \qquad [8.10]$$

$$\text{Right-hand moment} = Wa^2b/L^2 \qquad [8.11]$$

For a uniformly distributed load over part of the span as illustrated in Fig. 8.2 the fixed end moments are given by:

$$\text{Left-hand moment} = -w[c^3(4L-3c) - d^3(4L-3d)]/12L^2 \qquad [8.12]$$

$$\text{Right-hand moment} = w[b^3(4L-3b) - a^3(4L-3a)]/12L^2 \qquad [8.13]$$

where $c = (L-a)$ and $d = (L-b)$. The sign convention used is clockwise positive.

The end shears resulting from the moments at the fixed ends can be expressed, see Fig. 8.3, as:

$$R_1 = -(M_1 + M_2)/L \qquad [8.14]$$

$$R_2 = +(M_1 + M_2)L \qquad [8.15]$$

These equate to zero for symmetrical loading since M_1 and M_2 would be equal and opposite in value.

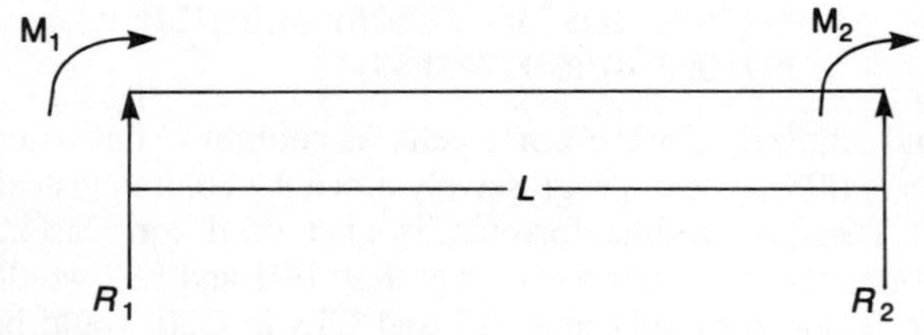

Figure 8.3 Sign convention, fixed beam

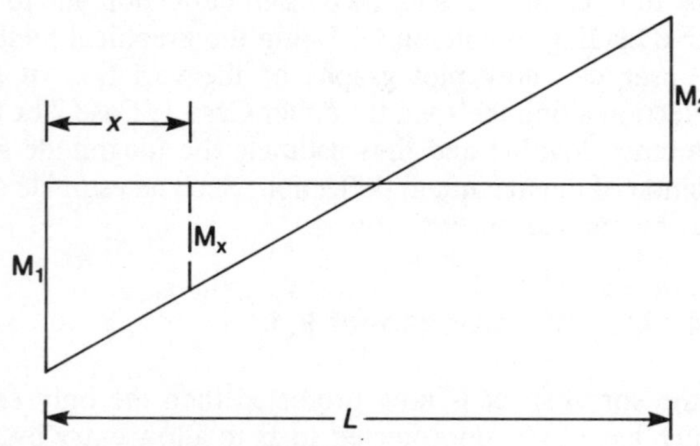

Figure 8.4 Additional moment due to fixity

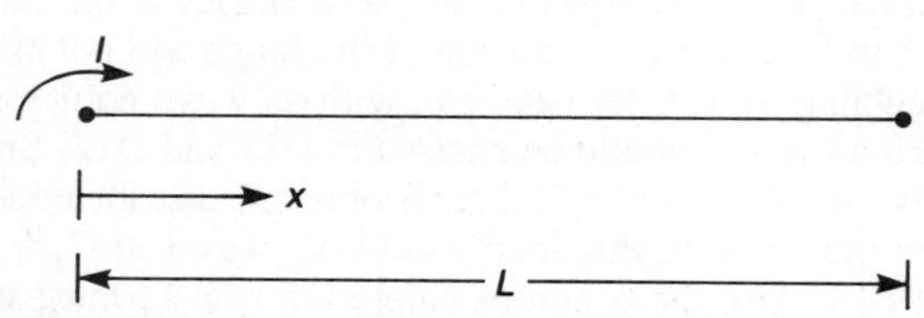

Figure 8.5 Unit moment on simply supported beam

The additional fixed-end moment at any section x, Fig. 8.4, is given by:

$$M_{xx} = M_1 - x(M_2+M_1)/L \qquad [8.16]$$

and this moment must be added algebraically to the simply supported moment resulting from the load.

To obtain the deflection for the fixed ended beam it is convenient to consider the effects of unit moment applied to one end of a simply supported beam, Fig. 8.5. For unit moment applied at the left-hand support the deflection at any section is given by:

$$\delta_x = (L-x)(2L-x)x/6EIL \qquad [8.17]$$

and for unit moment applied to the right-hand support:

$$\delta_x = (x^2/L^2 - 1)Lx/6EI \qquad [8.18]$$

where x is measured from the left-hand support. Using the deflections resulting from unit moment the actual deflection due to the fixing moments can be determined by multiplication. The overall deflection can then be found by combining the deflection due to the loading system with those due to the end moments. Note that for obtaining discrete deflections at particular nodes the value of x in Eqns 8.17 and 8.18 would be changed to ui, where u represents the nodal interval and i the nodal number.

8.6 Spreadsheet layout for fixed beams

The layout used in Sheet 8.1, for simply supported beams, can be modified so as to allow for the effects of end fixity and in what follows only the modifications to Sheet 8.1 will be described.

Fixed-end moments

The fixed-end moments resulting from the point loads are calculated in rows 13 and 14 using formulae based on Eqns 8.10 and 8.11. Typically these formulae are represented by:

```
D13: -(D11*D12*(($B$7-D12)^2))/($B$7^2)
D14: +(D11*(D12^2)*($B$7-D12))/($B$7^2)
```

and the totals collected in cells I13 and I14.

Similarly using Eqns 8.12 and 8.13 the fixed end moments for the distributed loads are calculated in rows 23 and 24. Typically:

```
C23: -(C20/(12*($B$7^2)))*((($B$7-C21)^3)
     *(4*$B$7-3
     *($B$7-C21))-(($B$7-C22)^3)
     *(4*$B$7-3*($B$7-C22)))
C24: +(C20/(12*($B$7^2)))*((C22^3)
     *(4*$B$7-3*C22)-(C21^3)
     *(4*$B$7-3*C21))
```

with the total fixed-end moments collected in H23 and H24.

Shear forces

For the point loads there are two values of the shear at each support and at each load point and these have been calculated in the block E30 to F36. Apart from the formula in F30, which is based on Eqn 8.14 the other formulae are similar to those used previously in the block C26 to D32 of Sheet 8.1 and typically:

```
E30: 0.0
F30: +D30-($I$14+$I$13)/B7
E31: @IF(B31>0,F30," ---")
F31: @IF(B31=B$7,0,@IF(B31>0
     #AND#B31<B$7,
     (E31-D$11)," ---")
```

and similarly for the lower cells. Note that the last term with the dollar sign in F31 will change to E$11 for cell F32 and similarly for the other cells. Cell F36 contains:

```
F36: @IF(B36>0,0," ---")
```

The shear force in the fixed-end beam due to distributed load is calculated in the range H41 to H47 using formulae of the type:

```
H41: +C41-($H$24+$H$23)/$B$7
H42: @IF(B42>=0,C42-($H$24
     +$H$23)/$B$7," ---")
```

and similarly for the cells H43 to H47 but without the = sign. The above formulae are based on Eqn 8.16.

Sheet 8.2 (Page 1)

1/A		B	C	D	E	F	G	H	I
2				Single Span Fixed Beams					
3				------------------------					
4	Initial Data								
5	------------								
6	Span (m)			Mod.(E) N/mm2		I mm4			
7		**9**		**2.0E+05**		**1.4E+08**			
8									
9	PointLoads			1	2	3	4	5	
10	------------			------------	------------	------------	------------	------------	------------
11	Load (kn)			**10**	**8**	**6**	**4**	**2**	
12	Position (m)			**1**	**3**	**4**	**6**	**8**	Total
13	Fix-MomentLt			-7.90	-10.67	-7.41	-2.67	-0.20	-28.84
14	Fix-MomentRt			0.99	5.33	5.93	5.33	1.58	19.16
15									
16	Uniformly Distributed Loads								
17	------------------------------								
18	CASE 1				CASE 2				
19	------				------				
20	load		**10** kn/m		load	**3** kN/m			
21	StartPos		**0** m		StartPos	**1** m			
22	LastPos		**6** m		LastPos	**5** m		Total	
23	Fix-MomLt		-60.00			-14.22		-74.22	
24	Fix-MomRt		40.00			8.00		48.00	
25									
26	Results for Shear Force								
27	------------------------						Distributed		Location
28		Point Loads			Fixed Ends		Loads		of
29		Dist	Shear (kN)		Point Load Shear		Dist	Shear	Zero
30		0.00	0.00	19.11	0.00	20.19	0.00	48.00	Shear
31		1.00	19.11	9.11	20.19	10.19	0.00	48.00	0.00
32		3.00	9.11	1.11	10.19	2.19	1.00	38.00	0.00
33		4.00	1.11	-4.89	2.19	-3.81	5.00	-14.00	3.92
34		6.00	-4.89	-8.89	-3.81	-7.81	6.00	-17.00	0.00
35		8.00	-8.89	-10.89	-7.81	-9.81	9.00	-17.00	0.00
36		9.00	-10.89	0.00	-9.81	0.00			
37		------------	------------	------------	------------	------------	------------	------------	------------
38							Fixed Ends		
39		Distributed Loads			Useful Extras		Resulting Shear		
40		Dist	Shear (kN)		------------		Uniform Load		
41		0.00	48.00		U D Loads			50.91	
42		0.00	48.00		Dist	Shear		50.91	
43		1.00	38.00		1.00	38.00		40.91	
44		3.92	0.00		3.00	12.00		2.91	
45		5.00	-14.00		4.00	-1.00		-11.09	
46		6.00	-17.00		6.00	-17.00		-14.09	
47		9.00	-17.00		8.00	-17.00		-14.09	

Sheet 8.2 (Page 2)

```
52     B          C          D          E          F          G           H          I
53 Results for Bending Moment
54 ------------------------------                         Fixed Ends
55           Pt. Loads      Distributed Loads             Resulting Bending
56        Dist  Bending         Dist Bending (Knm)        PtLoads        U D.
57        0.00       0.00      0.00     0.00               -28.84      -74.22
58        1.00      19.11      0.00     0.00                -8.65      -74.22
59        3.00      37.33      1.00    43.00                10.64      -28.31
60        4.00      38.44      3.92    98.54                10.68       32.83
61        6.00      28.67      5.00    91.00                 1.98       19.92
62        8.00      10.89      6.00    75.50               -15.80        4.19
63        9.00       0.00      9.00    24.50               -27.76      -40.98
64
65 Additional Calculations          Useful Extras
66 ---------------------------------------
67 Load1                 60            Distributed Load Bending Moments
68 Load2                 12            At Point Load Locations
69 Dist1                  3       No.       Dist Moment (kNm)
70 Dist2                  3         1       1.00    43.00
71                                  2       3.00    93.00
72                                  3       4.00    98.50
73                                  4       6.00    72.00
74                                  5       8.00    24.00
75
76 Results for Point Load Deflection at 11 Nodes (mm)
77 ------------------------------------------------------------------
78 Nodal Interval =                0.9 m                             Total Point Load
79                       1          2          3          4        5  Deflection
80         Node  ------------------------------------------------------------------
81            0       0.00       0.00       0.00       0.00     0.00       0.00
82            1       0.77       1.26       0.99       0.51     0.09       3.62
83            2       1.34       2.39       1.88       0.98     0.18       6.78
84            3       1.68       3.23       2.61       1.39     0.26       9.17
85            4       1.82       3.67       3.07       1.69     0.32      10.57
86            5       1.78       3.70       3.20       1.85     0.36      10.88
87            6       1.60       3.37       2.97       1.84     0.36      10.14
88            7       1.30       2.77       2.47       1.62     0.34       8.50
89            8       0.91       1.96       1.76       1.19     0.27       6.10
90            9       0.47       1.02       0.92       0.63     0.15       3.19
91           10       0.00       0.00       0.00       0.00     0.00       0.00
```

Bending moments

The bending moments at the salient points for the system of point loads are calculated in the range G57 to G63 and for the distributed loads in H57 to H63. Note that the values in these two columns are not calculated at the same sections and therefore cannot be added together without further calculation. Such calculation has not been included since simple sketches of the bending moments due to the two load systems will readily yield the required information. Examples of the formulae used in the block G57 to H63 are:

G57: **+I13**
H57: **+H23**
G58: **@IF(B58>=,+C58+$I$13−(B58−B57)**
***($I$14+$I$13)/$B$7," ---")**
H58: **@IF(D58>=,+E58+$H$23−(D58−D57)**
***($H$24+$H$23)/$B$7," ---")**

and similarly for the other formulae in the range except that the = sign is omitted.

Sheet 8.2 (Page 3)

```
 97      B          C          D          E          F          G          H          I
 98
 99 Results for U.D.Load Deflection (mm)
100 ------------------------------------------------------
101                                     Case1      Case2
102             Reaction                   40          8
103 Constant of Integration =             240    54.6666
104
105                              Total U D L              Overall
106                      1       2 Deflection         Deflection (mm)
107         Node     ------------------------     Simply Supported
108            0       0.00       0.00       0.00             0.00
109            1       7.55       1.72       9.27            12.89
110            2      14.20       3.24      17.43            24.21
111            3      19.25       4.37      23.62            32.79
112            4      22.25       5.01      27.26            37.83
113            5      22.98       5.12      28.09            38.97
114            6      21.45       4.72      26.17            36.31
115            7      17.91       3.90      21.81            30.31
116            8      12.81       2.78      15.58            21.69
117            9       6.66       1.44       8.10            11.29
118           10       0.00       0.00       0.00             0.00
119   -------------------------------------------------------------------
120
121             Fixed End Moment Deflections                      Overall
122                             Point        U.D                  Deflection
123              Distance       Loads      Loads      Total       Fixed Beam
124         Node   ------------------------------------------    ----------
125            0       0.00       0.00       0.00       0.00         0.00
126            1       0.90      -3.29      -8.41     -11.70         1.19
127            2       1.80      -5.78     -14.75     -20.53         3.68
128            3       2.70      -7.49     -19.09     -26.58         6.21
129            4       3.60      -8.44     -21.52     -29.96         7.87
130            5       4.50      -8.68     -22.10     -30.78         8.19
131            6       5.40      -8.22     -20.91     -29.13         7.18
132            7       6.30      -7.09     -18.03     -25.13         5.18
133            8       7.20      -5.33     -13.54     -18.87         2.82
134            9       8.10      -2.96      -7.50     -10.46         0.84
135           10       9.00       0.00       0.00       0.00         0.00
     ------------------------------------------------------------------
```

Deflections

The deflections due to moments, of magnitude equal to the fixed-end moments, applied to a simply supported beam can be obtained by multiplying the deflections resulting from unit moments applied at the ends by the value of the fixed-end moments. Deflections for end unit moments are given by Eqns 8.17 and 8.18 and the value of the fixed-end moments for the two loadings have been calculated previously. The required spreadsheet formulae for the deflections due to moments equal to the fixed-end moments are calculated in cells D125 to D135, for the point loads, and cells E125 to D135 for the distributed loads. For convenience the nodal points given in B125 to B135 are converted to distances in C125 to C135. Typical formulae used are:

```
C125: (B125*$D$78)
D125: ((($I$13*$D$78*B125)*($B$7 – ($D$78*
      B125))*(2*$B$7 – (B125*$D$78)))
      +((($I$14*$D$78*B125*$B$7)
      *(((($D$78*B125)/$B$7)^2) – 1)))*$B$7)
      *(1000^4)/(6*$D$7*$F$7*$B$7)
```

E125: ((($H$23*$D$78*B125)*($B$7 – ($D$78 *B125))*(2*$B$7 – (B125*$D$78))) +((($H$24*$D$78*B125*$B$7) *(((($D$78*B125)/$B$7)^2) – 1)))*$B$7) *(1000^4)/(6*$D$7*$F$7*$B$7)

and similarly for the lower rows.

The overall deflection for the beam with fixed ends is the sum of the deflections due to the loading (G108 to G118) and those due to the fixed end moments (F125 to F135). This addition has been carried out in the range H125 to H135 and if required the user could use the graphical facility to plot a graph of deflection against distance from the left support. Such a graph would indicate the approximate magnitude and position of the maximum deflection. However a reasonable estimate of these values can be determined by inspection.

8.7 Use of spreadsheet 8.2

If the spreadsheet is now protected then the only cells which have to be unprotected to allow entry by the user are: B7, D7, F7, D11 to H11, D12 to H12, C20 to C22, and F20 to F22.

Spreadsheet 8.2 is similar in use to Sheet 8.1 and could be used instead of Sheet 8.1 since the calculations used for a simply supported beam have not been changed. The same restrictions exist and it has been assumed that any zero entries in the range D11 to H12 appear at the end of the range and not at the beginning so that, for example, with only two point loads the load values would be entered in D11 and D12. Similarly for C20 and F20, if there is only one distributed load then the value of this load would be placed in C20 and not F20. Also the positions entered in row 22 must start with the one nearest to the left support and increase progressively along the beam to the right.

8.8 Continuous beams

For this example a continuous beam of up to five spans is considered. Each span can be loaded with five point loads acting at any specified positions and two uniformly distributed loads of any length. In practice one of these uniformly distributed loads could represent the self weight of the beam and the other an applied uniform load. If either end of the beam overhangs the support then the known moment resulting from this overhang is included in the analysis.

No new principles are involved for the development of the spreadsheet for this case and therefore only the procedural differences are explained. The rotation at each support is determined by multiplying the support moment vector by the inverse of the stiffness matrix, as shown previously in Chapter 7. Once the rotations are known the moments can be determined from the slope deflection equations and the reactions calculated from these moments. In Chapter 7 the matrix to be inverted had a maximum size of 4 × 4 but for a continuous beam with up to 5 spans this matrix has to be increased to a 6 × 6.

8.9 Spreadsheet layout for continuous beams

Initial data. The initial data for all spans has been gathered together on the first page. The span length, modulus of elasticity and second moment of inertia of each span is entered in rows 7 to 11. The example used on Sheet 8.3 refers to a three-span continuous beam with an overhang at the right support as shown in Fig. 8.6.

Since there are only three spans the entries for the span lengths of spans 4 and 5 are zero. In general if there are less than five spans then information about the spans must start from span 1 and proceed to the next span in sequence. Additionally any residual entries, remaining from a previous problem, in cells which relate to spans of zero length do not affect the result so that there is no requirement to enter a zero value in these cells. For example the modulus is given as 1 in cells E10 and E11 but since the span length for these rows is zero these values for the modulus do not affect the results. Similarly with the information left over in rows 28 to 36, since this information refers to spans of zero length.

If the real values of E and I are entered in columns E and G then the true values of the support rotations will be calculated but if relative values are used then the rotations will be in the correct ratio to each other but will not be absolute values. The values of the reactions and bending moments will be correct for either case. In the example shown on Sheet 8.3 the value of E has been entered as 1 for all spans and the values of I as 2:6:3 representing the relative values of the second moment of area of each span.

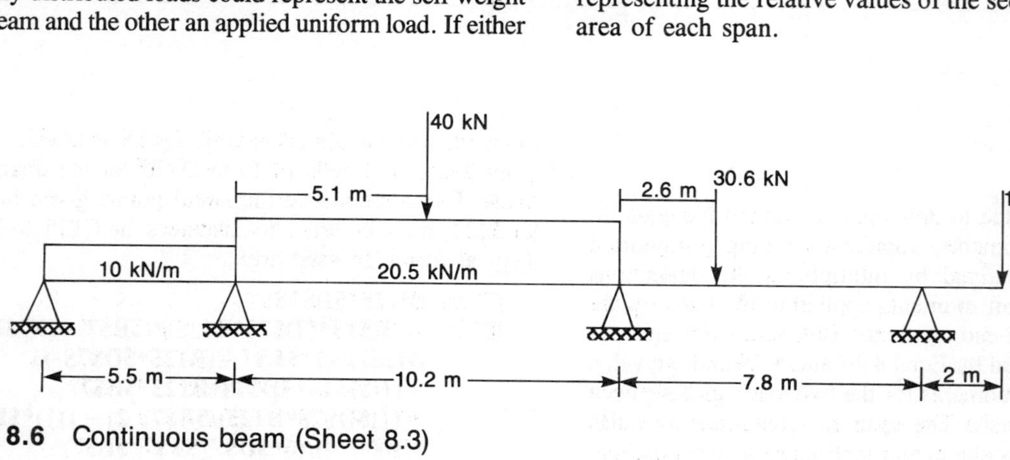

Figure 8.6 Continuous beam (Sheet 8.3)

Sheet 8.3 (Page 1)

1/A	B	C	D	E	F	G	H	I
2			Continuous Beams of up to 5 SPANS					
3			------------------------------------					
4 Initial Data								
5 -------------								
6		m		E kN/m2		I m4		
7 Span 1		5.5		1.0E+00		2.0E+00		
8 Span 2		10.2		1.0E+00		6.0E+00		
9 Span 3		7.8		1.0E+00		3.0E+00		
10 Span 4		0		1.0E+00		1.5E+00		
11 Span 5		0		1.0E+00		5.0E+00		
12			Loads on Span 1					
13 Load (kn)			0	0	0	0	0	
14 Position (m)			3	3	4	6	8	
15 UDLoad 1	(kNm)		10	StartPos	0	LastPos	5.5	m
16 UDLoad 2	(kNm)		0	StartPos	1	LastPos	5	m
17			Loads on Span 2					
18 Load (kn)			40	0	0	0	0	
19 Position (m)			5.1	3	4	6	8	
20 UDLoad 1	(kNm)		20.5	StartPos	0	LastPos	10.2	
21 UDLoad 2	(kNm)		0	StartPos	1	LastPos	5	
22			Loads on Span 3					
23 Load (kn)			30.6	0	0	0	0	
24 Position (m)			2.6	3	4	6	8	
25 UDLoad 1	(kNm)		0	StartPos	0	LastPos	9	
26 UDLoad 2	(kNm)		0	StartPos	1	LastPos	5	
27			Loads on Span 4					
28 Load (kn)			0	0	0	0	0	
29 Position (m)			4.5	3	4	6	8	
30 UDLoad 1	(kNm)		0	StartPos	0	LastPos	9	
31 UDLoad 2	(kNm)		0	StartPos	1	LastPos	5	
32			Loads on Span 5					
33 Load (kn)			0	0	0	0	0	
34 Position (m)			4.5	3	4	6	8	
35 UDLoad 1	(kNm)		2	StartPos	0	LastPos	50	
36 UDLoad 2	(kNm)		0	StartPos	1	LastPos	5	
37								
38			Boundary Conditions					
39			--------------------					
40		Simply Supported = 0, Fixed = 1, Overhang = 2						
41 Span 1		Left Support =		0 Right =		0		
42			(Overhang in kNm)		0		0	
43 Span 2				Right Support =		0		
44			(Overhang Moment in kNm if Any		=		0	
45 Span 3				Right Support =		2		
46			(Overhang Moment in kNm if Any		=		28.4	
47 Span 4				Right Support =		0		
48			(Overhang Moment in kNm if Any		=		0	
49 Span 5				Right Support =		0		
50			(Overhang Moment in kNm if Any		=		20	

Information about the loading on each active span has been entered in rows 13 to 36 as appropriate. These entries are similar to those explained for the single span beam described in Sheets 8.1 and 8.2.

Boundary conditions. In order to allow for different support conditions the numerals 0, 1 and 2 are used to represent the three support conditions simply supported, fixed and overhanging respectively. For the first span it is necessary to enter the support condition at both the left and right supports but for any remaining spans only the right support condition must be defined. These entries are made in rows 41, 43, 45, 47, and 49. If the support overhangs then the value of the resulting bending moment is required and this would be placed in rows 42, 44, 46, 48 or 50. In Fig. 8.4 there is an overhang at the fourth

Sheet 8.3 (Page 2)

54	B	C	D	E	F	G	H	I
55			Member Stiffness (kNm)					
56			------------------------					
57		1	2	3	4	5		
58		3.6E-01	5.9E-01	3.8E-01	0.0E+00	0.0E+00		
59	--							
60	SPAN 1 Fixed End Moments and Applied Load Reactions							
61	--							
62	Load No	1	2	3	4	5	Total	
63	Fix-MomLt	0.00	0.00	0.00	0.00	0.00	0.00	
64	Fix-MomRt	0.00	0.00	0.00	0.00	0.00	0.00	
65		Case 1			Case 2		Total	
66	Fix-MomLt	-25.21			0.00		-25.21	
67	Fix-MomRt	25.21			0.00		25.21	
68				Left	Right			
69	Point Load Reactions			0	0			
70	Uniform Load Reactions			27.5	27.5			
71	--							
72	SPAN 2 Fixed End Moments and Applied Load Reactions							
73	--							
74	Load No	1	2	3	4	5	Total	
75	Fix-MomLt	-51.00	0.00	0.00	0.00	0.00	-51.00	
76	Fix-MomRt	51.00	0.00	0.00	0.00	0.00	51.00	
77		Case 1			Case 2		Total	
78	Fix-MomLt	-177.74			0.00		-177.74	
79	Fix-MomRt	177.74			0.00		177.74	
80				Left	Right			
81	Point Load Reactions			20	20			
82	Uniform Load Reactions			104.55	104.55			
83	--							
84	SPAN 3 Fixed End Moments and Applied Load Reactions							
85	--							
86	Load No	1	2	3	4	5	Total	
87	Fix-MomLt	-35.36	0.00	0.00	0.00	0.00	-35.36	
88	Fix-MomRt	17.68	0.00	0.00	0.00	0.00	17.68	
89		Case 1			Case 2		Total	
90	Fix-MomLt	0.00			0.00		0.00	
91	Fix-MomRt	0.00			0.00		0.00	
92				Left	Right			
93	Point Load Reactions			20.4	10.2			
94	Uniform Load Reactions			0	0			
95	--							
96	SPAN 4 Fixed End Moments and Applied Load Reactions							
97	--							
98	Load No	1	2	3	4	5	Total	
99	Fix-MomLt	0.00	0.00	0.00	0.00	0.00	0.00	
100	Fix-MomRt	0.00	0.00	0.00	0.00	0.00	0.00	
101		Case 1			Case 2		Total	
102	Fix-MomLt	0.00			0.00		0.00	
103	Fix-MomRt	0.00			0.00		0.00	
104				Left	Right			
105	Point Load Reactions			0	0			
106	Uniform Load Reactions			0	0			
107	--							

support with a clockwise moment of 28.4 kN m. On the spreadsheet a value of 2 has been entered in G45 and a value of 28.4 in H46.

Member stiffness. Values of the member stiffnesses are not required until the stiffness matrix is set up but for convenience the stiffness of each span, defined as EI/L, is calculated in the cells C58 to G58 using formulae of the type:

C58: @IF(C7>0,(E7*G7/C7),0)

Sheet 8.3 (Page 3)

110	B	C	D	E	F	G	H	I
111	SPAN 5 Fixed End Moments and Applied Load Reactions							
112	--							
113	Load No	1	2	3	4	5	Total	
114	Fix–MomLt	0.00	0.00	0.00	0.00	0.00	0.00	
115	Fix–MomRt	0.00	0.00	0.00	0.00	0.00	0.00	
116		Case 1			Case 2		Total	
117	Fix–MomLt	0.00			0.00		0.00	
118	Fix–MomRt	0.00			0.00		0.00	
119				Left	Right			
120	Point Load Reactions			0	0			
121	Uniform Load Reactions			0	0			
122								
123								
124			Reactions, Moments and Thetas at Supports					
125			------------------------------------					
126			Reactions	Moments	Theta/EI			
127			kN	kNm	(with EI in kN m2)			
128	Span1	Left	3.87	0.00	−2.5E+01			
129		Right	51.13	129.97	8.4E+01			
130	Span2	Left	124.73	−129.97	8.4E+01			
131		Right	124.37	128.09	−8.5E+01			
132	Span3	Left	33.18	−128.09	−8.5E+01			
133		Right	−2.58	28.40	4.9E+01			
134	Span4	Left	0.00	0.00	4.9E+01			
135		Right	0.00	0.00	0.0E+00			
136	Span5	Left	0.00	0.00	0.0E+00			
137		Right	0.00	0.00	0.0E+00			

Fixed-end moments and applied load reactions

The fixed-end moments and the simply supported reactions resulting from the applied load for each active span are calculated in rows 62 to 121 using formulae based on Eqns 8.1, 8.2, 8.4, 8.5 and 8.10 to 8.13. The fixed-end moments are calculated in a simlar manner to that explained for Sheet 8.2, D13, D14 for point loads and C23, C24 for uniformly distributed loads. The applied load reactions are given by:

E69: **@IF(C7>0,((D13*(C7−D14))+(E13 *(C7−E14))+(F13*(C7−F14))+(G13 *(C7−G14))+(H13*(C7−H14))/C7,0)**

for point loads and:

E70: **@IF(C7>0,(((H15−F15)*D15) *((H15−F15)/2)+(C7−H15))) +((H16−F16)*D16)*(((H16−F16)/2) +(C7−H16)))/C7,0)**

for uniformly distributed loads.

Stiffness matrix

Prior to setting up the stiffness matrix a count is made of the number of active support rotations and the number of active spans as defined by the input on the first page of Sheet 8.3. This is achieved by placing formulae in cells L4 to L9 for the rotations and N4 to N8 for the spans. Typically:

L4: **@IF(E41< >1,1,0)**
N4: **@IF(C7>0,1,0)**

so that a 1 is output if the rotation or span is active and 0 if they are not. Then by summing L4 to L9 the number of active rotations can be found and by summing N4 to N8 the number of active spans. Cells Q4 and Q5 have been used for these summations.

The stiffness matrix, which can be 6 × 6, is housed in the block K11 to P16 and the required cell formulae are:

K11: **@IF(L4=1,4*C58,0)**
L11: **@IF(L4=1#AND#L5=1,2*C58,0)**
K12: **@IF(L4=1#AND#L5=1,2*C58,0)**
L12: **@IF(L5=1,4*(C58+@IF(L6=1,D58,0)),0)**
M12: **@IF(L5=1#AND#L6=1,2*D58,0)**
L13: **@IF(L6=1,2*D58,0)**
M13: **@IF(L6=1,4*(D58+@IF(L7=1,E58,0)),0)**
N13: **@IF(L6=1#AND#L7=1,2*E58,0)**
M14: **@IF(L7=1,2*E58,0)**
N14: **@IF(L7=1,4*(E58+F58),0)**
O14: **@IF(L7=1#AND#L8=1,2*F58,0)**
N15: **@IF(L8=1,2*F58,0)**
O15: **@IF(L8=1,4*(F58+G58),0)**
P15: **@IF(L8=1#AND#L9=1,2*G58,0)**
O16: **@IF(L9=1,2*G58,0)**
P16: **@IF(L9=1,4*G58,0)**

Sheet 8.3 (Page 4)

1/J	K	L	M	N	O	P	Q	R
2			Matrix Inversion and Multiplication					
3			------------------------------------					
4	Theta1	1	Span1	1	No of Spans =		3	
5	Theta2	1	Span2	1	No of Theta =		4	
6	Theta3	1	Span3	1				
7	Theta4	1	Span4	0				
8	Theta5	0	Span5	0				
9	Theta6	0		Initial Matrix				
10				---------------				
11	1.5E+00	7.3E−01	0.0E+00	0.0E+00	0.0E+00	0.0E+00	1.0E+00	0.0E+00
12	7.3E−01	3.8E+00	1.2E+00	0.0E+00	0.0E+00	0.0E+00	0.0E+00	1.0E+00
13	0.0E+00	1.2E+00	3.9E+00	7.7E−01	0.0E+00	0.0E+00	0.0E+00	0.0E+00
14	0.0E+00	0.0E+00	7.7E−01	1.5E+00	0.0E+00	0.0E+00	0.0E+00	0.0E+00
15	0.0E+00	0.0E+00	0.0E+00	0.0E+00	0.0E+00	0.0E+00	0.0E+00	0.0E+00
16	0.0E+00	0.0E+00	0.0E+00	0.0E+00	0.0E+00	0.0E+00	0.0E+00	0.0E+00
17								
18	1.0E+00	5.0E−01	0.0E+00	0.0E+00	0.0E+00	0.0E+00	6.9E−01	0.0E+00
19	0.0E+00	3.4E+00	1.2E+00	0.0E+00	0.0E+00	0.0E+00	−5.0E−01	1.0E+00
20	0.0E+00	1.2E+00	3.9E+00	7.7E−01	0.0E+00	0.0E+00	0.0E+00	0.0E+00
21	0.0E+00	0.0E+00	7.7E−01	1.5E+00	0.0E+00	0.0E+00	0.0E+00	0.0E+00
22	0.0E+00	0.0E+00	0.0E+00	0.0E+00	0.0E+00	0.0E+00	0.0E+00	0.0E+00
23	0.0E+00	0.0E+00	0.0E+00	0.0E+00	0.0E+00	0.0E+00	0.0E+00	0.0E+00
24								
25	1.0E+00	0.0E+00	−1.7E−01	0.0E+00	0.0E+00	0.0E+00	7.6E−01	−1.5E−01
26	0.0E+00	1.0E+00	3.4E−01	0.0E+00	0.0E+00	0.0E+00	−1.5E−01	2.9E−01
27	0.0E+00	0.0E+00	3.5E+00	7.7E−01	0.0E+00	0.0E+00	1.7E−01	−3.4E−01
28	0.0E+00	0.0E+00	7.7E−01	1.5E+00	0.0E+00	0.0E+00	0.0E+00	0.0E+00
29	0.0E+00	0.0E+00	0.0E+00	0.0E+00	0.0E+00	0.0E+00	0.0E+00	0.0E+00
30	0.0E+00	0.0E+00	0.0E+00	0.0E+00	0.0E+00	0.0E+00	0.0E+00	0.0E+00
31								
32	1.0E+00	0.0E+00	0.0E+00	3.8E−02	0.0E+00	0.0E+00	7.7E−01	−1.6E−01
33	0.0E+00	1.0E+00	0.0E+00	−7.5E−02	0.0E+00	0.0E+00	−1.6E−01	3.2E−01
34	0.0E+00	0.0E+00	1.0E+00	2.2E−01	0.0E+00	0.0E+00	4.9E−02	−9.8E−02
35	0.0E+00	0.0E+00	0.0E+00	1.4E+00	0.0E+00	0.0E+00	−3.8E−02	7.5E−02
36	0.0E+00	0.0E+00	0.0E+00	0.0E+00	0.0E+00	0.0E+00	0.0E+00	0.0E+00
37	0.0E+00	0.0E+00	0.0E+00	0.0E+00	0.0E+00	0.0E+00	0.0E+00	0.0E+00
38								
39	1.0E+00	0.0E+00	0.0E+00	0.0E+00	0.0E+00	0.0E+00	7.7E−01	−1.6E−01
40	0.0E+00	1.0E+00	0.0E+00	0.0E+00	0.0E+00	0.0E+00	−1.6E−01	3.3E−01
41	0.0E+00	0.0E+00	1.0E+00	0.0E+00	0.0E+00	0.0E+00	5.5E−02	−1.1E−01
42	0.0E+00	0.0E+00	0.0E+00	1.0E+00	0.0E+00	0.0E+00	−2.8E−02	5.5E−02
43	0.0E+00	0.0E+00	0.0E+00	0.0E+00	0.0E+00	0.0E+00	0.0E+00	0.0E+00
44	0.0E+00	0.0E+00	0.0E+00	0.0E+00	0.0E+00	0.0E+00	0.0E+00	0.0E+00
45								
46	1.0E+00	0.0E+00	0.0E+00	0.0E+00	0.0E+00	0.0E+00	7.7E−01	−1.6E−01
47	0.0E+00	1.0E+00	0.0E+00	0.0E+00	0.0E+00	0.0E+00	−1.6E−01	3.3E−01
48	0.0E+00	0.0E+00	1.0E+00	0.0E+00	0.0E+00	0.0E+00	5.5E−02	−1.1E−01
49	0.0E+00	0.0E+00	0.0E+00	1.0E+00	0.0E+00	0.0E+00	−2.8E−02	5.5E−02
50	0.0E+00	0.0E+00	0.0E+00	0.0E+00	0.0E+00	0.0E+00	0.0E+00	0.0E+00
51	0.0E+00	0.0E+00	0.0E+00	0.0E+00	0.0E+00	0.0E+00	0.0E+00	0.0E+00

All the remaining cells of the stiffness matrix contain a zero.

Following the approach outlined in Chapter 7 the unit matrix is now placed adjacent to the stiffness matrix, in block Q11 to V16, and the process of inversion carried out by operating on each row in turn so that the original stiffness matrix becomes a unit matrix and the original unit matrix becomes the inverse matrix. This inversion process, which is entered in the block K18 to V63, follows a similar procedure to that given in Chapter 7 for the inversion of a 4 × 4 matrix. (Only the first two columns of the six column unit matrix are shown on page 4 of Sheet 8.3.)

The load vector resulting from the point loads has been placed in the range K67 to K72 and the vector from the uniformly distributed loads in L67 to L72 using:

Sheet 8.3 (Page 5)

56								
57	K	L	M	N	O	P	Q	R
58	1.0E+00	0.0E+00	0.0E+00	0.0E+00	0.0E+00	0.0E+00	7.7E−01	−1.6E−01
59	0.0E+00	1.0E+00	0.0E+00	0.0E+00	0.0E+00	0.0E+00	−1.6E−01	3.3E−01
60	0.0E+00	0.0E+00	1.0E+00	0.0E+00	0.0E+00	0.0E+00	5.5E−02	−1.1E−01
61	0.0E+00	0.0E+00	0.0E+00	1.0E+00	0.0E+00	0.0E+00	−2.8E−02	5.5E−02
62	0.0E+00	0.0E+00	0.0E+00	0.0E+00	0.0E+00	0.0E+00	0.0E+00	0.0E+00
63	0.0E+00	0.0E+00	0.0E+00	0.0E+00	0.0E+00	0.0E+00	0.0E+00	0.0E+00
64				Support Theta's				
65	PT Loads	UD Loads		PT Loads	UD Loads			Inversion
66	--------	--------		--------	--------		----	----
67	0.0E+00	2.5E+01		−8.7E+00	−1.6E+01		7.7E−01	−1.6E−01
68	5.1E+01	1.5E+02		1.7E+01	6.7E+01		−1.6E−01	3.3E−01
69	−1.6E+01	−1.8E+02		−7.8E+00	−7.7E+01		5.5E−02	−1.1E−01
70	−1.8E+01	2.8E+01		−7.6E+00	5.7E+01		−2.8E−02	5.5E−02
71	0.0E+00	0.0E+00		0.0E+00	0.0E+00		0.0E+00	0.0E+00
72	0.0E+00	0.0E+00		0.0E+00	0.0E+00		0.0E+00	0.0E+00
73								
74								
75				Final Momemts from				
76				Pt Loads	U D Loads	Total		Shears
77		1		0.00	0.00	0.00		−2.4E+01
78		2		19.06	110.91	129.97		2.4E+01
79		3		−19.06	−110.91	−129.97		1.8E−01
80		4		53.20	74.89	128.09		−1.8E−01
81		5		−53.20	−74.89	−128.09		1.3E+01
82		6		0.00	28.40	28.40		−1.3E+01
83		7		0.00	0.00	0.00		0.0E+00
84		8		0.00	0.00	0.00		0.0E+00
85		9		0.00	0.00	0.00		0.0E+00
86		10		0.00	0.00	0.00		0.0E+00

K67: **−H63**
K68: **−(H64+H75)**
K69: **−(H76+H87)**
L67: **@IF(E41=2,F42,0)−H66**
L68: **@IF(G41=2,H42,0)−(H67+H78)**
L69: **@IF(G43=2,H44,0)−(H79+H90)**

and similarly for the other cells in the block.

Determination of support rotations

The rotation at each support due to the point loads has been calculated in the range N67 to N72 and for the uniformly distributed loads in O67 to O72. This separate calculation of the rotation resulting from the two types of loading has only been introduced for convenience. The actual rotation will be the sum of these two rotations. Examples of the required formulae are:

N67: **(Q67*$K$67)+(R67*$K$68)+(S67*$K$69)+(T67*$K$70)+(U67*$K$71)+(V67*$K$72)**
N68: **(Q68*$K$67)+(R68*$K$68)+(S68*$K$69)+(T68*$K$70)+(U68*$K$71)+(V68*$K$72)**
O67: **(Q67*$L$67)+(R67*$L$68)+(S67*$L$69)+(T67*$L$70)+(U67*$L$71)+(V67*$L$72)**
O68: **(Q68*$L$67)+(R68*$L$68)+(S68*$L$69)+(T68*$L$70)+(U68*$L$71)+(V68*$L$72)**

and similarly for the other cells.

Calculation of the support moments and reactions

Once the values of rotation have been calculated the bending moment at the supports can be determined from the standard slope deflection equations. On Sheet 8.3 the calculations for the two types of loading have been kept separate and the results for the point loads are placed in the range N77 to N86 and for the distributed loads in O77 to O86. The total moment at each support is placed in P77 to P86. Examples of the formulae required to determine these values are:

N77: **+H63+4*C58*N67+2*C58*N68**
N78: **+H64+2*C58*N67+4*C58*N68**
O77: **+H66+4*C58*O67+2*C58*O68**
O78: **+H67+2*C58*O67+4*C58*O68**

The support reactions are calculated in cells R77 to R86 using:

R77: **@IF($C$7>0,−(P77+P78)/$C$7,0)**
R78: **@IF($C$7>0,(P77+P78)/$C$7,0)**

and similarly for the remaining cells.

Finally the values of the support reactions, bending moments and rotations are grouped together in block D128 to F137 so that the user can view the results without having to move off the screen to the left-hand set of calculations. Typically these cells contain:

D128: **+R77+E69+E70**
E128: **+P77**
F128: **(N67+O67)**

8.10 Use of spreadsheet 8.3

The unprotected cells are:

C7 to C11	Span lengths
E7 to E11	Values of E
G7 to G11	Values of I
D13 to H14	Point loads, values and positions, Span 1
D15,F15,H15	Location and magnitude of UDL 1 on Span 1
D16,F16,H16	Location and magnitude of UDL 2 on Span 1
D18 to H19	Point loads, values and positions, Span 2
D20,F20,H20	Location and magnitude of UDL 1 on Span 2
D21,F21,H21	Location and magnitude of UDL 2 on Span 2
D23 to H24	Point loads, values and positions, Span 3
D25,F25,H25	Location and magnitude of UDL 1 on Span 3
D26,F26,H26	Location and magnitude of UDL 2 on Span 3
D28 to H29	Point loads, values and positions, Span 4
D30,F30,H30	Location and magnitude of UDL 1 on Span 4
D31,F31,H31	Location and magnitude of UDL 2 on Span 4
D33 to H34	Point loads, values and positions, Span 5
D35,F35,H35	Location and magnitude of UDL 1 on Span 5
D36,F36,H36	Location and magnitude of UDL 2 on Span 5
E41,G41	Boundary conditions, Span 1
F42,H42	Values of any overhang moment, Span 1
G43,G45,G47,G49	Boundary conditions, Spans 2 to 5
H44,H46,H48,H50	Values of any overhang moment, Spans 2 to 5

Cell entries for the above are quite straightforward except for values of E and I. True values of the rotation will only be found if the true values of E and I are input. If relative values are used then relative values of rotation will be calculated but this will not affect the values found for the support reactions and bending moments.

9 Composite action

WALL BEAMS

9.1 Introduction

When a wall is supported on a reinforced beam then the stresses in the wall are reduced because of the stiffening effect of the beam on the wall. The beam acts as a tie resulting in an arching effect in the wall which in turn reduces the stresses.

Prior to 1952 it was common practice to assume that if a wall and its supporting beam acted as a composite unit then the beam could support a triangular wedge of brickwork in the wall above. The load from this wedge could be calculated using the span of the beam as the base of an equilateral triangle.

Since that period experimental and theoretical studies have led to a better understanding of the problem so that a more realistic modelling of the composite action is possible.

It has been shown that, assuming composite action between the wall and beam exists, the action of a uniformly distributed load along the top of the wall results in:

(1) A vertical stress distribution at the wall/beam interface which is mainly compressive. This stress is a maximum at the supports and then reduces in value towards the centre becoming either a minimum compressive stress or a small maximum tensile stress at the centre.

(2) A horizontal stress distribution at mid-span which is compressive in the wall and tensile in the beam so that the beam acts as a tie.

Typical stress distributions are shown in Fig. 9.2 for the wall beam shown in Fig. 9.1.

Composite action cannot be achieved unless there is sufficient bond at the interface to support shear stresses. The large compressive stresses near the supports result in large frictional forces along the interface and it is generally accepted that the depth/span ratio must be > 0.6 for the frictional forces to be large enough to supply the required shear capacity.

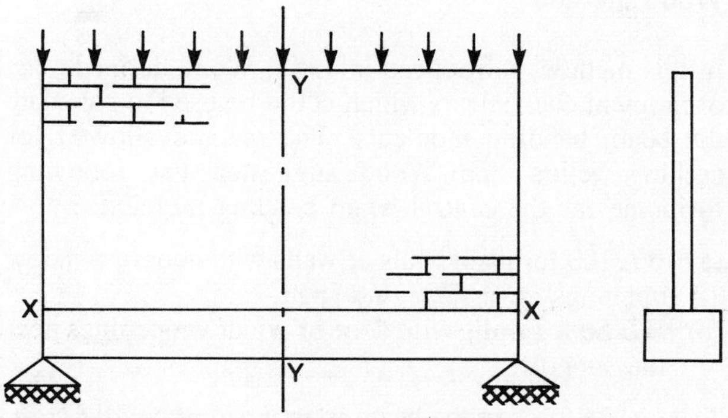

Figure 9.1 Simply supported wall beam

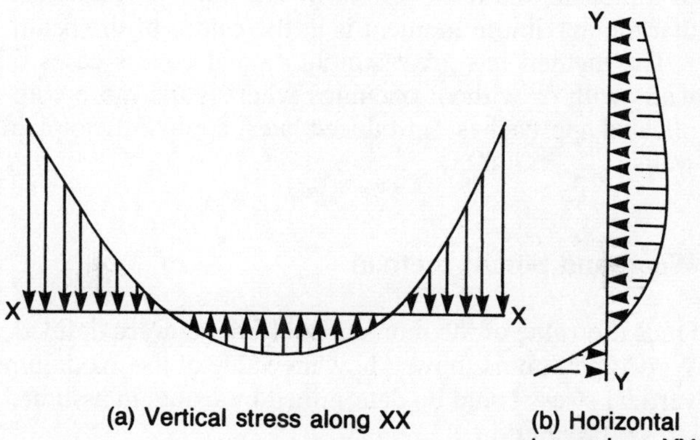

Figure 9.2 Stress distribution

9.2 Design objectives

The initial purpose of any design approach is to determine the following:

(a) the maximum vertical stress in the wall;
(b) the axial force in the beam;
(c) the maximum shear stress along the interface;
(d) the central bending moment in the beam;
(e) the maximum bending moment in the beam.

Once these have been determined additional quantities,

such as the required amount of beam reinforcement, can be determined using the relevant code of practice.

Many different approaches have been proposed for calculating the five basic quantities and brief details of some of these methods are given below. Spreadsheet solutions of the selected methods are then outlined and an illustration given of how they compare.

The methods selected for the comparison are:

(1) Wood (1952);
(2) Wood and Simms (1969);
(3) Stafford-Smith and Riddington (1977);
(4) Davies and Ahmed (1978).

In effect methods 1 and 2 are combined into a single approach.

9.3 Design methods

Wood method

In this method, introduced in 1952, Wood defined a set of moment coefficients which could be used to calculate the beam bending moments. The method allowed for arching action and Wood suggested the following formulae for the central beam bending moment:

(a) $WL/100$ for plain walls or walls with door or window openings at or near mid-span.
(b) $WL/50$ for walls with door or window openings near the support.

The tensile force in the beam is assumed to be $3WL/16h$.

In these formulae W represents the total load, L the span of the beam and h the height of the wall. It is assumed that the maximum moment is at the centre of the beam.

The method has great simplicity and covers cases of walls with or without openings whereas the more complicated approaches, introduced later, apply only to plain walls.

Wood and Simms method

Once the value of the moment coefficients were defined, Wood and Simms showed how the value of the maximum vertical stress could be determined by using an assumed stress distribution.

They assumed that the vertical stresses near the supports formed a rectangular stress block which extended a distance x into the span from each end of the beam. Using

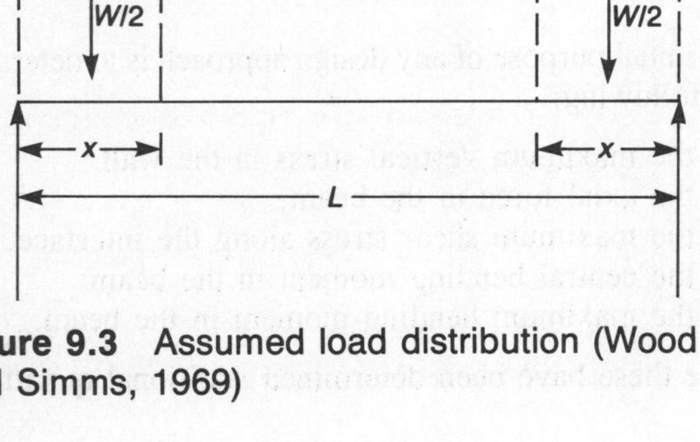

Figure 9.3 Assumed load distribution (Wood and Simms, 1969)

this assumption the central beam bending moment is given by $Wx/4$ and comparing this with the formulae given by Wood above, in the form WL/k, enables the value of x to be found from the value used for k. See Fig. 9.3.

For example if k is taken as 100 then x would be $L/25$ and if k is taken as 50 then x would be $L/12.5$. This approach could also be used for other values of k. Once x has been found then the value of the maximum vertical stress can be determined on the basis that each stress block carries half the load so that the maximum vertical stress is given by:

$$f_m = W/2x \qquad [9.1]$$

for example if $k = 100$ then $f_m = 12.5W/L$.

Equation 9.1 can be expressed as:

$$f_m = Wk/8L \qquad [9.2]$$

by eliminating x and thus giving a general relationship for the maximum vertical stress in terms of the assumed values of k.

In summary using the method advanced by Wood and amended by Wood and Simms results in methods for determining four out of the five required quantities listed in section 9.2. The maximum shear stress along the interface is not determined since it is assumed that the depth/span ratio will be > 0.6 and the shear stresses will be small for this case.

Stafford-Smith and Riddington method

These authors illustrated that the wall beam problem was similar to the problem of a beam on an elastic foundation in that the distribution of stresses between the wall and beam depends on their relative stiffness. They introduced a relative stiffness parameter K_s based on this principle and also assumed that a ratio of wall height to span equal to 0.6 was representative of walls for which this ratio was greater. The parameter K_s was defined as:

$$K_s = (E_w tL^3/E_b I)^{1/4} \qquad [9.3]$$

where $E_b I$ is the flexural rigidity of the beam, E_w the elastic modulus of the wall and t the wall thickness.

The authors developed a finite element program using an element which was selected to represent the required characteristics of the problem and then used this program to conduct an extensive study over a wide range of wall beam parameters. The results are shown in Fig. 9.4.

Using these results the authors produced the following equations to represent the problem:

$$M_m = WL/[4(E_w tL^3/E_b I)^{0.33}] \qquad [9.4]$$

$$f_m = 1.63(W/Lt)(E_w tL^3/E_b I)^{0.28} \qquad [9.5]$$

$$T = W/3.4 \qquad [9.6]$$

where M_m represents the maximum beam moment, f_m the maximum vertical stress and T the tie force in the beam. It was assumed that the maximum beam moment occurs at the centre of the beam and that the depth to span ratio is > 0.6 so that calculation of the shear force along the interface was not required.

If the value of 0.6 is substituted for h/L in the equation

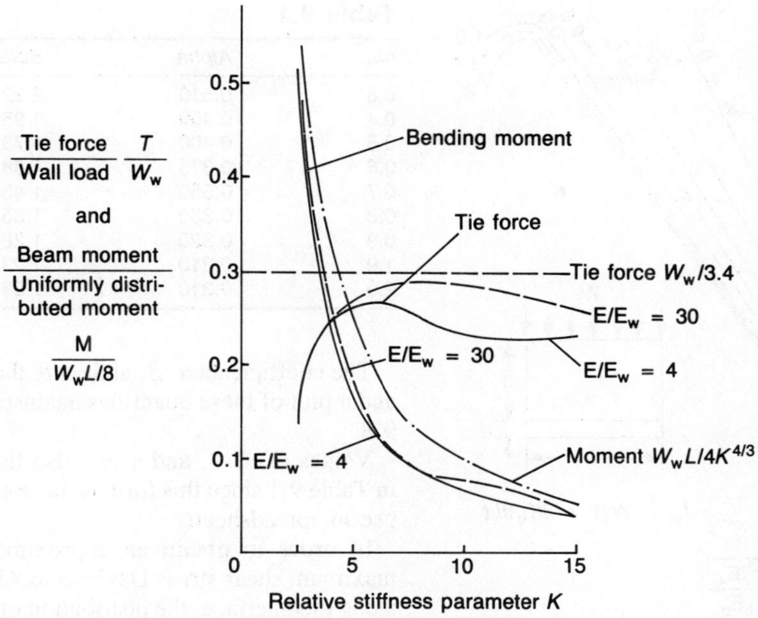

Figure 9.4 Maximum bending moments and tie force in beam (Stafford-Smith and Riddington, 1977)

given by Wood for the tie force then the equation would become:

$$T = W/3.2$$

which is very similar to the value suggested by Stafford-Smith and Riddington in Eqn 8.6.

Davies and Ahmed method

The approach used by Davies and Ahmed in 1978 was similar to that used by Stafford-Smith and Riddington in that an extensive finite element study was carried out but they produced a different set of design equations to represent the problem.

Two relative stiffness parameters were defined. One was similar to that used by Stafford-Smith and Riddington, except that the depth of the wall h was used in the equation instead of the span, and the other was an axial stiffness parameter. These parameters were defined as:

$$K_d = (E_w th^3)/E_b I)^{1/4} \quad [9.7]$$

$$K_a = E_w th/E_b A \quad [9.8]$$

Two further assumptions were made by Davies and Ahmed. The first is illustrated in Fig. 9.5.

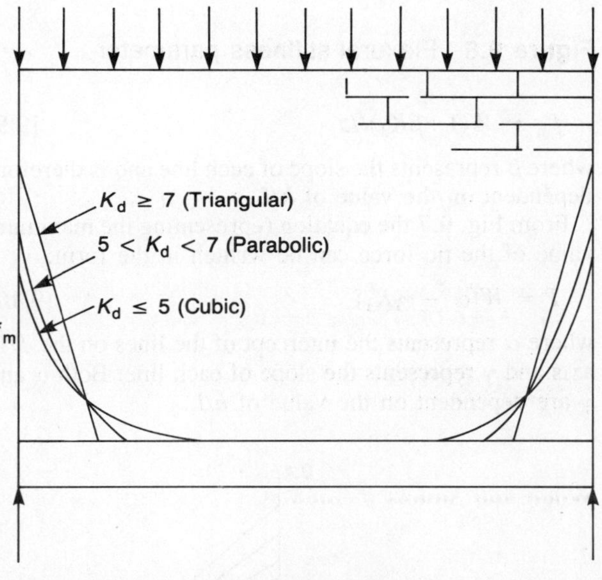

Figure 9.5 Vertical stress distribution

This figure shows that the vertical stress distribution was assumed to be related to the value of the relative stiffness parameter K_d. For values of K_d greater than 7 the distribution was assumed to be linear and for values less than 5 it was assumed to be cubic. For values of K_d between 5 and 7 the distribution was assumed to be parabolic.

The second assumption was that the axial force in the beam was assumed to vary linearly from zero at the supports to a maximum value at the centre.

Using these assumptions equations can be derived, for each of the three cases illustrated in Fig. 9.5, for the five required quantities listed in Section 9.2. To obtain these equations Davies and Ahmed plotted graphs of the results from the finite element study for:

(a) The ratio of the maximum vertical stress to the average vertical stress against the flexural stiffness parameter K_d.
(b) The ratio T/W against the axial stiffness parameter K_a.

It was found that these plots could each be represented by a series of straight lines which were dependent on the value of h/L as shown in Figs 9.6 and 9.7.

From Fig. 9.6 the equation for the maximum vertical stress can be written in the form:

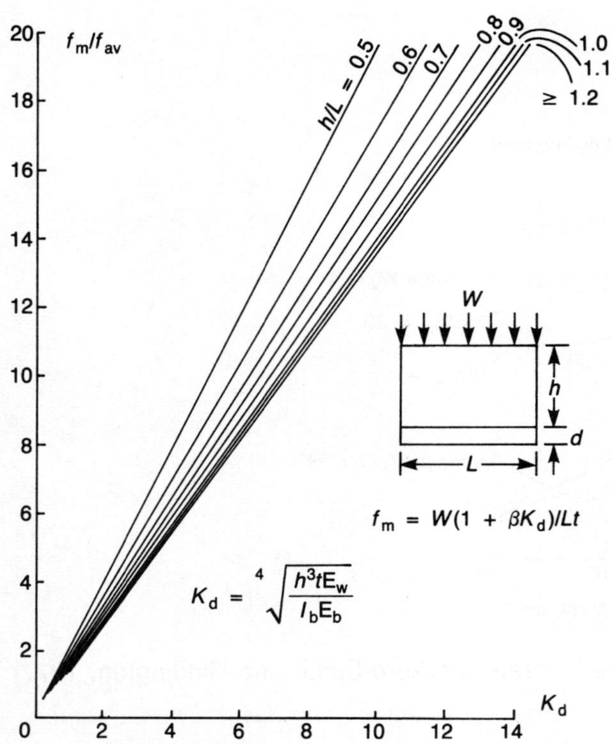

Figure 9.6 Flexural stiffness parameter

$$f_m = W(1+\beta K_d)/Lt \qquad [9.9]$$

where β represents the slope of each line and is therefore dependent on the value of h/L.

From Fig. 9.7 the equation representing the maximum value of the tie force can be written in the form:

$$T = W(\alpha - \gamma K_a) \qquad [9.10]$$

where α represents the intercept of the lines on the T/W axis and γ represents the slope of each line. Both α and γ are dependent on the value of h/L.

Table 9.1

h/L	*Alpha*	*Beta*	*Gamma*
0.3	0.510	2.22	0.155
0.4	0.450	1.95	0.115
0.5	0.400	1.73	0.090
0.6	0.375	1.58	0.075
0.7	0.350	1.45	0.062
0.8	0.330	1.35	0.050
0.9	0.320	1.28	0.045
1.0	0.310	1.23	0.040
2.0	0.310	1.23	0.040

The coefficients α, β, and γ are the key to the solution and a plot of these quantities against h/L is given in Fig. 9.8.

Values of α, β, and γ are also listed in tabular form in Table 9.1 since this form is more convenient for direct use in spreadsheets.

In order to obtain an approximate estimate of the maximum shear stress Davies and Ahmed assumed that, along the interface, the distribution of the vertical stresses was linear and the shear stress distribution was bi-linear in each half of the wall beam as shown in Fig. 9.9.

In their analysis the authors introduced two 'contact lengths', l_v and l_s, as shown in Fig. 9.9. Then, using the results from the finite element study, showed that the ratio l_s/l_v was between 2 and 3 and that a conservative value of 2 could be used to derive an equation for the maximum shear stress. This equation was expressed as:

$$\tau_m = W(\alpha - \gamma K_a)(1 + \beta K_d)/Lt \qquad [9.11]$$

To calculate expressions for the bending moments it is necessary to combine the moment due to the vertical loading with the moment due to the horizontal forces in the beam. Using the assumption that the tie force in the beam varies linearly from zero at the supports to a maximum value at the centre the authors showed that the maximum and central bending moment could be expressed by the equations:

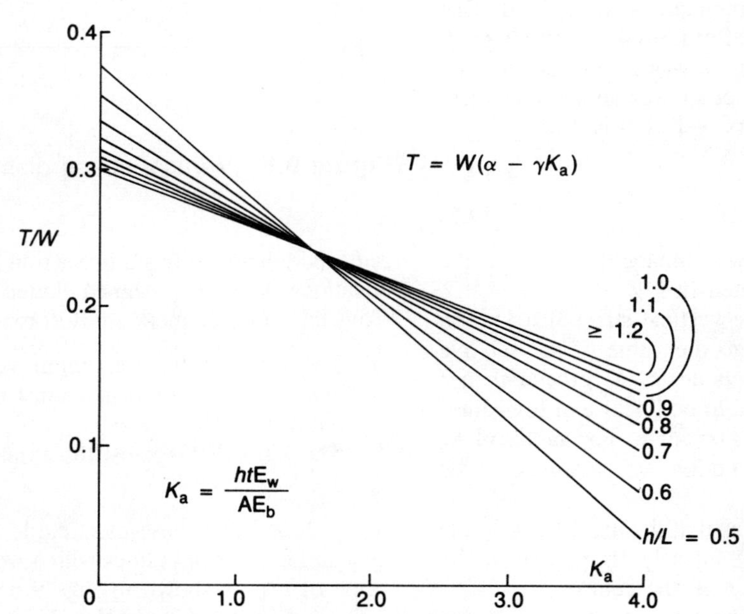

Figure 9.7 Axial stiffness parameter

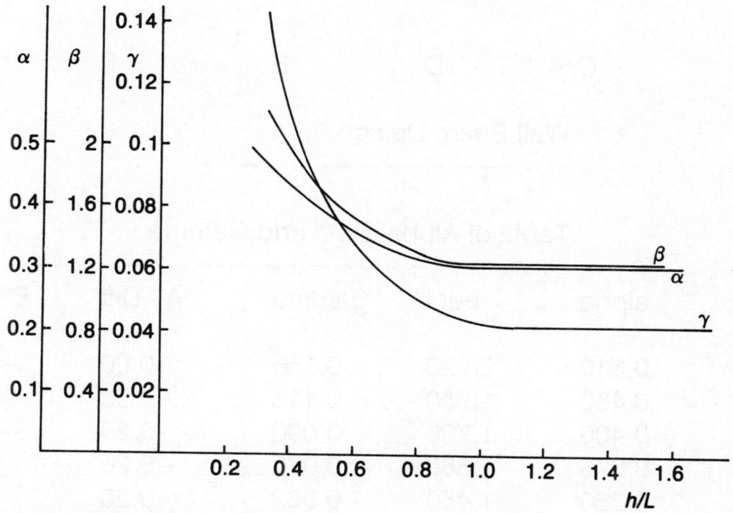

Figure 9.8 Variation of α, β and γ with h/L

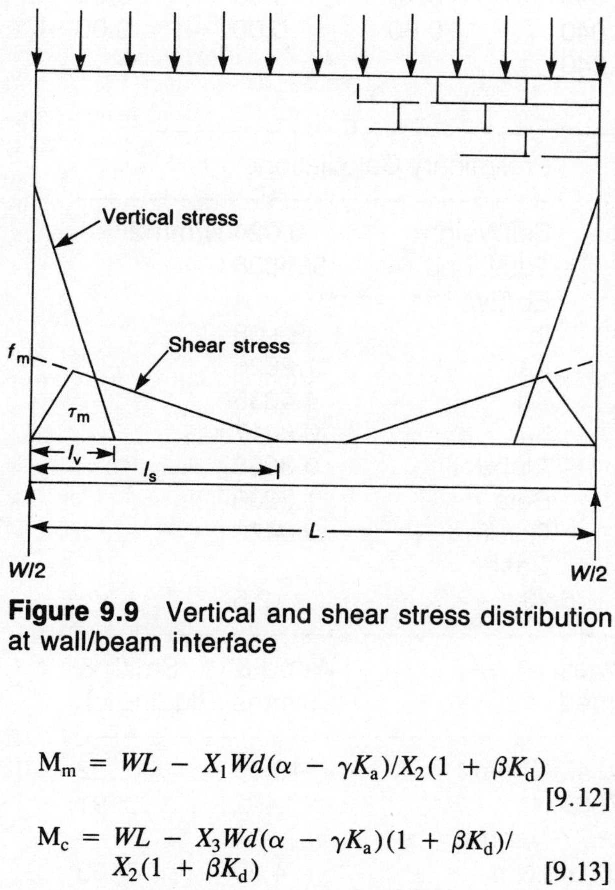

Figure 9.9 Vertical and shear stress distribution at wall/beam interface

$$M_m = WL - X_1Wd(\alpha - \gamma K_a)/X_2(1 + \beta K_d) \quad [9.12]$$

$$M_c = WL - X_3Wd(\alpha - \gamma K_a)(1 + \beta K_d)/X_2(1 + \beta K_d) \quad [9.13]$$

where X_1, X_2 and X_3 are coefficients related to the value of the relative stiffness parameter K_d. Values of these coefficients are given in Table 9.2.

Table 9.2

K_d	X_1	X_2	X_3
≤ 5	10	5	2.5
$5<K_d<7$	8	5.33	2.66
≥ 7	6	6	3

9.4 Spreadsheet layout

The design methods outlined in the previous section can all be incorporated into the same spreadsheet so that a comparison between the results of the different approaches is immediately available. Additionally it will be shown in the next section that the spreadsheet can be used as a research tool so that more general comparisons between the different methods can be made.

Referring to Sheet 9.1 the tabular values of α, β, and γ shown in Table 9.1 are entered on the spreadsheet in the block B10 to E18 and to these are added the three difference columns which allow interpolated values of α, β, and γ to be calculated for any intermediate value of h/L. These additional columns are entered in block F10 to H17 and typical examples of the contents of the cells in these additional columns are:

F10: **(C11−C10)/(B11−B10)**
F11: **(C12−C11)/(B12−B11)**
G10: **(D11−D10)/(B11−B10)**
G11: **(D12−D11)/(B12−B11)**
H10: **(E11−E10)/(B11−B10)**
H11: **(E12−E11)/(B12−B11)**.

Initial data. Initial values of the wall density, the applied load, the dimensions and elastic moduli of the wall and the beam are entered in cells D23 to D31. The corresponding labels are placed alongside in cells B23 to B31.

Note that in this particular example the wall modulus has been entered as 1 and the beam modulus as 4. This is permissible since only the relative values are required in the equations.

Preliminary calculations

Formulae for calculating the quantities indicated by the labels in cells F23 to F33 are entered in cells G23 to G33 which contain:

G23: **+D23*(D25+D26)**

to calculate the self weight.

Sheet 9.1

1/A	B	C	D	E	F	G	H
2							
3			Wall Beam Design				
4			----------------				
5							
6			Table of Alpha, Beta and Gamma				
7			------------------------------------				
8	h/L	alpha	beta	gamma	A−Diff	B−Diff	G−Diff
9							
10	0.3	0.510	2.220	0.155	−0.60	−2.70	−0.40
11	0.4	0.450	1.950	0.115	−0.50	−2.20	−0.25
12	0.5	0.400	1.730	0.090	−0.25	−1.50	−0.15
13	0.6	0.375	1.580	0.075	−0.25	−1.30	−0.13
14	0.7	0.350	1.450	0.062	−0.20	−1.00	−0.12
15	0.8	0.330	1.350	0.050	−0.10	−0.70	−0.05
16	0.9	0.320	1.280	0.045	−0.10	−0.50	−0.05
17	1.0	0.310	1.230	0.040	0.00	0.00	0.00
18	2.0	0.310	1.230	0.040			
19							
20	--------------------	--------------------	--------------------	--------------------	--------------------	--------------------	--------------------
21 Initial Values					Preliminary Calculations		
22 ----------------					----------------------		
23 Wall Density			0.00002 N/mm3		SelfWeight	0.024 N/mm2	
24 Applied Load			1.500 N/mm2		TotalLoad	365808	
25 Wall Height			1010 mm		Eb/Ew	4	
26 Beam Depth			200 mm		Ib	1.3E+08	
27 Wall Thickness			200 mm		Ka	1.2625	
28 Beam Thickness			200 mm		Kd	4.4335	
29 Wall & Beam Span			1200 mm		h/L	0.8417	
30 Wall Modulus			1		Alpha	0.3258	
31 Beam Modulus			4		Beta	1.3208	
32					Gamma	0.0479	
33					Case	1	
34 x1		10 x2		5 x3		2.5	
35	--------------------	--------------------	--------------------	--------------------	--------------------	--------------------	--------------------
36 Final Calculations				Davies		Wood &	Smith &
37				& Ahmed		Simms	Riddington
38	--------------------	--------------------	--------------------	--------------------	--------------------	--------------------	--------------------
39 Max Vertical Sress		=		10.45 N/mm2		19.05	15.22
40 Max Axial Force		=		97063 N		81492	107591
41 Max Shear Stress		=		2.77 N/mm2		---	---
42 Max Bending Moment		=		7.14 kNm		4.39	12.68
43 Central Moment		=		3.10 kNm		4.39	12.68

G24: **((D24+G23)*D29*D27)**

to calculate the total load.

G25: **(D31/D30)**

for the modular ratio.

G26: **(D28*(D26^3)/12)**

to calculate the second moment of area of the beam using $td^3/12$. No allowance has been made for the contribution of the reinforcement although there would be no difficulty in including this if required.

G27: **(D25*D27)/(D28*D26*G25)**

to calculate K_a in accordance with Eqn 9.8.

G28: **((D27*(D25^3))/(G26*G25))^0.25**

to calculate K_d in accordance with Eqn 9.7.

G29: **(D25/D29)**

G30: **@VLOOKUP(G29,B10..H18,1)+(G29−(@VLOOKUP(G29,B10..H18,0)))*(@VLOOKUP(G29,B10..H18,4))**

to calculate the interpolated value of α from the value of h/L determined in cell G29.

G31: **@VLOOKUP(G29,B10..H18,2)+(G29−(@VLOOKUP(G29,B10..H18,0)))*(@VLOOKUP(G29,B10..H18,5))**

to calculate the interpolated value of β from the value of h/L determined in cell G29.

G32: **@VLOOKUP(G29,B10..H18,3)+(G29−(@VLOOKUP(G29,B10..H18,0)))*(@VLOOKUP(G29,B10..H18,6)**

to calculate the interpolated value of γ from the value of h/L determined in cell G29.

G33: **@IF(G28<=5,1,@IF(G28>=7,3,2))**

is used to determine which case, in accordance with the limits shown in Table 9.2, applies. If K_d from cell G28 is less than or equal to 5 then a value of 1 results, if it is greater than or equal to 7 then the value would be 3, otherwise the value would be 2. Thus cell G28 will contain one of the values 1, 2 or 3 depending on which case applies to the present set of parameters. This output is for information only and is not used in subsequent calculations.

It was shown in Section 9.3 that, in order to calculate the maximum and central bending moments, Davies and Ahmed introduced three coefficients X_1, X_2 and X_3 the values of which depend on the case that applies in accordance with Table 9.2. In order to inform the spreadsheet as to which set of the coefficients are to be used formulae are placed in cells C34, E34 and G34 as follows:

C34: **@IF(G28<=5,10,@IF(G28>=7,6,8))**
E34: **@IF(G28<=5,5,@IF(G28>=7,6,5.33))**
G34: **@IF(G28<=5,2.5,@IF(G28>=7,3,2.66))**

so that the values which appear in these cells will be in accordance with the limits specified in Table 9.2.

Final calculations

Using the initial values and the values obtained in the preliminary calculations block it is now possible to complete the calculations and determine the five required quantities listed in Section 9.2.

The method proposed by Davies and Ahmed results in two distinct values for the maximum and central moments whereas for the other methods the maximum moment is assumed to occur at the mid-section.

The necessary equations are included in cells E39 to E43 for the method proposed by Davies and Ahmed, cells G39 to G43 for the Wood and Simms method and H39 to H43 for the Stafford-Smith and Riddington approach.

(1) Davies and Ahmed.

E39: **(G24*(1+(G31*G28)))/(D29*D28)**

from Eqn 9.9.

E40: **+G24*(G30−(G32*G27))**

from Eqn 9.10.

E41: **+E39*E40/G24**

from Eqn 9.11.

E42: **((G24*D29)−(C34*E40*D26))/(E34*(1+(G31*G28))*(10^6)**

from Eqn 9.12.

E43: **((G24*D29)−(G34*E40*D26*(1+(G31*G28))))/(E34*(1+(G31*G28))*1000000)**

from Eqn 9.13.

(2) Wood and Simms.

G39: **(12.5*G24/(D29*D27))**

from Eqn 9.1

G40: **(3*G24*D29/(16*D25))**

using $3WL/16$.

G42: **(G24*D29/100000000)**

using $WL/100$.

G43: **(G24*D29/100000000)**

using $WL/100$.

(3) Stafford-Smith and Riddington.

H39: **1.63*(G24/(D29*D27))*(((D27*(D29^3))/(G26*G25))^0.28)**

from Eqn 9.5.

H40: **+G24/3.4**

from Eqn 9.6.

H42: **(G24*D29)/(4000000*(((D27*(D29^3))/(G26*G25))^0.3333))**
H43: **(G24*D29)/(4000000*(((D27*(D29^3))/(G26*G25))^0.3333))**

Use of sheet 9.1

If the spreadsheet is now protected then the only cells which have to be unprotected in order to allow user input are: D23 to D31. So that with a very simple spreadsheet results can be obtained for the various quantities by three different methods. In Hendry (1981) the theoretical results obtained by the above three methods are compared with a limited number of test results and concludes that the Davies and Ahmed method appears to give the most consistent agreement with the experimental results.

Sheet 9.2

1/A	B	C	D	E	F	G	H
2							
3		Comparison of Wall Beam Design					
4		--					
5							
6		Table of Alpha, Beta and Gamma					
7		--					
8		h/L	alpha	beta	gamma		
9							
10		0.3	0.510	2.220	0.155		
11		0.4	0.450	1.950	0.115		
12		0.5	0.400	1.730	0.090		
13		0.6	0.375	1.580	0.075		
14		0.7	0.350	1.450	0.062		
15		0.8	0.330	1.350	0.050		
16		0.9	0.320	1.280	0.045		
17		1.0	0.310	1.230	0.040		
18		2.0	0.310	1.230	0.040		
19		--					
20 Stafford–Smith and Riddington							
21 --							
22 d/l			0.03	0.04	0.05	0.06	0.07
23 M/WL(4)			0.0052	0.0069	0.0087	0.0104	0.0121
24 M/WL(30)			0.0102	0.0136	0.0170	0.0203	0.0237
25 Davies and Ahmed							
26 --------------------							
27	h/L	d/L	0.03	0.04	0.05	0.06	0.07
28	0.4	h/d	13.333	10.000	8.000	6.667	5.714
29	x1		6	6	8	8	10
30	x2		6	6	5.33	5.33	5
31	0.6		20.000	15.000	12.000	10.000	8.571
32	x1		6	6	6	6	8
33	x2		6	6	6	6	5.33
34	0.8		26.667	20.000	16.000	13.333	11.429
35	x1		6	6	6	6	6
36	x2		6	6	6	6	6
37	1		33.333	25.000	20.000	16.667	14.286
38	x1		6	6	6	6	6
39	x2		6	6	6	6	6
40	1.2		40.000	30.000	24.000	20.000	17.143
41	x1		6	6	6	6	6
42	x2		6	6	6	6	6
43 M/WL(4)		0.4	0.0087	0.0104	0.0130	0.0141	0.0153
44		0.6	0.0081	0.0097	0.0110	0.0122	0.0145
45		0.8	0.0076	0.0092	0.0105	0.0117	0.0128
46		1	0.0071	0.0086	0.0099	0.0110	0.0120
47		1.2	0.0063	0.0076	0.0088	0.0098	0.0107
48							
49 Wood		0.010	0.010	0.010	0.010	0.010	0.010

Sheet 9.2A

1/A		B	C	D	E	F	G	H
2								
3			Comparison of Wall Beam Design					
4			--					
5								
6			Table of Alpha, Beta and Gamma					
7			--					
8			h/L	alpha	beta	gamma		
9								
10			0.3	0.510	2.220	0.155		
11			0.4	0.450	1.950	0.115		
12			0.5	0.400	1.730	0.090		
13			0.6	0.375	1.580	0.075		
14			0.7	0.350	1.450	0.062		
15			0.8	0.330	1.350	0.050		
16			0.9	0.320	1.280	0.045		
17			1.0	0.310	1.230	0.040		
18			2.0	0.310	1.230	0.040		
19	--							
20	Stafford–Smith and Riddington							
21	--							
22	d/l		0.00	0.01	0.02	0.03	0.04	0.05
23	M/WL(4)		0.0000	0.0017	0.0035	0.0052	0.0069	0.0087
24	M/WL(30)		0.0000	0.0034	0.0068	0.0102	0.0136	0.0170
25	Davies and Ahmed							
26	--------------------------							
27		h/L	d/L	0.03	0.04	0.05	0.06	0.07
28		0.4	h/d	13.333	10.000	8.000	6.667	5.714
29		x1		8	10	10	10	10
30		x2		5.33	5	5	5	5
31		0.6		20.000	15.000	12.000	10.000	8.571
32		x1		6	8	8	10	10
33		x2		6	5.33	5.33	5	5
34		0.8		26.667	20.000	16.000	13.333	11.429
35		x1		6	6	8	8	10
36		x2		6	6	5.33	5.33	5
37		1		33.333	25.000	20.000	16.667	14.286
38		x1		6	6	6	8	8
39		x2		6	6	6	5.33	5.33
40		1.2		40.000	30.000	24.000	20.000	17.143
41		x1		6	6	6	6	8
42		x2		6	6	6	6	5.33
43	M/WL(30)		0.4000	0.0144	0.0172	0.0189	0.0201	0.0208
44			0.6	0.0122	0.0158	0.0178	0.0196	0.0206
45			0.8	0.0116	0.0139	0.0172	0.0188	0.0204
46			1	0.0109	0.0130	0.0149	0.0178	0.0192
47			1.2	0.0096	0.0115	0.0132	0.0146	0.0170
48								
49	Wood		0.010	0.010	0.010	0.010	0.010	0.010

9.5 Comparison of design methods by spreadsheet

A more general comparison can be made between the results obtained from the three design methods by use of a simple spreadsheet and this is illustrated below and in Sheet 9.2. The value of the maximum bending moment is determined, by all three methods, using a common parameter range.

Since the method used by Davies and Ahmed uses a different definition for the relative stiffness parameter, compared with the Stafford-Smith and Riddington approach, it is first necessary to rewrite the equations using a common parameter and for this the ratio d/L is selected. Additionally the comparison has been made for only two values of the modular ratio, namely 4 and 30.

(a) Stafford-Smith and Riddington.

Taking $I = td^3/12$, Eqn 9.4 can be rewritten as:

$$M/WL = 0.173d/L \qquad [9.14]$$

for $E_b/E_w = 4$
and:

$$M/WL = 0.339d/L \qquad [9.15]$$

for $E_b/E_w = 30$

(b) Davies and Ahmed.

Taking $I = td^3/12$ and $A = td$ the stiffness parameters can be rewritten as:

$$K_d = 1.316\ (h/d)^{0.75} \qquad [9.16]$$

$$K_d = 0.795\ (h/d)^{0.75} \qquad [9.17]$$

for modular ratios of 4 and 30 respectively.

The limits shown in Table 9.2 relate to values of K_d but these limits can be changed, using Eqns 9.16 and 9.17, to functions of h/d by writing:

$$1.316\ (h/d)^{0.75} \leq 5$$

so that the limit becomes: $h/d \leq 5.9286$ for case 1 and $E_b/E_w = 4$. Following a similar approach the limit $K_d \geq 7$ can be converted to a limit $h/d \geq 9.285$.

The corresponding limits for a modular ratio of 30 are given by $h/d \leq 11.609$ and $h/d \geq 18.182$. The new limiting conditions are summarised in Table 9.3.

The equation for the maximum bending can now be written in the forms:

$$M/WL = (1 - X_1(d/L)(\alpha - \gamma 0.25(h/d))/X_2(1 + \beta 1.316((h/d)^{0.75}) \qquad [9.18]$$

$$M/WL = (1 - X_1(d/L)(\alpha - \gamma 0.033(h/d))/X_2(1 + \beta 0.795((h/d)^{0.75}) \qquad [9.19]$$

for modular ratios of 4 and 30 respectively.

Table 9.3

Ratio 4	*Ratio 30*	X_1	X_2	X_3
$h/d \leq 5.93$	$h/d \leq 11.61$	10	5	2.5
$5.93 < h/d < 9.285$	$11.61 < h/d < 18.18$	8	5.33	2.66
$h/d \geq 9.285$	$h/d \geq 18.18$	6	6	3

Comparison of results

To effect the comparison a graph can now be drawn using a plot of M/WL against d/L.

For the approach used by Wood the graph will be a horizontal line with $M/WL = 0.01$.

For the approach used by Stafford-Smith and Riddington the graph will be a straight line with a slope of 0.173 for a modular ratio of 4 and a slope of 0.339 for a ratio of 30.

For the approach used by Davies and Ahmed the equation varies with both d/L and h/d and there will be a set of graphs — one for each value of h/d. However, Eqns 9.18 and 9.19 are used to convert the expression for M/WL into a function of d/L and h/L since the parameter h/L is considered to be more convenient for the spreadsheet approach.

Spreadsheet layout

The design of the required spreadsheet is illustrated in Sheet 9.2. In this sheet block C10 to F18 contains the table of values of α, β, and γ. The three difference columns for interpolating are not required for the present comparison since only tabular values are used.

The formulae in block D23 to H24 are used to calculate the values of the ratio M/WL according to the method proposed by Stafford-Smith and Riddington. Only the first five columns of results are shown on Sheet 9.2 although the actual calculations extend to column M. Values of d/L are given in row 22 and M/WL calculated in row 23 for a modular ratio of 4 and row 24 for a ratio of 30. As an example cells D23 and D24 contain:

D23: **0.173*D22**
D24: **0.339*D22**

In the method of calculation for the Davies and Ahmed approach a value of h/L is assumed (say 0.4 in cell B28) and from this, the value of h/d is determined for each d/L using $(h/L)/(d/L)$. These are recorded in row 28 in which, for example:

D28: **+$B$28/D27**
E28: **+$B$28/E27**

Rows 29 and 30 are used to determine the case which applies and the corresponding coefficients X_1 and X_2. Examples of the required formulae are:

D29: **@IF(D28< =5.9286,10,@IF(D28 > =9.285,6,8))**
D30: **@IF(D28< =5.9286,5,@IF(D28 > =9.285,6,5.33))**
E29: **@IF(E28< =5.9286,10,@IF(E28 > =9.285,6,8))**
E30: **@IF(E28< =5.9286,5,@IF(E28 > =9.285,6,5.33))**

The above calculations, for $h/L = 0.4$, are repeated for the required range of h/L values and, in Sheet 9.2, these calculations appear in rows 31 to 42.

Finally the values of M/WL for the range of h/L values are calculated in rows 43 to 47. Typical examples of the required formulae are given below:

D43: (1 – (D29*D$27*($D$11 – ($F$11*0.25 *D28))))/(D30*(1 + ($E$11*1.316 *(D28^0.75))))
E43: (1 – (E29*E$27*($D$11 – ($F$11*0.25 *E28))))/(E30*(1 + ($E$11*1.316 *(E28^0.75))))
D44: (1 – (D32*D$27*($D$13 – ($F$13*0.25 *D31))))/(D33*(1 + ($E$13*1.316 *(D31^0.75))))
E44: (1 – (E32*E$27*($D$13 – ($F$13*0.25 *D31))))/(E33*(1 + ($E$13*1.316 *(E31^0.75))))

Values of 0.01 have been placed in row 49, for the full range of *d/L* values, to represent the results obtained by Wood's method.

The graphical function of the spreadsheet can now be used to compare the results. Typical plots are shown in Graph 9.1 for $E_w/E_b = 4$ and Graph 9.2 for a ratio of 30. Note that the results of the comparisons are shown on Sheet 9.2 for a modular ratio of 4 and on Sheet 9.2A for a ratio of 30.

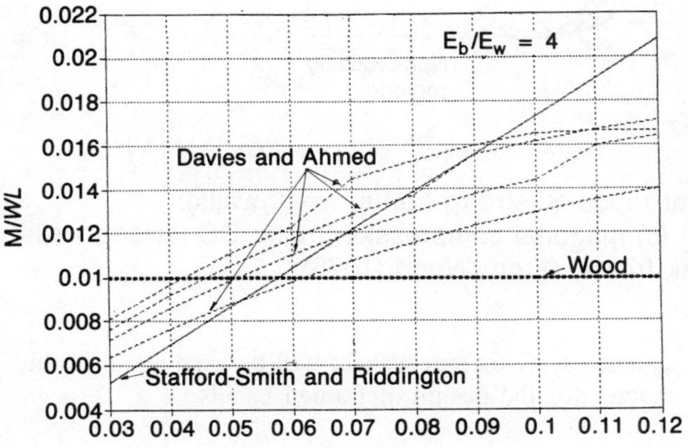

Graph 9.1 Comparison of maximum moments

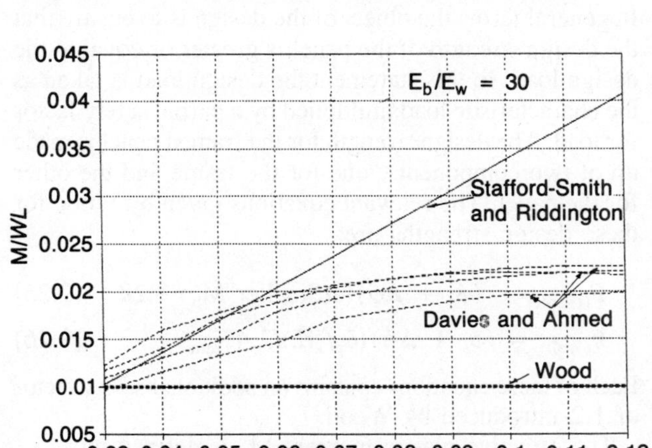

Graph 9.2 Comparison of maximum moments

FRAMED WALL PANELS

9.6 Introduction

If a wall panel is built into a framework of steel or reinforced concrete in such a way that they can interact then the overall stiffness of the combination will be increased. A number of researchers have considered this problem in an attempt to produce a theory which will predict the increase in stiffness for a given set of dimensions but the result of their studies does not appear to have filtered down to practical design. The degree of bonding of the brickwork to the framework may be uncertain and unless the designer can be sure of this then there would be a natural reluctance to use any increase in stiffness derived on the assumption that such bonding exists. Additionally the panel may be removed from the frame at some future date and thus negate any calculated increase stiffness.

In spite of these uncertainties there may be circumstances when the designer might wish to allow for the combined effect of the frame and wall. The method of design described in this chapter is based on the approach developed by Wood (1978). For the development of his theory Wood assumed idealised plastic failure modes and then applied a correction factor to allow for the fact that masonry is not ideally plastic. The method is normally cumbersome for practical design purposes, in that the correction factor has to be read from a series of graphs requiring three interpolations. This problem was illustrated in Chapter 2 where it was shown that a spreadsheet could be designed to apply the three interpolations automatically.

In this chapter one of the spreadsheets developed in Chapter 2 will be incorporated into the Framed Panel Design spreadsheet.

9.7 Failure modes

By studying the failure of a large number of framed panels Wood came to the conclusion that there were four idealised plastic failure modes and these are illustrated in Fig. 9.10.

9.8 Design procedure

Initially a parameter m_d is introduced which is related to the lowest plastic moment of either the beams or the columns of the frame and the characteristic strength of the panel:

$$m_d = 8M_p\gamma_{mm}/f_k t L^2 \qquad [9.20]$$

This parameter represents a frame/wall strength ratio and as such is related to the mode of failure. Wood suggests the following limits for m_d:

(a) If $m_d < 0.25$ collapse will be by 'diagonal compression' or 'corner crushing'.
(b) If $0.25 < m_d < 1$ collapse will be 'shear rotation'.
(c) If $m_d > 1$ collapse will be by 'shear'.

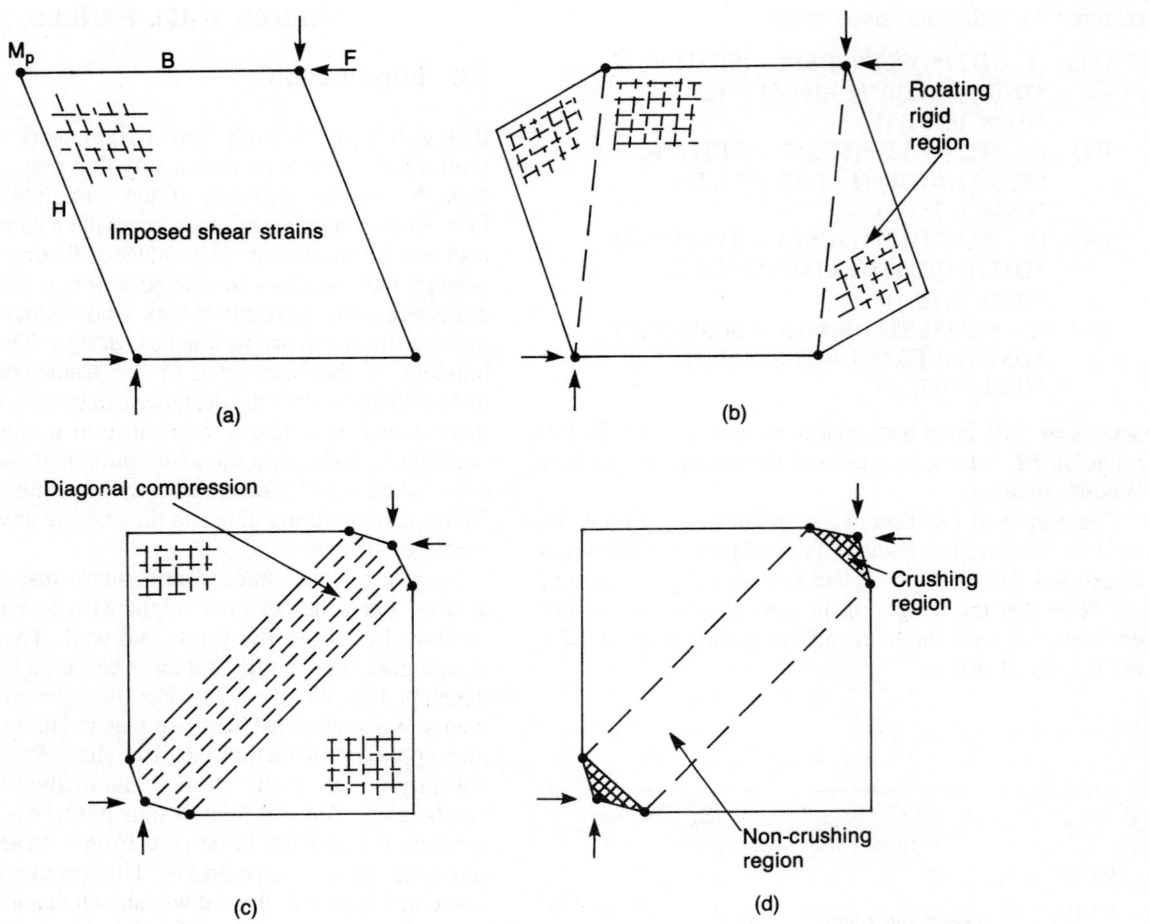

Figure 9.10 Idealised plastic failure modes: (a) shear mode S (strong frame, weak walls); (b) shear rotation mode SR (medium strength walls); (c) diagonal compression mode DC (strong walls, weak frame); (d) corner crushing mode CC (very weak frame). From Wood (1978).

The value of m_d determined from Eqn 9.20 is corrected to m_e using the equation:

$$m_e = m_d/\delta_p \qquad [9.21]$$

where δ_p is defined by:

$$\delta_p = 2.66m_d^2 - 1.37m_d + 0.406 \qquad [9.22]$$

Using the corrected parameter m_e a non-dimensional parameter ϕ_s can be found using:

$$\phi_s = 2/(\sqrt{m_e} + 1/\sqrt{m_e}) \qquad [9.23]$$

The non-dimensional parameter ϕ_s was derived for square panels with identical beams and columns and a correction factor $\Delta\phi$ is introduced for rectangular panels with unequal beams and columns. In fact up to this point the design is fairly routine but the determination of $\Delta\phi$ requires some skill in interpolating the true value from a design curve in which $\Delta\phi$ varies with m_e, μ_p and L/h. The parameter μ_p is equal to the ratio of the plastic moments of beams to the columns as shown in Eqn 9.24:

$$\mu_p = M_{pb}/M_{pc} \qquad [9.24]$$

The design chart is shown in Fig. 2.2 and, in Chapter 2, a spreadsheet approach was developed which enabled the value of $\Delta\phi$ to be found from any given values of m_e, μ_p and L/h. This spreadsheet will be used again in this chapter for the design of framed panels.

9.9 Design objectives

In general terms the object of the design is to ensure that the design strength of the panel is greater or equal to the design load. In this statement the design load is taken as the characteristic load multiplied by a partial safety factor for load. The design strength for the framed panel is made up of two components, one for the frame and the other for the panel. The relevant equations given by Wood for these design strengths are:

$$F_{(Frame)} = (\phi_s + \Delta\phi)(4(\text{Smaller } M_p)/1.2h \qquad [9.25]$$

$$F_{(Wall)} = (\phi_s + \Delta\phi)(\delta_p f_k tL)/2.4\gamma_{mm} \qquad [9.26]$$

Each of these equations contains an additional safety factor of 1.2 introduced by Wood.

The effective plastic moment M_p is defined by:

$$M_p = Zf_k/\gamma_{ms} \qquad [9.27]$$

where Z is section modulus of the beams or columns.

In order to check for the shear force in the beams and

columns it is first necessary to determine the limiting sliding friction, F_o, contributed by the panel. A lower limit of F_o can be defined for the wall, if composite action fails or m_d is very low, by:

$$F_o = Ltf_v/\gamma_{mv} \qquad [9.28]$$

where γ_{mv} is the partial safety factor for the shear strength of masonry and f_v, the characteristic shear strength of masonry, is defined by:

$$f_v = 0.35 + 0.6g_a \text{ N mm}^{-2} < 1.75 \qquad [9.29]$$

for mortar designation (i), (ii) or (iii) and

$$f_v = 0.15 + 0.6g_a \text{ N mm}^{-2} < 1.4 \qquad [9.30]$$

for mortar designation (iv) (BS 5628).

Once F_o has been determined the value of the shear in the columns and beams can be calculated from:

$$\text{Shear in columns} = 0.5(F-F_o) \qquad [9.31]$$

$$\text{Shear in beams} = 0.5(h/L)(F-F_o) \qquad [9.32]$$

9.10 Spreadsheet layout for design

Initial data. Values of the panel height, length and thickness are entered in block C6 to C8 of Sheet 9.3. The characteristic strengths and partial safety factors for the frame and panel material are entered in range C9 to C12 and the section modulus for the column and beam in cells C13 and C14.

To calculate the sliding friction, values of γ_{mv} and the mortar designation are required and these are placed in cells C15 and D16. Finally the value of the superimposed load is entered in cell D17.

Preliminary calculations

Using Eqn 9.27 values of the plastic moment for the columns and beams are calculated in cells F6 and F7 which contain:

```
F6: (C13*C10/(C12*1000000))
F7: (C14*C10/(C12*1000000))
```

The ratio of the plastic moments is entered in cell F8 and the value of L/h in cell F9:

```
F8: +F7/F6
F9: +C7/C6
```

Values of m_d, m_e, δ_p, and ϕ_s are determined from Eqns 9.20 to 9.23 using the formulae:

```
F10: 8*(@MIN(F6..F7))*C11*1000000/C9*C8
     *(C7^2))
F11: (2.663*(F10^2))-(1.371*F10)+0.406
F12: +F10/F11
F13: 2/((F12^0.5)+(1/(F12^0.5)))
```

The modes of failure, as defined by Wood, depend on the value of m_d calculated in cell F10 and an indication of the mode of failure, for the specified parameters, can be obtained from cell G11 using:

```
G11: @IF(F10<=0.25,"Diag. compression",
     @IF(F10>=1,"Shear",
     "Shear Rotation"))
```

Sheet 9.3

```
 1/A    B        C          D          E            F            G            H
  2          Design of Framed Wall Panels
  3          ---------------------------------
  4 Initial Data                      Preliminary Calculations
  5 -------------                     ---------------------------
  6 Height           2000 mm          Mpcol        130.435 kNm
  7 Length           4000 mm          Mpbeam       173.913 kNm
  8 Thickness         110 mm          mup             1.33
  9 Fk                 10 N/mm2       L/h             2.00
 10 Fy                250 N/mm2       md             0.184 Mode of Failure
 11 Gmm               3.1             deltap         0.244 Diag.Compression
 12 Gms              1.15             me            0.7533    or
 13 Zcol           600000 mm3         Phi           0.9901 Corner Crushing
 14 Zbeam          800000 mm3         DeltaPhi      0.0049
 15 Gmv              2.00             Fv              0.95 N/mm2
 16 Mortar Grade                 3
 17 Superimposed Load         1.00 N/mm2
 18
 19 -----------------------------------------------------------------
 20 Design Strength (F)                       Sliding Friction (Fo)
 21            Frame         216.29 kN              Fo =          209 kN
 22            Wall          143.56 kN        Design for Shear
 23            Total         359.85 kN        Col Shear          75.42 kN
 24                                           BeamShear          37.71 kN
```

According to Wood if the frame is very weak then corner crushing can occur and this possibility is included in the limit $m_d < 0.25$. This can be indicated on the spreadsheet, in the form of a reminder, by entering in cells G12 and G13:

G12: **@IF(F10<=0.25," or","")**
G13: **@IF(F10<=0.25,"Corner crushing","")**

which would output the appropriate wording if the condition was true and leave the cells blank if the condition was false.

Values of m_e, μ_p and L/h have now been calculated and it is necessary to determine the correction factor Δ_ϕ from the design chart shown in Fig. 2.2. The block of cells J3 to Z52 contains the necessary interpolation formulae for effecting this calculation. This block, which is not shown on Sheet 9.3, takes the values of m_e, μ_p and L/h from cells F12, F8 and F9 respectively, calculates the value of the correction factor Δ_ϕ and returns it to cell F14 which contains the formula:

F14: **+O28**

The method of interpolating from the design chart using a spreadsheet was illustrated in Chapter 2.

To calculate the value of the characteristic shear strength of the masonry an @IF function, which includes two @MIN functions, is used in cell F15:

F15: **@IF(D16<=3,(@MIN((0.35+(0.6*D17)), 1.75)),(@MIN((0.15+(0.6*D17)),1.4)))**

In this formula a check is first made if the mortar designation is 1 to 3 and, if this is found to be the case, then the minimum value from Eqn 9.29 is used, otherwise the minimum value from Eqn 9.30 is used.

Final design

All the necessary parameters have now been calculated and the lateral force carried by the wall and frame can be calculated using Eqns 9.25 and 9.26. These are placed in cells D21 and D22.

D21: **(F13+F14)*4*(@MIN(F6..F7)) *1000/(1.2*C6)**
D22: **((F13+F4)/1.2)*(0.5*F11*C9*C8 *C7/C11)/1000**

The total force is placed in cell D23 using **+D21+D22**.

Equation 9.28 is used to calculate the value of F_o in cell G21 which contains:

G21: **+C7*C8*F15/(C15*1000)**

and this value is used in cells G23 and G24 to determine the value of the shear force in the column and beam respectively:

G23: **(D23−G21)/2**
G24: **(D23−G21)*C6/(2*C7)**

9.11 Use of spreadsheet 9.3

If the spreadsheet is now protected then the only cells which have to be unprotected in order to allow user input are: C6 to C15, D16 and D17.

Appendix 1

Application to research

The spreadsheet techniques developed in this text can be used to simplify many of the routine calculations and comparisons required during research investigations and two examples are given in this Appendix. The topics selected are from investigations with which the author was involved. It is not the purpose of this text to explain in full the actual research topics or the conclusions reached but rather to illustrate how spreadsheets were used during the studies. As a result details about some of the parameters used may not be explained as fully as would be the case for a research publication and readers should consult the appropriate references for more details. Sufficient information is included to explain the use of the spreadsheets.

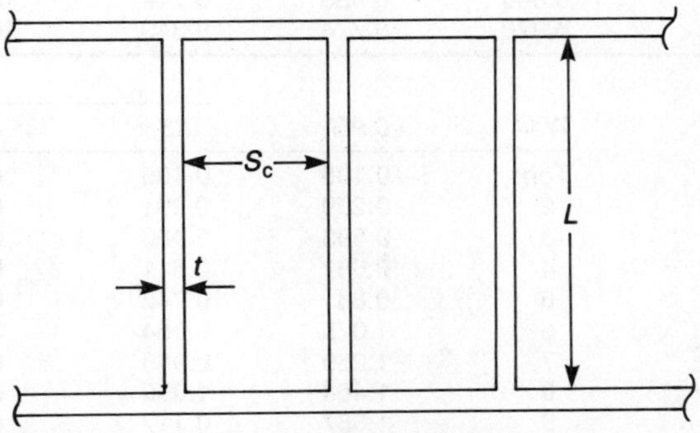

Figure A.1 Plan of simple crosswall

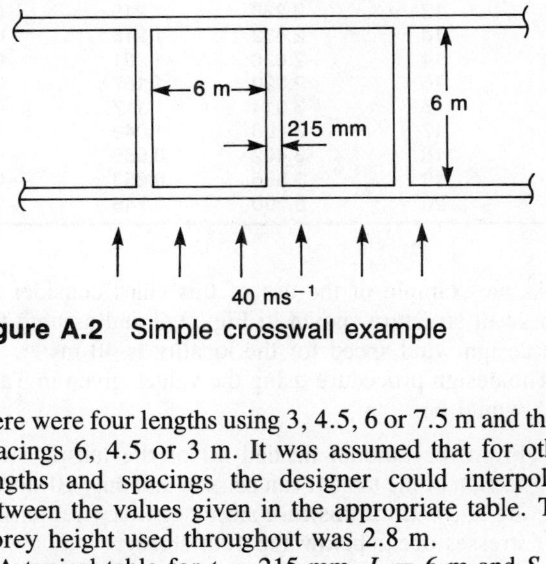

Figure A.2 Simple crosswall example

Example 1. Stresses in simple crosswall structures

Historical background

During the period 1968 to 1977 the Structural Clay Products consortium produced various editions of two booklets which were designed to assist engineers with the preliminary design of multi-storey buildings of simple crosswall construction. Initially the texts used imperial units but this was changed in later editions to the metric system. The Code of Practice in use at the time was CP111 and the tables and graphs in the SCP booklets were based on the requirements of this Code (SCP, 1976 and 1977).

In 1981 the author was asked to prepare a similar set of design tables, and/or charts, to those given in SCP 2 but based on the requirements of the code in BS 5628. This task was completed using a spreadsheet and a series of 24 design tables and 24 design charts produced to cover the range of the variables that had to be considered.

The thickness and length of the wall (t and L) and the spacing between walls (S_c) are important parameters and solutions were obtained for a range of different values of these quantities.

The thickness was taken as either 215 mm or 102.5 mm; there were four lengths using 3, 4.5, 6 or 7.5 m and three spacings 6, 4.5 or 3 m. It was assumed that for other lengths and spacings the designer could interpolate between the values given in the appropriate table. The storey height used throughout was 2.8 m.

A typical table for t = 215 mm, L = 6 m and S_c = 6 is shown in Table A.1.

Each table consisted of two parts, one which gave the values of the maximum compressive stress and the other the minimum value of the tensile stress at each storey level. They contained sufficient information to check for no tension and also to calculate the required value of f_k/γ_{mm} for each storey allowing for different values of the wind speed.

Table A.1
Wall thickness = 215.0 mm; Spacing = 6.0 m and Length = 6.0 m

Wall	*1.4G_k*	*1.4G_k + 1.6Q_k*	Max compressive stress (N mm⁻²) V_s (ms⁻¹) *25*	*30*	*35*	*40*	*45*	*50*
Top	0.239	0.306	0.306	0.306	0.306	0.306	0.306	0.306
2	0.541	0.668	0.668	0.668	0.668	0.668	0.668	0.676
3	0.842	1.023	1.023	1.023	1.023	1.037	1.088	1.146
4	1.144	1.372	1.372	1.372	1.408	1.489	1.581	1.684
5	1.446	1.713	1.713	1.749	1.859	1.985	2.129	2.289
6	1.747	2.055	2.055	2.184	2.342	2.524	2.731	2.961
7	2.049	2.397	2.462	2.644	2.859	3.107	3.388	3.702
8	2.350	2.739	2.890	3.128	3.408	3.732	4.099	4.509
9	2.652	3.081	3.335	3.636	3.991	4.401	4.865	5.384
10	2.954	3.422	3.797	4.168	4.606	5.112	5.686	6.327
11	3.255	3.758	4.276	4.725	5.255	5.867	6.561	7.337
12	3.557	4.093	4.771	5.305	5.937	6.665	7.491	8.414
13	3.859	4.428	5.284	5.911	6.652	7.507	8.476	9.559
14	4.160	4.763	5.813	6.540	7.400	8.391	9.515	10.771
15	4.462	5.098	6.359	7.194	8.180	9.319	10.609	12.051
16	4.763	5.433	6.922	7.872	8.994	10.290	11.758	13.398
17	5.065	5.768	7.502	8.574	9.842	11.304	12.961	14.813
18	5.367	6.103	8.099	9.301	10.722	12.361	14.219	16.295
19	5.668	6.438	8.712	10.052	11.635	13.461	15.531	17.845
20	5.970	6.774	9.343	10.827	12.581	14.605	16.898	19.462

Min tensile stress N mm⁻²

Wall	*0.9G_k*	*25*	*30*	*35*	*40*	*45*	*50*
Top	0.105	0.105	0.105	0.105	0.105	0.105	0.105
2	0.299	0.291	0.287	0.283	0.278	0.272	0.266
3	0.493	0.460	0.445	0.427	0.407	0.384	0.358
4	0.687	0.611	0.578	0.538	0.493	0.441	0.384
5	0.881	0.746	0.687	0.617	0.536	0.444	0.341
6	1.075	0.864	0.771	0.662	0.535	0.392	0.232
7	1.269	9.965	0.832	0.674	0.492	0.285	0.055
8	1.463	1.050	0.868	0.653	0.405	0.124	−0.190
9	1.657	0.117	0.880	0.599	0.275	−0.092	−0.502
10	1.851	1.168	0.867	0.512	0.102	−0.362	−0.882
11	2.044	1.201	0.830	0.392	−0.114	−0.688	−1.329
12	2.238	1.218	0.769	0.238	−0.374	−1.068	−1.843
13	2.432	1.218	0.684	0.052	−0.676	−1.502	−2.425
14	2.626	1.201	0.574	−0.167	−1.022	−1.991	−3.074
15	2.820	1.167	0.440	−0.419	−1.411	−2.535	−3.791
16	3.014	1.117	0.282	−0.705	−1.843	−3.133	−4.575
17	3.208	1.049	0.099	−1.023	−2.319	−3.786	−5.427
18	3.402	0.965	−0.108	−1.375	−2.837	−4.494	−6.346
19	3.596	0.863	−0.339	−1.759	−3.399	−5.257	−7.333
20	3.790	0.745	−0.594	−2.177	−4.004	−6.074	−8.387

As an example of the use of this chart consider the crosswall structure shown in Fig. A.2 and assume that the design wind speed for the locality is 40 ms^{-1}.

The design procedure using the values given in Table A.1 would be:

(1) From the column headed 40 of the tensile stress section of the table it can be seen that only 10 storeys are allowed for this case since negative (i.e. tensile) stresses occur below the 10th storey.
(2) The required design compressive stress at each level is given by the value in the column headed 40 of the compressive stress section of the table.
(3) To calculate the required values of f_k the compressive stresses must be multiplied by γ_{mm}/β where β is a capacity reduction factor which allows for the effects of slenderness and eccentricity. Values of β are given in tabular form in BS 5628.

Note that Table A.1 contains values for other wind speeds but these values are not required for the present example.

Basic theory

The stresses at the top and bottom of each wall result from the roof load, R, the floor load, F, the imposed load, IL, and the wall weight, W.

If the walls are numbered from the top as shown in Fig A.3 then the basic (or characteristic) stresses at the top and bottom of the Ith wall are given by the following equations:

Dead load stresses

$$G_k(\text{top}) = [S_c R + (I-1)S_c F + (I-1)W(h-t_s)]/t \qquad [A.1]$$

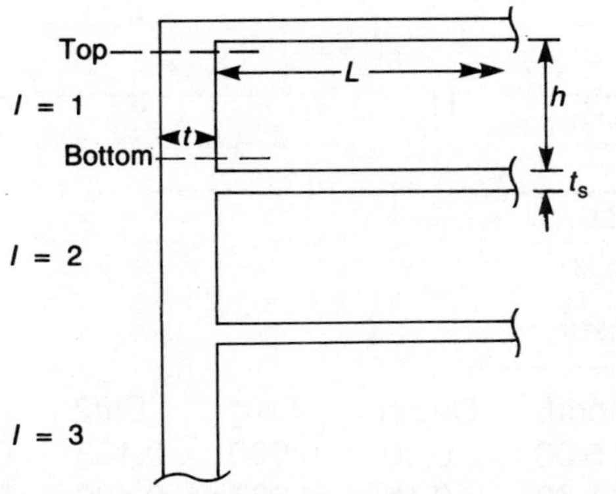

Figure A.3 Cross-section of simple crosswall

$$G_k(\text{bot}) = [S_cR + (I-1)S_cF + IW(h-t_s)]/t \quad [A.2]$$

where h is the storey height and t_s the thickness of the slab.

Imposed stresses (at top and bottom)

$$Q_k = [S_cIL \times X_{(m)}]/t \quad [A.3]$$

where $X_{(m)}$ is a load reduction factor for the imposed load. In the British code this reduction factor is given as 90%, 80%, 70% and 60% of the total imposed load at levels 2 to 5 respectively and 60% for each storey below the 5th.

Wind stresses

$$W_k(\text{top}) = [1.839C_fV_s^2h^2S_c(I-1)^2]/[10^3C_iL^2t] \quad [A.4]$$

$$W_k(\text{bot}) = [1.839C_fV_s^2h^2S_cI^2]/[10^3C_iL^2t] \quad [A.5]$$

where V_s is the wind speed for the locality in metres per sec and C_f is a force correction factor which compensates for the fact that the wind force is shared between more than one wall. Values of this coefficient for particular cases are given in CP3: Chapter V: Part 2 (1972).

C_i is a correction factor to allow for non-rectangular rectangular sections and is the ratio of the second moment of area of the section to the second moment area of a rectangle. The results obtained for non-rectangular sections, although useful for preliminary design, are only approximate since no allowance is made for the true sectional area in calculating the dead and imposed stresses.

Design loads

The basic stresses calculated above must now be multiplied by the partial safety factors and combined in different ways in order to determine the most severe conditions. These conditions must consider both tension and compression. No combination of factored loads must produce tension and the combination that produces the maximum compressive stresses must be determined.

As far as the no tension condition is concerned the British Standard code lists four combinations of factored loads that must be checked to ensure that they do not produce tension. These load combinations are applied to the top of the wall since the dead loads will be smaller and the no tension condition more severe. The load combinations to be checked are:

(a) $0.9G_k - 1.4W_k$
(b) $1.2G_k + 1.2Q_k - 1.2W_k$
(c) $0.95G_k + 0.35Q_k - 0.35W_k$
(d) $1.2G_k + 1.2Q_k - 0.015G_k$

but it can be shown that for the range of values being considered the governing condition for no tension will be combination (a). In a previous study the author has shown that although condition (d) will sometimes govern at the very top of the building the error in using condition (a) instead of (d) is less than 2%.

The combinations to be used to determine the most severe compressive stresses are listed in the British Standard code as:

(a) $1.4G_k + 1.6Q_k$
(b) $1.4G_k + 1.4W_k$
(c) $1.2G_k + 1.2Q_k + 1.2W_k$
(d) $1.05G_k + 0.35Q_k + 0.35W_k$
(e) $1.4G_k + 0.015G_k$
(f) $1.2G_k + 1.2Q_k + 0.015G_k$

but it can be shown that for the range of values being considered the governing condition will be (a), (b), or (c) and the calculations can be limited to these three conditions.

Once the design load has been determined then the design can be completed by ensuring that: Design strength $\geq$ Design load. Using the BS code, this would become:

$$\beta f_k/\gamma_{mm} \geq \text{Design load} \quad [A.6]$$

or

$$f_k \geq (\text{Design load})\gamma_{mm}/\beta \quad [A.7]$$

Note that the design loads, as calculated above, have already been divided by the thickness and unit length and therefore represent stresses rather than loads.

Limiting values of the parameters

As stated above, limits were placed on the value of certain parameters in the original study in order to limit the number of tables and/or graphs required for the final publication. It was assumed that for the preliminary design of simple crosswalls with dimensions different from those given in the publication the user would be able to interpolate between the values used in the study.

Using a spreadsheet eliminates the requirement for any interpolation and generally solutions can be obtained for any set of values. However, in the examples given in the following sections the wall thicknesses are confined to either 215 mm or 102.5 mm and the wall weights for these thicknesses taken as 4.8 and 2.7 respectively as given in the original document. In a similar way, although any value of the spacing between 0 and 7 m can be entered, the values of the roof and floor loading and the thickness of the floor slab are calculated by interpolation from the values given in the original document for spacings of 3, 4.5 and 6 m. See Table 1 on Sheet A.1.

Sheet A.1

1/A	B	C	D	E	F	G	H	I	J	K
2			Design of Simple Crosswall Structures							
3		-------	-------	-------	-------	-------	-------	-------		
4 Initial Data										
5 -------	-------	-------								
6 Thickness		**215**				Table 1				
7 Spacing		**6.00**				-------				
8 Length		**6.00**	Long	Span	RoofL	FloorL	Depth	Diff1	Diff2	Diff3
9 Length		**0.00**	Short	3.00	3.80	5.00	0.10	0.000	0.133	0.033
10 StoreyHt		**2.60**		4.50	3.80	5.20	0.15	0.267	0.400	0.033
11 WindSp		**40.00**		6.00	4.20	5.80	0.20	0.000	0.000	0.000
12 Cf		**1.00**		7.00	4.20	5.80	0.20			
13 Ci		**1.00**								
14 Gmm		**2.80**		Table 2	(Beta against S.R)					
15 -------	-------	-------	-------	-------	-------	-------	-------	-------	-------	-------
16 S.R.		0	6	8	10	12	14	16	18	20
17 Beta		1.00	1.00	1.00	0.97	0.93	0.89	0.83	0.77	0.70
18 Diff		0.000	0.000	−0.005	−0.007	−0.007	−0.010	−0.010	−0.012	−0.013
19 -------	-------	-------	-------	-------	-------	-------	-------	-------	-------	-------
20 Preliminary Calculations										
21 -------	-------	-------	-------	-------	-------	-------				
22 RoofLd		4.200		WallWt	4.800					
23 FloorLd		5.800		S.R	9.070					
24 SlabTh		0.200		Beta	0.995					
25 Imposed		1.500		WindSp	40.000					
26 -------	-------	-------	-------	-------	-------	-------	-------	-------		
27 Check for Tension				Final Design						
28 -------	-------	-------	-------	-------	-------	-------	1.2Gk	Fk	Fk	
29		No			1.4Gk	1.4Gk	1.2Qk	Long	Short	
30	I	Tension		Storeys	1.6Qk	1.4Wk	1.2Wk	Wall	Wall	
31	1	1	Top	10	0.306	0.261	0.274	0.86	0.00	
32	2	1		9	0.661	0.627	0.628	1.86	0.00	
33	3	1		8	1.003	1.037	1.009	2.92	0.00	
34	4	1		7	1.331	1.489	1.417	4.19	0.00	
35	5	1		6	1.647	1.985	1.852	5.59	0.00	
36	6	1		5	1.988	2.524	2.345	7.11	0.00	
37	7	1		4	2.330	3.107	2.874	8.75	0.00	
38	8	1		3	2.672	3.732	3.440	10.51	0.00	
39	9	1		2	3.014	4.401	4.043	12.39	0.00	
40	10	1	Ground	1	3.356	5.112	4.683	14.39	0.00	
41	11	0		0						
42	12	0		0						
43	13	0		0						
44	14	0		0						
45	15	0		0						
46	16	0		0						
47	17	0		0						
48	18	0		0						
49	19	0		0						
50	20	0		0						

Solution by spreadsheet A.1

Initial Data. Values of the thickness, spacing, length, storey height, wind speed, force coefficient, correction factor for 2nd moment of area and the partial safety factor for the compressive strength of masonry are entered in cells C6 to C14. Note that two values have been entered for length in order to allow for the preliminary design of crosswalls made up of two lengths. It is assumed that, for two walls of unequal length, the length of the longer wall is entered in cell C8 and of the shorter wall in cell C9. If there is only one length then this value would be entered in cell C8 with 0.0 in cell C9.

Tables 1 and 2 (Sheet A.1). Table 1 in Sheet A.1 is used in order to determine the values to be used for the roof and floor loads and also the depth of the slab. As stated above, Table 1 is constructed from values given in SCP (1977) but there would be no difficulty in constructing a table with other values or ignoring the table and using known values of slab depth, roof load and floor load. Appropriate values of the span, roof and floor loads and slab depth are entered in the block E9 to H12 and the difference ratios required for interpolation entered in block I9 to K11.

Table 2 of Sheet A.1 is used to calculate the value of β from the slenderness ratio and to achieve this, values of β and appropriate slenderness ratio together with the difference ratios required for interpolation are entered in block C16 to O18. On Sheet A.1 only columns A to K are shown. The values required for this table have been taken from BS 5628: Part 1: (Table 7) on the assumption that the eccentricity is less than $0.05t$. Other eccentricities could be allowed for by extending the table in a similar manner to that shown previously. This example illustrates the different ways available for the construction of interpolation tables. Since Table 1 is a vertical table the function used to reference the values will be @VLOOKUP. Table 2 is a horizontal table and the reference function is therefore @HLOOKUP.

Preliminary calculations

This section is used to determine the values of the loading to be used in subsequent calculations. These values have been determined from Table 1 or 2 or from the predefined wall weights quoted previously for wall thicknesses of 215 mm or 102.5 mm. The value of the imposed load in cell C25 has been entered directly without any preliminary calculation and has only been included in this section to keep all the loads together.

The roof load, floor load and slab depth are entered in cells C22 to C24 which contain the interpolation formulae:

```
C22: @VLOOKUP($C$7,$E$9..$K$12,1)
     +($C$7-(@VLOOKUP
     ($C$7,$E$9..$K$12,0)))*(@VLOOKUP
     ($C$7,$E$9..$K$12,4))
C23: @VLOOKUP($C$7,$E$9..$K$12,2)
     +($C$7-(@VLOOKUP
     ($C$7,$E$9..$K$12,0)))*(@VLOOKUP
     ($C$7,$E$9..$K$12,5))
C24: @VLOOKUP($C$7,$E$9..$K$12,3)
     +($C$7-(@VLOOKUP
     ($C$7,$E$9..$K$12,0)))*(@VLOOKUP
     ($C$7,$E$9..$K$12,6))
```

The wall weight is calculated in cell F22 using:

```
F22: @IF(C6=215,4.8,2.7)
```

and, to calculate the value of β the slenderness ratio is first calculated in cell F23, assuming that the effective length is $0.75L$ and then β determined from Table 2 using the formula in cell F24:

```
F23: 0.75*(C10*1000/C6)
F24: @HLOOKUP($F$23,$C$16..$O$18,1)
     +($F$23-(@HLOOKUP($F$23,
     $C$16..$O$18,0)))*(@HLOOKUP($F$23,
     $C$16..$O$18,2))
```

Although the value of the wind speed has been entered in cell C11 a correction is applied to this value when the wall is divided into two lengths. The formula for this correction factor has been derived in SCP (1977) as:

$$V_2/V_s = \sqrt{1/(1+(L_1/L_2)^3)} \qquad [A.8]$$

where V_2 is the corrected value of the wind speed V_s to be used for the longer wall. L_1 and L_2 are the lengths of the short and long walls respectively. This corrected value is calculated in cell F25 using the condition that if there is only one length (C9 = 0), then use V_s, otherwise use V_2 from Eqn A.8. The formula in cell F25 is:

```
F25: @IF(C9=0,C11,C11*@SQRT
     (1/(1+((C9/C8)^3))))
```

It is also shown in SCP (1977) that the stresses in the shorter wall can be determined from the stresses in the longer wall by multiplying by the ratio L_1/L_2 and this equation will be used later to calculate the value of f_k for the short wall.

No tension design

To determine how many storeys are permitted for the given parameters the condition $0.9G_k - 1.4W_k \geq 0$ is used in cells C31 to C50 in such a way that 1 is output if the condition is satisfied and 0 if it is not. In Sheet A.1 the values entered are such that the condition is satisfied for 10 storeys. The numbers 1 to 20 entered in cells B31 to B50 act as counters and give the value of I required in the basic equations A.1 to A.5. Examples of the formulae in cells C31 to C50 are:

```
C31: @IF((0.9*($C$7*$C$22+(B31-1)
     *$C$7*$C$23+(B31-1)*$F$22
     *($C$10-$C$24))/$C$6)-(1.4*1.839
     *$C$12*($F$25^2)*(($C$10^2)*$C$7
     *((B31-1)^2))/(1000*$C$13*($C$8^2)
     *$C$6))>0,1,0)
C32: @IF((0.9*($C$7*$C$22+(B32-1)
     *$C$7*$C$23+(B32-1)*$F$22
```

```
*($C$10-$C$24))/$C$6)-(1.4*1.839
*$C$12*($F$25^2)*(($C$10^2)*$C$7
*((B32-1)^2))/(1000*$C$13*($C$8^2)
*$C$6))>0,1,0)
```

The string value 'Top' is entered in cell D31 to represent the top storey and a search is made, by the spreadsheet, for the lowest storey by placing formulae in cells D32 to D50 as represented by:

```
D32: @IF(E32=1,"Ground","")
```

so that the string 'Ground' will be entered for ground floor storey and the null string, "", for the other storeys.

This is followed by arranging for the ouput to row E below cell E30 to contain an integer representing the storey level. In the example shown on Sheet A.1 the integer 10 appears adjacent to the top storey and the integer 1 adjacent to the ground floor. These values are purely cosmetic and are not used in subsequent operations although they indicate at a glance how many storeys are allowed under the no tension rule. The formulae required to achieve this output is of the form:

```
E31: @IF(@SUM(C31..$C$50)>0),
     @SUM(C31..$C$50),0)
E32: @IF(@SUM(C32..$C$50)>0),
     @SUM(C32..$C$50),0)
```

in which the @SUM function is used to count the number of storeys without tension. To achieve this use is made of a variable summation as illustrated in the above formulae. If this @SUM becomes zero then the @IF function would output a zero.

This completes the no tension check and the user is informed of how many storeys are permitted for the given parameters. If more storeys were required then this would require a change in some of the initial data or the use of reinforcement in the lower parts of the building.

Design for compression

As discussed above under Design Loads, the maximum compressive stress results from one of the following load combinations:

(a) $1.4G_k + 1.6Q_k$
(b) $1.4G_k + 1.4W_k$
(c) $1.2G_k + 1.2Q_k + 1.2W_k$

and for the range of parameters normally used it will be found that condition (a) will govern at the top of the building and condition (b) at lower levels. However, the decision as to which condition governs at any particular level can be left to formulae in the spreadsheet and this is the method used in block I31 to J50. First the value of each of the above load combinations are evaluated at each level, in block F31 to H50 using formulae of the type:

```
F31: @IF(E31>0,(1.4*($C$7*$C$22)+(B31-1)
     *$C$7*$C$23+B31*$F$22
     *($C$10-$C$24))/$C$6)+(1.6*$C$7
     *$C$25)/$C$6,"")
F32: @IF(E32>0,(1.4*($C$7*$C$22+(B32-1)
     *$C$7*$C$23+B32*$F$22
     *($C$10-$C$24))/$C$6)+(1.6*$C$7
     *$C$25*1.8)/$C$6,"")
```

The value 1.8 used in the formula in cell F32 represents the reduction of 90 per cent allowed for the imposed load. There are two floors so that the reduction is calculated using 2 × 0.9. In a similar way the equivalent constant for the imposed load reduction in cells F33 to F35 are 2.4, 2.8 and 3. The reduction for cells below F35 can be represented by a formula which can be copied to the lower cells. The content of cell F36 is:

```
F36: @IF(E36>0,(1.4*($C$7*$C$22+(B36-1)
     *$C$7*$C$23+B36*$F$22
     *($C$10-$C$24))/$C$6)+(1.6*$C$7
     *$C$25*(3+0.6*(B36-5)))/$C$6,"")
```

and this formula can be copied to cells F37 to F50 in the normal way. The imposed load term appearing in cells H31 to H50 is treated in the same way. Cells G31 and G32 contain:

```
G31: @IF(E31>0,(1.4*($C$7*$C$22+(B31-1)
     *$C$7*$C$23+B31*$F$22
     *($C$10-$C$24))/$C$6)+(1.4*1.839
     *$C$12*($F$25^2)*($C$10^2)
     *$C$7*(B31^2))/(1000*$C$13*($C$/8^2)
     *$C$6),"")
G32: @IF(E32>0,(1.4*($C$7*$C$22+(B32-1)
     *$C$7*$C$23+B32*$F$22
     *($C$10-$C$24))/$C$6)+(1.4*1.839
     *$C$12*($F$25^2)*($C$10^2)
     *$C$7*(B32^2))/(1000*$C$13*($C$/8^2)
     *$C$6),"")
```

and cells H31 and H32:

```
H31: @IF(E31>0,1.2*(($C$7*$C$22+(B31-1)
     *$C$7*$C$23+B31*$F$22
     *($C$10-$C$24))+($C$7*$C$25)+((1.839
     *$C$12*($F$25^2)*($C$10^2)*$C$7
     *(B31^2))/(1000*$C$13
     *($C$8^2))))/$C$6,"")
H32: @IF(E32>0,1.2*(($C$7*$C$22+(B32-1)
     *$C$7*$C$23+B32*$F$22
     *($C$10-$C$24))+($C$7*$C$25
     *1.8)+((1.839*$C$12*($F$25^2)*($C$10^2)
     *$C$7*(B32^2))/(1000*$C$13
     *($C$8^2))))/$C$6,"")
```

Note that the constant 1.8 used in the formula in cell H32 would be changed in the lower cells in accordance with the change in the reduction factor for the imposed load as described previously.

Finally the required values of f_k are determined in the block I31 to J50 and the appropriate formulae are represented by:

```
I31: @IF(E31>0,(@MAX(F31..H31))
     *$C$14/$F$24,"")
J31: @IF(E31>0,(@MAX(F31..H31))
     *$C$14*$C$9/$F$24*$C$8,"")
```

Example 2. Draft Eurocode/BS 5628 comparison

In 1987 a draft of a new *European Code for Masonry Structures (EC6)* was published and, in order to comment on the draft, comparisons were made between the results obtained from the new and existing codes. Comparisons were made for different types of structures with different loading cases and some of the results of these comparisons were published in *Masonry* (4), *The Proceedings of the 2nd Masonry Conference (1990)*.

The author was involved in the comparative study of a 10-storey building and since the results were obtained using a spreadsheet a description of the layout used is given below. In this description only general layout details are given since the equations in the draft code are of secondary importance and subject to revision.

The study included both brickwork and blockwork, solid and cavity walls, walls of different thicknesses, three different mortar strengths, three different loading cases, ten storey levels and the recommendations of the two design codes. This represents a large number of designs which would be very tedious without the aid of a computer program. To write a computer program which includes all the possible variations and all the input and output routines would be time consuming but the results can be obtained by spreadsheet using only 120 rows, most of which are repetitive and can be copied.

Spreadsheet layout

The four top storeys of a ten-storey cavity wall structure are shown in Fig. A.4. Levels are numbered downwards from the top of the building.

Given the dimensions, the dead and imposed loading and the density of the wall material for this building the dead and imposed loading at each floor level can be determined as described in Example 1 above. The values calculated from the initial data given in cells C13 to C20 are shown in F13 to F22 (for G_k) and G13 to G22 (for Q_k) in Sheet A.2. The wall density is shown in C20, the wall thicknesses in C15 and C16 and the reduction factors for the imposed loading in D13 to D22.

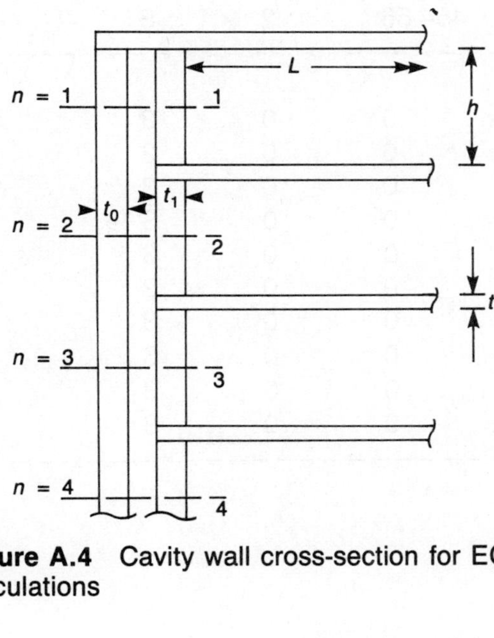

Figure A.4 Cavity wall cross-section for EC6 calculations

To calculate G_k in accordance with BS 5628 the top of each wall, immediately below the floor slab, has been used to define each level. Using EC6 it was more convenient to use the mid-height of each wall for the reference level so that it was necessary to calculate alternative values of G_k (shown in Sheet A.2, Page 1, Column I and named EGk).

To determine the wind loading at each level arbitrary wind stresses (σ) were selected at the base of the building and the wind distribution assumed to be parabolic so that at any level I the wind stress could be taken as $\sigma I^2/100$. The value of σ shown on Sheet A.2 is 1.2 N mm^{-2} in cell C19 but for the complete study comparative results were also obtained for other values such as 0.1 and 0.7 in order to cover a range of wind loadings from light to heavy.

On the basis of the above, the formulae entered in cells F13 to I13 were:

F13: **(E13*$C$13*$C$18/2)+((E13−1)*$C$20*C$17*$C$16/1000)+(0*$C$20*3*$C$15*$C$17/1000)**
G13: **(E13*$C$14*$C$18/2)*D13**
H13: **((E13/10)^2)*$C$19*$C$16**
I13: **(E13*$C$13*$C$18/2)+((((2*E13)−1)/2)*$C$20*$C$17*$C$16/1000)+(0*$C$20*3*$C$15*$C$17/1000)**

The zero values at the beginning of the last bracketed terms of the formulae in F13 and I13 are due to the fact that, for the construction shown in Fig. A.4, the external wall does not contribute to the loadings for the first three storeys. By entering the formulae in the form shown they can be copied to the lower levels and only a minimal change to the formulae is required to allow for the weight of the external wall. The changes to the zero value required at other levels are:

Use 1 for levels 4, 5 and 6;
Use 2 for levels 7, 8 and 9;
Use 3 for level 10.

The three loading cases considered are shown in rows 5 to 10 and the calculation for each case and both codes are shown in the block B26 to G35 using for example:

B26: **1.4*F13+1.6*G13**
C26: **1.35*I13+1.5*G13**

The maximum value for each code and at each level are shown in block B38 to C47 using:

B38: **@MAX(B26,D26,F26)**
C38: **@MAX(C26,E26,G26)**

In order to determine if the loading case, which governs for design, changes between codes, some additional entries were made in block D38 to I47. The first three columns of this block were used to determine which loading case of BS 5628 had been used to calculate the maximum value in the range B38 to B47. If the Case 1 value was the same as the maximum value then a 1 would be output in column D, otherwise a 0 would appear.

Sheet A.2 (Page 1)

A/1	B	C	D	E	F	G	H	I	J
2	COMPARISON BETWEEN BRITISH AND EUROCODE FOR MASONRY								
3	CAVITY WALL								
4	--								
5			CASE 1	BS5268 = 1.40*Gk + 1.6*Qk					
6			-----	E.6 = 1.35*Gk + 1.5*Qk					
7			CASE 2	BS5628 = 1.40*Gk + 1.4*Wk					
8			-----	E.6 = 1.35*Gk + 1.5*Wk					
9			CASE 3	BSB628 = 1.20*(Gk+Qk+Wk)					
10			-----	E.6 = 1.35*(Gk+Qk+Wk)					
11					Gk	Qk	Wk	EGk	
12		Data	Red	Level	kn/m	kn/m	kn/m		
13	gk–load	**4.00**	1.00	1	9.00	4.50	1.23	11.61	
14	qk–load	**2.00**	0.90	2	23.23	8.10	4.92	25.84	
15	Ext wall	**102.50**	0.80	3	37.46	10.80	11.07	40.07	
16	Int Wall	**102.50**	0.70	4	67.37	12.60	19.68	69.98	
17	Height	**2.55**	0.60	5	81.59	13.50	30.75	84.21	
18	Span	**4.50**	0.60	6	95.82	16.20	44.28	98.43	
19	Sigma	**1.20**	0.60	7	125.73	18.90	60.27	128.34	
20	Rho	**20.00**	0.60	8	139.96	21.60	78.72	142.57	
21			0.60	9	154.19	24.30	99.63	156.80	
22			0.60	10	184.10	27.00	123.00	186.71	
23	--								
24	CASE1		CASE2		CASE3		MAX		
25	BS	E6	BS	E6	BS	E6	BS	E6	
26	19.80	22.43	14.32	17.52	17.68	23.41	1	3	
27	45.48	47.04	39.41	42.27	43.50	52.46	1	3	
28	69.72	70.29	67.93	70.70	71.19	83.62	3	3	
29	114.47	113.37	121.86	123.99	119.57	138.05	2	3	
30	135.83	133.93	157.28	159.80	151.01	173.42	2	3	
31	160.07	157.19	196.14	199.31	187.56	214.53	2	3	
32	206.26	201.61	260.40	263.67	245.88	280.14	2	3	
33	230.50	224.87	306.15	310.55	288.33	327.90	2	3	
34	254.74	248.13	355.34	361.12	333.74	378.98	2	3	
35	300.93	292.56	429.93	436.56	400.91	454.56	2	3	
36	--								
37	Max	Max							
38	19.80	23.41	1	0	0	0	0	3	
39	45.48	52.46	1	0	0	0	0	3	
40	71.19	83.62	0	0	3	0	0	3	
41	121.86	138.05	0	2	0	0	0	3	
42	157.28	173.42	0	2	0	0	0	3	
43	196.14	214.53	0	2	0	0	0	3	
44	260.40	280.14	0	2	0	0	0	3	
45	306.15	327.90	0	2	0	0	0	3	
46	355.34	378.98	0	2	0	0	0	3	
47	429.93	454.56	0	2	0	0	0	3	
48	--								

Sheet A.2 (Page 2)

52	B	C	D	E	F	G	H	I	J
53	SPAN	**4.50**	Height	**2.55**	Ext wall	102.50	qk−load	2.00	
54	GAM−M.EC	**3.00**	ES/EW	**2.00**	Int wall	102.50	gk−load	4.00	
55	sigma	1.20	Kbrk	**0.45**					
56	GAM−M.BS	**3.50**	Kblk	**0.45**					
57			−−−−−−−−	−−−−−−−−	−−−−−−−−	−−−−−−−−	−−−−−−−−		
58	Reaction	18.90	ALPHA	5.17	TO2/3	136.67	Wall−in	7.06	
59	T−slab	170.00	N	167.59	T−EFF	136.67	Wall−out	21.17	
60	MTS	1.30	EA/T	0.04		BRICK		BLOCK	
61	MBS	0.56	SI−Ratio	13.99	DELTA	0.70	DELTA	1.40	
62	MD	1.67			D^0.75	0.77	D^0.75	1.29	
63	MTOP	2.97	MBOT	−1.11					
64	M11	0.68	M11	1.14					
65	M4.5	0.61	M4.5	1.03					
66	M2.5	0.51	M2.5	0.87					
67	−−−−−−−−	−−−−−−−−	−−−−−−−−	−−−−−−−−	−−−−−−−−	−−−−−−−−			
68	Reaction	19.80							
69	Wall−in	7.32							
70	Wall−out	21.96		LOAD	EX/T	EA/T			
71	19.80		1	19.80	0.22	0.07			
72	19.08		2	45.48	0.10	0.07			
73	18.36		3	69.72	0.06	0.07			
74	17.64		4	92.52	0.05	0.07			
75	16.92		5	113.87	0.04	0.07			
76	16.92		6	138.11	0.03	0.07			
77	16.92		7	162.35	0.03	0.07			
78	16.92		8	186.59	0.02	0.07			
79	16.92		9	210.83	0.02	0.07			
80	16.92		10	235.07	0.02	0.07			
81	−−−−−−−−	−−−−−−−−	−−−−−−−−	−−−−−−−−	−−−−−−−−	−−−−−−−−	−−−−−−−−	−−−−−−−−	−−−−−−−−
82									
83					BS 5628 with GAM−M =			3.50	
84							Fb(2A)		
85		LOAD	EM/T	BETA	FK	1	2	3	1
86	1	19.80	0.22	0.61	1.11	15.40	15.40	15.40	3.60
87	2	45.48	0.12	0.83	1.88	15.40	15.40	15.40	3.60
88	3	71.19	0.10	0.87	2.79	15.40	15.40	15.40	3.60
89	4	121.86	0.10	0.89	4.67	15.40	15.40	15.40	5.46
90	5	157.28	0.09	0.90	5.95	15.40	15.95	17.41	7.16
91	6	196.14	0.09	0.91	7.35	17.86	21.50	24.27	9.25
92	7	260.40	0.08	0.92	9.69	25.43	31.93	37.77	13.31
93	8	306.15	0.08	0.92	11.33	31.19	40.10	48.78	16.58
94	9	355.34	0.08	0.93	13.10	37.81	49.67	62.02	20.48
95	10	429.93	0.08	0.93	15.81	48.75	65.84	85.02	27.21
96	−−−−−−−−	−−−−−−−−	−−−−−−−−	−−−−−−−−	−−−−−−−−	−−−−−−−−	−−−−−−−−	−−−−−−−−	−−−−−−−−

Sheet A.2 (Page 3)

102	B	C	D	E	F	G	H	I	J
103					EC.6/BS5628 Ratios				
104		LOAD	EM/T	PHI	FKratio	1	2	3	1
105	1	23.41	0.45	0.00	0.00	1.00	1.00	1.00	1.00
106	2	52.46	0.23	0.33	2.50	1.00	1.00	1.24	1.84
107	3	83.62	0.17	0.47	1.86	1.00	1.12	1.41	2.10
108	4	138.05	0.12	0.58	1.48	1.44	1.65	2.08	2.03
109	5	173.42	0.11	0.61	1.40	1.83	2.03	2.34	1.97
110	6	214.53	0.10	0.63	1.35	2.00	1.90	2.12	1.93
111	7	280.14	0.09	0.66	1.28	1.89	1.72	1.83	1.80
112	8	327.90	0.08	0.67	1.26	1.86	1.66	1.72	1.75
113	9	378.98	0.08	0.68	1.24	1.83	1.59	1.61	1.69
114	10	454.56	0.07	0.69	1.21	1.76	1.50	1.46	1.58
115									

116 For EC.6 calculations delta = 0.7 for brick and 1.4 for block
117 For BS 5628 Dry/Wet Strengths = 1.1 for brick and 1.2 for block
118
119 Mortar strengths EC.6 1 = 15N/mm2 BS5628 1 = 1/0.25/3
120 2 = 10N/mm2 2 = 1/0.5/4.5
121 3 = 5N/mm2 3 = 1/1/6

Similarly if the Case 2 value was the same as the maximum value then a 2 would be output in column E otherwise a 0 would appear. Similarly for Case 3 and also for the three cases for EC6. Then by finding the maximum value for the set of three in each row the loading case used can be determined. These calculations for BS 5628 were repeated in columns G, H and I for the determination of the governing case for EC6. Finally this information was gathered together into the block H26 to I35 so that the governing cases can be compared. On Sheet A.2 the governing case for EC6 was Case 3 throughout whilst for BS 5628 it was mainly Case 2.

The formulae entered in some of the above cells are:

D38: **@IF(B38=B26,1,0)**
E38: **@IF(B38=D26,2,0)**
H26: **@MAX(D38,E38,F38)**

Having calculated the value of the design load at each level it is now possible to proceed with the determination of the compressive strength of the units required at each level following the recommendations of BS 5628.

These calculations, which are shown on page 2 of Sheet A.2 in rows 84 to 95, first require the determination of the eccentricity, the value of β and f_k. These preliminary calculations are performed in blocks B71 to G80 and C86 to C95. Typical examples of the required formulae are:

B71: **((1.6*1*$I$53)+(1.4*$I$54))*$C$53/2**

representing the reaction from the loaded slab and allowing for the reduction in the imposed load so that the 1 in the first bracket becomes 0.9, 0.8, 0.7 and 0.6 for lower levels.

E71: **(D71*B71)+((D71−1)*$C$69)**

representing the total load on the inner leaf at each level.

F71: **($G$59/6)/$G$54**
F72: **+$C$68*$F$71/E72**

in which the formulae in F72 to F80 use the value calculated in cell F71. See Chapter 5 of Hendry, *et al.* (1987).

G71: **((($E$53*1000*0.75/$G$59)^2)/2400) −0.015)**
D86: **@MAX(F71,(0.6*F71)+G71)**
E86: **@IF(D86−0.05<=0,1,1.1*(2*D86)))**

using equations 1, 2 and 4 of Appendix B of BS 5628.

F86: **($C$56*C86)/(E86*$G$54)**

using equation A.7.

Cells G85, H85 and I85 are marked 1, 2, and 3 to represent the mortar designations used for the calculations and these mortar designations are defined in block H119 to H121. The three columns are collectively labelled Fb(2A) to show that the values are derived from Table 2A of BS 5628. The values given in Table 2A were first expressed as a 2nd degree polynomial so that given a value of f_k and a mortar designation the value of f_b could be calculated. For the type of construction being considered solid walls would be classed as narrow (Clause 23.1.2) so a factor of 1.15 has to be introduced if the formula is to allow for both solid and cavity walls. Additionally the values must be multiplied by a factor of 1.1 to allow for the dry/wet strength ratio. Typically the formula in G86 is:

G86: **(@MAX(14,@IF(G53>0,(−0.9+(2.16 *F86/1.15)+(0.08*((F86/1.15)^2))), (−0.9+(2.16*F86)+(0.08*(F86^2))))))*1.1**

where the numerical values represent the coefficients of

the polynomial representing the values in Table 2A of BS 5628 for mortar of designation (i). Note that an upper limit of **14*1.1** has been imposed on the value of f_b. The calculations in block G86 to I95 are repeated in block J86 to L95 using the values from Table 2d of BS 5628 so that both brickwork and blockwork results are available on the same spreadsheet. Columns K and L are not shown on Sheet A.2.

To calculate comparative values using the draft of EC6 a number of equations given in the code are required. These equations are given below as equations A.9 to A.21, without derivation since they are used only to illustrate their use in obtaining a spreadsheet solution.

$$\alpha = (h/L)(E_s/E_w)(t_s/t_i)^3 \quad [A.9]$$

where t_s refers to the thickness of the slab which is taken as 150 mm for a span of 3.5 m and 170 mm for a span of 4.5 m.

$$N_s = 4(\alpha + 2)^2 - (\alpha + 1)^2 \quad [A.10]$$

$$S = (0.75h/t_e)1000 \quad [A.11]$$

where t_e refers to the effective thickness calculated as the maximum value of the thickness of the inner wall or two-thirds the sum of the thicknesses of both walls.

$$\text{MTS} = (\alpha + 2)1.5q_k L^2/(2N_s) \quad [A.12]$$

$$\text{MBS} = (\alpha + 1)1.5q_k L^2/(4N_s) \quad [A.13]$$

$$\text{MD} = (5.4L^2)/(8\alpha + 24) \quad [A.14]$$

$$\text{MTop} = \text{MTS} + \text{MD} \quad [A.15]$$

$$\text{MBot} = \text{MBS} - \text{MD} \quad [A.16]$$

Equations A.12 to A.16 are used to determine the bending moments at the top and bottom of a wall.

The cell formulae required for these equations are fairly straightforward and have been entered on Sheet A.2 in rows 58 to 62.

$$M_{fm} = \Delta^{0.75} K_b f_{ms}{}^{0.25} \quad [A.17]$$

where K_b is taken as 0.45, the mortar designation f_{ms} as either 15, 10 or 5 and Δ for brickwork as 0.7 and for blockwork as 0.77. For the comparison this equation is used six times, three for brickwork and three for blockwork using each of the mortar strengths in turn. The calculations are made in cells C64 to C66 and E64 to E66 using values of $\Delta^{0.75}$ previously calculated in cells G62 and I62. Examples of the required entries are:

C64: **+G62*E55*(15^0.25)**

where 15 = mortar strength.

G62: **(G61^0.75)**.

Finally to determine the EC6 values it is necessary to calculate the eccentricity ratio e_m/t and hence the value of ϕ. Using ϕ the value of f_k can be determined and hence the required value of f_b. The equations required for these steps are:

$$e_m/t = 1000(\text{MTop}+\text{MBot})/(2\text{Load(n)}t_i + 750h/(450t_i) \quad [A.18]$$

$$\phi = \text{Max}[0,(1.14(1-2e_m/t))-0.02S_R)] \quad [A.19]$$

$$f_k = \gamma_{mm}\text{Load(n)}/\phi t_i \quad [A.20)$$

$$f_k = 0 \text{ if } \phi = 0$$

$$f_b = \text{Max}[15.4,(f_k/M_{fm})^{1.333}] \quad [A.21]$$

The values of the load at each level, Load(n), have been determined previously in C38 to C47 and these can now be transferred to C105 to C114. Examples of the entries for the block D105 to F121 are:

D105: **(1000*($C$63+$E$63)/(2*C26*$G$54))+$E$60**

E105: **0**

E106: **@MAX(0,((1.14*(1−(2*D106)))−(0.02*$E$61)))**

F105: **(@IF(E105=0,0,($C$54*C105)/(E105*$G$54)))/F86**

Equation A.21 can now be used to determine the required values of f_b for each material, at each level and for the three mortar strengths. These calculations are performed in block G105 to L114 where they are expressed as ratios by dividing each value of f_b by the equivalent value obtained using BS 5628. Columns K and L are not shown on Sheet A.2. The required formulae are represented by:

G105: **(@MAX(15.4,((F105*F86/$C$64)^1.333)))/G86**

A ratio of 1 in any cell of block G105 to L114 would indicate that the two codes give the same result for that cell. The comparison shows that the use of the EC6 code is generally more onerous than BS 5628 with a maximum difference of 2.34 in cell I109, i.e. the brick strength required at this level using EC6 would be 2.34 times greater than that required using BS 5628 for the parameters used in this example.

Note. The EC6 equations were taken from a draft code and these might be subject to change in the final publication.

A spreadsheet layout similar to the one developed in this example would be useful for determining the effect of any changes and their influence on the final results.

Appendix 2

Disk information

A2.1 Contents of Disk

The 3½″ disk supplied with this textbook is a high density, 1.44 MB, disk which contains 44 files as listed below. It is formatted for use with IBM compatibles but can also be used with an Apple Macintosh that has an Apple File Exchanger. Readers who use disks of other density and size will have to arrange to have the disk copied into a suitable format for their hardware. It is recommended that readers should first make a working copy of the disk and then store the original for use as a backup.

For convenience the file names on the disk have been abbreviated so that, for example, SH11.WK1 is the file name associated with Sheet 1.1 in the text, and similarly for the other file names.

File (WK1)	*Size*	*Purpose*
SH11	5583	Interpolation of stress for given strain (BS8110).
SH12	4417	Interpolation of stress for given strain (BS5628).
SH13	3909	Solution of up to 5 simultaneous equations.
SH21	16579	Triple interpolation (framed panels).
SH31	12793	Design of reinforced concrete beams (Method 1).
SH32	12804	Design of reinforced concrete beams (Method 1).
SH33	8573	Design of reinforced concrete beams (Method 2).
SH34	8560	Design of reinforced concrete beams (Method 2).
SH34R	6718	Design of reinforced concrete beams allowing for redistribution.
SH35	8401	Stress block coefficients for reinforced masonry beams.
SH36	10045	Design for bending. Reinforced masonry beams.
SH37	12097	Design for bending and shear. Reinforced masonry beams.
SH38	12073	Design for bending and shear. Reinforced masonry beams.
SH39	12046	Design for bending and shear. Reinforced masonry flanged beams.
SH41	24896	Reinforced masonry. Short columns, uniaxial bending.
SH42	26945	Reinforced masonry. Short columns, uniaxial bending.
SH43	33523	Reinforced masonry. Short columns, biaxial bending.
SH44	33523	Reinforced masonry. Short columns, uniaxial or biaxial bending.
SH45	34013	Reinforced masonry. Slender columns, uniaxial or biaxial bending.
SH46	21424	Reinforced concrete. Short columns, uniaxial bending.
SH47	36658	Reinforced concrete. Short columns, uniaxial bending (alternative method).
SH48	43580	Reinforced concrete. Short columns, uniaxial or biaxial bending.
SH49	45655	Reinforced concrete. Short/slender columns, uniaxial or biaxial bending.
SH49EX	45910	Reinforced concrete. Short/slender columns, uniaxial or biaxial bending.
SH51	16561	Reinforced concrete. Simply supported slabs.
SH52	12577	Moments in r.c. slabs. Johansen, Hillerborg and BS8110 methods.
SH61	17707	Prestressed concrete beams at ultimate load.
SH62	9703	Prestressed concrete beams at service load.

SH63	24383	Prestressed concrete beams at transfer, service and ultimate loads.
SH71	15046	Beam analysis of multi storey braced frames.
SH72	14922	Beam analysis of braced sub frames.
SH73	9524	Beam and column analysis of unbraced frames.
SH74	9783	Matrix multiplication and inversion.
SH75	23041	Analysis of braced frames with automatic matrix inversion.
SH76	14111	Analysis of braced sub frame for column design.
SH81	35456	Simply supported beams. Shear, moment and deflection.
SH82	44893	Fixed beams. Shear, moment and deflection.
SH83	78472	Continuous beams. Rotation, shear and moment at each support.
SH91	8834	Design of wall–beams by three methods.
SH92	21876	Comparison of methods of wall–beam design with $E_{beam}/E_{wall} = 4$.
SH92A	24634	Comparison of methods of wall–design with $E_{beam}/E_{wall} = 30$.
SH93	15860	Design of framed wall panels.
SHA1	30649	Design of simple crosswall structures.
SHA2	39578	Comparison of British and Eurocode for design of masonry walls.

Total = 948,335 bytes (44 files).

A2.2 Details of special design files.

Some of the files are duplicates in the sense that the same spreadsheet has been used to illustrate two examples. For example SH31 and SH32 use the same spreadsheet to solve, in the first case, a singly reinforced concrete section and, in the second case, a doubly reinforced section. Other groups of files have been designed to show the progression from a simple spreadsheet, which can be used to solve one aspect of the problem, to more comprehensive spreadsheets which can cope with several aspects. For example for the design of reinforced concrete columns spreadsheet SH49 can be used for short or slender columns and uniaxial or biaxial bending whilst spreadsheets SH46 to SH48 illustrate the progressive development to SH49. This development process has been deliberately maintained throughout since one of the purposes of the text was to show how the method could be applied to a range of problems so that readers could then use similar procedures to solve other problems.

However, a by-product of this progression is that the final spreadsheet, for each set of problems, can be used directly for solving a number of routine but tedious designs from within that set. For example, as stated above, SH49 could be used directly for the design of a range of reinforced concrete columns which fall into different categories of length and loading. Similarly SH63 can be used for the design of prestressed concrete beams at both service and ultimate loading. A list of the other spreadsheeets which will be found useful for direct design is given below together with a brief description of the required input and the resulting output.

SH34R *Design of reinforced concrete beams allowing for redistribution.* Input of the dimensions of the section, the material properties, the design moment and the % redistribution, results in the value of the required areas of steel reinforcement which is then readily converted, using the spreadsheet, into acceptable bar diameters.

SH39 *Design for bending and shear. Reinforced masonry flanged beams.* First the stress block coefficients are calculated from the characteristic strength of the reinforcement and the ultimate masonry strain. Then input of the dimensions of the section, the material properties, the design moment and shear force, results in the value of the required areas of steel reinforcement which is then readily converted, using the spreadsheet, into acceptable bar diameters. Beams can be rectangular or flanged, simply supported, continuous or cantilevered.

SH44 *Reinforced masonry. Short columns, uniaxial or biaxial bending.* Input of the dimensions of the section, the material properties, the bar locations and diameters, the applied moments and axial force, results in a graph which indicates if the trial values are acceptable and economical.

SH45 *Reinforced masonry. Slender columns, uniaxial or biaxial bending.* As for SH44 for slender columns.

SH49 *Reinforced concrete. Short/slender columns, uniaxial or biaxial bending.* Input of the dimensions of the section, the material properties, bar diameters, applied moments at top and bottom of column, axial force and end conditions, results in a graph which indicates if the trial values are acceptable and economical. Up to 16 bars can be used.

SH51 *Reinforced concrete simply supported slabs with redistribution.* Input of the dimensions of the section, the material properties, the design moment, the shear force and the % redistribution, results in the required areas of steel reinforcement which is then readily converted, using the spreadsheet, into acceptable bar diameters. Checks are also made for link diameter and spacing, secondary reinforcement, cracking and fire resistance.

SH52 *Moments in reinforced concrete slabs. Johansen, Hillerborg and BS8110 methods.* Input of the dimensions, density, loading and support conditions, results in the values of the bending moments.

SH63 *Prestressed concrete beams at transfer, in service and at ultimate loads.* Input of the dimensions of the section, the material properties, the dead and imposed loading, the preload, and the % losses at transfer and in service, results in a check on the stresses for service, transfer and ultimate conditions. Checks are also made on the section moduli, the preload and the tendon stress.

SH75 *Analysis of braced frames with automatic matrix conversion.* The frame is assumed to consist of up to 3 bays with, or without, upper and lower columns at the end of each bay. Input of the member dimensions, the vertical dead and imposed load on each bay, the modulus of elasticity and load factor for each member, results in the values of the reactions and moments in the beams and columns.

SH81 *Simply supported beams. Shear, moment and deflection for different loading.* Input of the span, the second moment of area, the modulus of elasticity, up to five point loads and up to two distributed loads starting and ending at defined locations, results in the values of the shear forces and bending moments at the salient sections and the overall deflection at 10 locations along the beam.

SH82 *Fixed beams. Shear, moment and deflection for different loading.* The fixed beam equivalent of SH81.

SH83 *Continuous beams. Rotation, shear and moment at each support.* Up to five spans, where each span can have up to five point loads and two distributed loads acting over any defined range and the end supports can be simply supported, fixed or overhanging. The spreadsheet outputs the reactions, bending moments and rotations at each support.

A2.3 Method of using disk files.

To use the files a spreadsheet package such as Lotus 1-2-3, Quattro Pro or Excel is required. Other packages which can import Lotus 1-2-3 files, such as SUPERCALC, should also be suitable and the reader should consult their manual to check if this is possible.

Start the spreadsheet application, as recommended in the user manual, and then LOAD, or IMPORT, the required file. This should result in a screen output of the file similar to that shown in the text. The cells which contain values of the initial data are UNPROTECTED and the user will be able to change these values in order to solve a similar problem for a different set of initial data. Values in the PROTECTED cells cannot be changed unless the user deliberately UNPROTECTs these cells using the method recommended in the manual for the application in use. If the user wishes to modify the spreadsheet in order to incorporate additional, or alternative, equations or limits then it would be necessary to remove any cell protection before this could be achieved. It would be advisable to make a working copy of the appropriate files before embarking on such modifications.

A2.4 Troubleshooting

If the selected spreadsheet does not appear on the screen then check that:

(a) the correct file name has been used;
(b) the correct PATH has been specified;
(c) the file exists on the specified path;
(d) the package is not searching in a different drive such as the default drive.

If the file looks different from the example in the text then check if the widths of the columns have been transferred accurately. During transfer, from one application to another, a small number of packages output a standard default width for the columns which is different from the column widths used in the original file. For example a column width of 20 spaces might be reduced to a width of 9 spaces during transfer and if this occurs the screen output would be affected. To correct for this the user would have to alter the column widths as appropriate.

In a similar way a small number of packages remove the protection from PROTECTED cells during transfer. If this occurs then the user should follow the instructions in their manuals to re-protect the affected cells.

References

Armer, G.S.T. (1968) *The Strip Method: A New Approach to the Design of Slabs,* **CP81/68**, Building Research Station, Watford.

BSI (1972) Wind loads, *British Standards Code of Practice 3; Chapter V; Part 2*, British Standards Institute, London and Milton Keynes.

BSI (1985) Structural use of reinforced and prestressed masonry, *BS 5682: Part 2, British Standard Code of Practice for the Use of Masonry*, British Standards Institute, London and Milton Keynes.

BSI (1985) Structural use of reinforced concrete, *BS 8110: Part 2, British Standard Code of Practice for the Use of Concrete*, British Standards Institute, London and Milton Keynes.

BSI (1990) Structural use of steelwork in building, *BS 5950: Part 1, Code of Practice for Design in Simple and Continuous Construction*, London and Milton Keynes.

Davies, S.R. and Ahmed, A.E. (1978) An approximate method for analysing composite wall beams, *Proceedings of the British Ceramics Society,* **27**, pp 305–20.

Davies, S.R. and Hendry, A.W. (1986) Reinforced masonry beams, *Proceedings of the British Masonry Society,* **1**, pp 73–6.

Edgell, G.J. (1981) The ultimate strength of reinforced clay brickwork beams, *British Ceramic Research Association Seminar*, June, Edinburgh.

Hendry, A.W. (1981) *Structural Brickwork*, Macmillan Press Ltd., London.

Hendry, A.W. (Ed.) (1991) *Reinforced and Prestressed Masonry*, Longman Scientific & Technical, Harlow.

Hendry, A.W., Sinha, B.P. and Davies, S.R. (1987) *Load Bearing Brickwork Design*, 2nd edn, Ellis Horwood Series in Engineering Science, Chichester.

Hillerborg, A. (1982) The advanced strip method — a simple design tool, *Magazine of Concrete Research,* **34/121**, December, pp 175–81.

Hognestad, E. (1953) Yield-line theory for the ultimate flexural strength of reinforced concrete slabs, *Proc. ACI 49,* **7**, March, pp 637–56.

ISE/ICE joint committee (1985) *Manual for the Design of Reinforced Concrete Building Structures*, Institution of Structural Engineers, London.

Johansen, K.W. (1972) *Yield-line Formulae for Slabs*, Cement & Concrete Association, London.

Jones, L.L. and Wood, R.H. (1967) *Yield-line Analysis of Slabs*, Thames and Hudson with Chatto and Windus, London.

Kleinlogel, Prof. (1964), *Rahmenformeln unt Mehrstielige Rahmen*, Wilhelm Ernst & Sohn, Berlin.

Kong, F.K. and Evans, R.H. (1987) *Reinforced and Prestressed Concrete*, 3rd edn, Van Nostrand Reinhold, Wokingham.

Martin, L.H., Croxton, P.C.L. and Purkiss, J.A. (1989) *Concrete Design to BS 8110*, Edward Arnold, London.

Parsons, D.E., Strong, A.H. and McBurney, J.W. (1932) Shear tests of reinforced brick masonry beams, *Journal Res. Natl. Bur. Stand.,* **9**, pp 749–68.

Rathbone, A.J. (1981) The behaviour of reinforced concrete blockwork members, *British Ceramic Research Association Seminar*, June, Edinburgh.

Redpath Brown (1964) *Handbook of Structural Steelwork*, Redpath Brown & Co. Ltd., London.

Scheider, H. (1976) Tests on shear resistance of masonry, *Proceedings of the Fourth International Brick Masonry Conference*, paper 4.b.12, Brugge.

SCP (1976) *Calculated Brickwork Data Sheets, SCP 1–R*, Structural Clay Products Ltd., Hertford.

SCP (1977) *Calculated Brickwork Data Sheets, SCP 2–R*, Structural Clay Products Ltd., Hertford.

Stafford-Smith, B. and Riddington, J.R. (1977) The composite behaviour of elastic wall beam systems, *Proc. Inst. C. Engs,* **63**, pp 377–91.

Suter, G.T. and Hendry, A.W. (1975) Limit state shear design of reinforced brickwork beams, *Proceedings of the British Ceramics Society,* **24**, pp 191–6.

Thomas, F.G. and Simms, L.G. (1981) The strength of some reinforced brick masonry beams, *British Ceramic Research Association Seminar*, June, Edinburgh.

West, H.W.H. (Ed.) (1990) *Masonry (4): Proceedings of the 2nd Masonry Conference*, July, London, British Masonry Society, Stoke-on-Trent.

Withey, M.O. (1933) Tests on Brick Masonry Beams, *Proceedings Am. Test. Mat.,* **33**, pp 651–65.

Wood, R.H. (1952) Studies in composite construction, Part 1, the composite action of brick panel walls supported on reinforced concrete beams, *National Building Studies Research Paper 13*, HMSO, London.

Wood, R.H. (1978) Plasticity, composite action and collapse design of unreinforced shear wall panels in frames, *Proc. Inst. C. Engs, Part 2,* **65**, June, London, pp 381–411.

Wood, R.H. and Armer, G.S.T. (1968) The theory of strip method for design of slabs, *Proc. Inst. C. Engs,* **41**, October, pp 285–311.

Wood, R.H. and Simms, L.G. (1969) A tentative design method for the composite action of heavily loaded brick panel walls supported on reinforced concrete beams, **CP26/69**, Building Research Station, Watford.

Index of Spreadsheets

Index